AF182488

CISM COURSES AND LECTURES

Series Editors:

The Rectors of CISM
Sandor Kaliszky - Budapest
Mahir Sayir - Zurich
Wilhelm Schneider - Wien

The Secretary General of CISM
Giovanni Bianchi - Milan

Executive Editor
Carlo Tasso - Udine

The series presents lecture notes, monographs, edited works and proceedings in the field of Mechanics, Engineering, Computer Science and Applied Mathematics.
Purpose of the series is to make known in the international scientific and technical community results obtained in some of the activities organized by CISM, the International Centre for Mechanical Sciences.

INTERNATIONAL CENTRE FOR MECHANICAL SCIENCES

COURSES AND LECTURES - No. 380

ELECTROHYDRODYNAMICS

EDITED BY

ANTONIO CASTELLANOS
UNIVERSITY OF SEVILLA

Springer-Verlag Wien GmbH

Le spese di stampa di questo volume sono in parte coperte da
contributi del Consiglio Nazionale delle Ricerche.

This volume contains 137 illustrations

This work is subject to copyright.
All rights are reserved,
whether the whole or part of the material is concerned
specifically those of translation, reprinting, re-use of illustrations,
broadcasting, reproduction by photocopying machine
or similar means, and storage in data banks.
© 1998 by Springer-Verlag Wien
Originally published by Springer-Verlag Wien New York in 1998
SPIN 10686913

In order to make this volume available as economically and as rapidly as possible the authors' typescripts have been reproduced in their original forms. This method unfortunately has its typographical limitations but it is hoped that they in no way distract the reader.

ISBN 978-3-211-83137-3 ISBN 978-3-7091-2522-9 (eBook)
DOI 10.1007/978-3-7091-2522-9

PREFACE

This book deals with Electrohydrodynamics (EHD) which is an interdisciplinary science dealing with the interaction of fluids with electric fields. It is addressed to scientists, engineers and students possessing a basic knowledge of mathematics and physics. It may be particularly useful to the practising fluid dynamicists, chemical and electrical engineers who need to understand and apply the principles of electrohydrodynamics in their daily work.

The book is the outcome of a series of lectures on EHD delivered at the Seventh IUTAM (International Union of Theoretical and Applied Mechanics) Summer School held in Udine, Italy in July 1996. This Summer School was proposed to the General Assembly of IUTAM by the Centre International des Sciences Mécaniques (CISM). With this decision IUTAM has recognized the rapid development and increasing importance of Electrohydrodynmics both from the fundamental and applied point of view.

The book consists of five parts. In Part I the foundations of EHD are established for the case of homogeneous liquid dielectrics. In the first three chapters of Part II conduction phenomena are dealt in depth, followed by a discussion of surface waves on charged interfaces, and EHD pumping in the last two chapters. In the first chapter of Part III conduction is reviewed again for liquids used in applications. Then instabilities, pre-breakdown and breakdown phenomena, and electrospray are considered in the following four chapters. EHD motion driven by Coulomb forces acting in the bulk will constitute the object of Part IV. Part V is dedicated to multiphase flows, that are of paramount importance in industrial processes.

Each author has developed his part with autonomy, thus reflecting his own interest and approach. As a result some topics have been treated by more than one author. This is an enriching characteristic of the book, as the reader may appreciate the different points of view in some of the still unresolved problems in EHD. This is natural since the authors come from universities, research laboratories and industry. However an effort has been made to use SI units through the book, and to unify as much as possible the nomenclature.

I wish to take the opportunity to thank all the contributing authors of this volume as well as their institutions. My sincere thanks are also extended to Prof. Kaliszky and Prof. Tasso as well as to the other members of the staff of CISM for their hospitality and help that has made this book possible. I would also like to thank J.L. Lara and A. Mayer for the re-editing of most chapters of the book in order to make the format as uniform as possible.

Antonio Castellanos

CONTENTS

PART I

BASIC CONCEPTS AND EQUATIONS IN ELECTROHYDRODYNAMICS

A. Castellanos
University of Sevilla, Sevilla, Spain

The purpose of chapter 1 is to present a concise introduction to the electrical equations governing the electrohydrodynamics of dielectric liquids. First Maxwell's equations are stated, followed by a discussion of the Galilean limits of electromagnetism. Then, the electroquasistatic equations in moving media are derived. Finally, the electrical energy is computed and discussed in the context of thermodynamics.

When a d.c. electric field is applied to a fluid there are two responses: the charge carriers that are free to move through the fluid gives rise to the conduction, and those that are bounded or blocked gives rise to the polarization. In chapter 2 we will discuss these phenomena with the aim of understanding the laws of conductivity and permittivity.

In chapter 3 we present a concise introduction to the mechanical equations governing the electrohydrodynamics of dielectric fluids. The conservation laws for mass and momentum are presented, followed by a derivation of the electric force density, using thermodynamic arguments. Finally, some nondimensional numbers are introduced and discussed.

The temperature equation in Electrohydrodynamics is discussed in chapter 4. The conservation laws satisfied by each kind of macroscopic energy, i. e. kinetic, potential and electrical energies present in Electrohydrodynamics are derived, with the aim of deducing an equation for the internal energy. Using Gibbs relation and the conservation law for the internal energy the equation for the entropy is derived. Thermoelectric effects and other coupled phenomena arise naturally from Onsager reciprocity theorem. Finally, the temperature equation is formulated for both nonpolar and polar liquids.

I would like to acknowledge the contribution of F. Pontiga since the review of conduction phenomena that I presented in the first part of chapter 2, is based in a substantial part in the first chapter of his PhD thesis (in Spanish). Also I would like to thank A. Ramos, A. Gonzalez and A.T. Perez for the fruitful discussions I had with them on the question of the energy in electrohydrodynamics. Finally, thanks are due to N. Green that read carefully the manuscript and corrected my English.

Chapter 1

ELECTRICAL EQUATIONS

In this chapter we study the governing electrical equations in Electrohydrodynamics as a particular case of Electroquasistatics for fluids in nonrelativistic motion. First, we present in a very succint way Maxwell's equations. Secondly, we show how these equations may be simplified after choosing suitable scales to make them dimensionless, and the electroquasistatic and magnetoquasistatic limits to the Maxwell's equations are obtained. In these limits a nonrelativistic approach suffices, and using the Galilean principle of relativity the equations of Electroquasistatics and Magnetoquasistatics will be obtained for slowly moving fluids. Here slow means slow compared to the velocity of light, though the fluids could be moving at high speeds by ordinary standards. Galilean invariance imposes a transformation of the field quantities that is specific to this limits; it is important to note that, contrary to Mechanics, these limits are not obtained simply taken the factor $\gamma = [1 - (u/c)^2]^{-1/2} = 1$ in the relativistic transformations of the fields [1]. For an alternative approach see [2]. The boundary and jump conditions are considered. Finally, a discussion of the concept of electrical energy is done in order to place this concept in the appropriate thermodynamic context [3],[4].

1.1 Maxwell's equations

A fundamental property of elementary particles, which are the constituents of the material bodies, is the electrical charge that is responsible for the electromagnetic interaction. The elementary unit of charge is the electron charge $e = 1.6 \cdot 10^{-19}$C. In most conditions the number of charge carriers is enormous and the nearest neighbour separation is negligible by any macroscopic standards. It is then possible to take averages over local volumes δV, considered infinitesimal from the point of view of the measuring apparatus but still containing an extremely large number of particles. It is useful to distinguish between charge carriers that are free to move macroscopic distances from those that are bounded or blocked. We define the free (bounded) charge density as the ensemble (spatio-temporal) average of the free (bounded) charge carriers.

$$q = \frac{< \sum_{i=1}^{n} n_i e_i >_f}{\delta V}, \qquad q_p = \frac{< \sum_{i=1}^{n} n_i e_i >_p}{\delta V}, \tag{1.1}$$

where n_i, and e_i, are the number, and charge of particles of type i, contained in the volume δV, $< g >$ means the ensemble average of g, and the subscript $f(p)$ refers to free (bounded) charge carriers.

Similarly the electric and magnetic fields **E**, and **B**, are ensemble (spatio-temporal) averages of the fields due to the individual particles. The basic laws governing the behavior of these fields in material media have been the object of continued research and may be summarised in the celebrated Maxwell's equations. It is not possible to trace the development of the full equations, and we must refer the reader to classical texts in electromagnetism [1,4]. However, we shall state these laws and we will discuss those aspects of interest to EHD.

Gauss's law states that

$$\varepsilon_o \nabla \cdot \mathbf{E} = q + q_p, \tag{1.2}$$

with $\varepsilon_o = 8.8542 \cdot 10^{-12} \mathrm{Fm}^{-1}$ the permittivity of the free space, q the free charge density, and q_p represents the ensemble average density due to the bounded charges, usually known as the polarization charge density. Since for any body, inside volume V,

$$\int_V q_p dV = 0, \tag{1.3}$$

we may write for this body, $q_p = -\nabla \cdot \mathbf{P}$, with the only restriction for $\mathbf{P}$ of being zero outside the body. The vector $\mathbf{P}$ is called the polarization vector.

Defining the displacement vector, $\mathbf{D} = \varepsilon_o \mathbf{E} + \mathbf{P}$, Gauss's law is restated as

$$\nabla \cdot \mathbf{D} = q. \tag{1.4}$$

The nonexistence of magnetic charges implies that there are no scalar sources for the magnetic field, and therefore

$$\nabla \cdot \mathbf{B} = 0, \tag{1.5}$$

Faraday's law states that a time dependent magnetic field gives rise to an electric field through

$$\nabla \times \mathbf{E} = -\frac{\partial \mathbf{B}}{\partial t}. \tag{1.6}$$

Finally, the fourth law tell us that the vector sources of the magnetic field are the current density due to the motion of charge carriers, either free or bounded, as well as the magnetic field induced by a time varying electric field

$$\frac{1}{\mu_o} \nabla \times \mathbf{B} = < \sum_{i=1}^{n} n_i e_i \mathbf{u}_i > + \varepsilon_o \frac{\partial \mathbf{E}}{\partial t}, \tag{1.7}$$

where $\mu_o = 4\pi \cdot 10^{-7} \mathrm{Hm}^{-1}$ is the permeability of the vacuum, and $\mathbf{u}_i$ the velocity of particles of type i. The velocity of a charged particle is the sum of three terms: the average (macroscopic) velocity of the element of volume δV, $\mathbf{u}$; the velocity of the particle relative to the fluid, $\mathbf{u}_{ri}$ (see chapter 2); and a fluctuating velocity with zero ensemble average, $\mathbf{u'}_i$. The conduction current density, $\mathbf{J}$, is defined as

$$\mathbf{J} = q\mathbf{u} + \frac{< \sum_{i=1}^{n} n_i e_i \mathbf{u}_{ri} >_f}{\delta V} = q\mathbf{u} + \mathbf{J}', \tag{1.8}$$

with $\mathbf{J}'$ the current density measured by an observer comoving with the element of volume δV.

Formally, we have that the current due to all charge carriers is given by, $\mathbf{J} + q_p \mathbf{u}_p + < q'\mathbf{u}' >$, with $\mathbf{J}$ the conduction current density due to the motion of the free charge carriers relative to the liquid, $\mathbf{u}_p$ the velocity of the bounded charged density, and q' and $\mathbf{u}'$ are charge and velocity fluctuations with zero average. From

$$\frac{\partial q_p}{\partial t} + \nabla \cdot (q_p \mathbf{u}_p) = 0, \tag{1.9}$$

and $q_p = -\nabla \cdot \mathbf{P}$ we may put

$$q_p \mathbf{u}_p = \frac{\partial \mathbf{P}}{\partial t}. \tag{1.10}$$

On the other hand, since the term, $< q'\mathbf{u}' >$, gives a zero contribution to the electrical current across any section of a material body, $\nabla \cdot < q'\mathbf{u}' >= 0$, and a general solution for this term may be expressed as

$$< q'\mathbf{u}' >= \nabla \times \mathbf{M}. \tag{1.11}$$

The vector $\mathbf{M}$ is called the magnetization vector, and a precise understanding of this term was not possible until the advent of quantum mechanics. It should be noticed that magnetism is absent in classical physics [5] (Miss van Leeuwen's theorem), and therefore any simple picture in terms of elementary circuit currents is doomed to fail. Defining the magnetic intensity field, $\mathbf{H} = {\mu_0}^{-1}\mathbf{B} - \mathbf{M}$, the fourth Maxwell equation may be written as

$$\nabla \times \mathbf{H} = \mathbf{J} + \frac{\partial \mathbf{D}}{\partial t}. \tag{1.12}$$

Now $\mathbf{J}$ is due only to the the free charge carriers that are convected and drift through the fluid under the action of the electromagnetic field.

1.2 Nondimensional Maxwell's equations

Consider a physical system of characteristic size, l, in the presence of an electromagnetic field. In general, and depending on the phenomena we are interested, we may have a set of different characteristic times corresponding to thermal, chemical, mechanical, electrical and magnetic subprocesses taking place in the system. For simplicity we assume that there is only one characteristic time, τ, and consequently a characteristic velocity, u. It will be also assumed for the same reasons of simplicity that the constitutive relations, $\mathbf{D} = \varepsilon\mathbf{E}$, $\mathbf{B} = \mu\mathbf{H}$, and $\mathbf{J} = \sigma \mathbf{E}$, are valid with ε, μ, and σ, the permittivity, the permeability and the electrical conductivity, considered constants.

The scales l and τ will then be used to nondimensionalize the coordinates and time. Since we are interested in electromagnetic phenomena there is always another characteristic velocity, c, which is the velocity of light in the system, $c = 1/\sqrt{\varepsilon\mu}$. Associated with this velocity there is a characteristic time given by the transit time of an electromagnetic wave across the system, $\tau_{em} = l/c$.

To choose a scale to nondimensionalize the EM fields we need to compare the relative magnitude of the electric to the magnetic field. For radiation fields, such as plane waves, it is well known that $E = cB$ with E the magnitude of the electric field, c the velocity of light, and B the magnitude of the magnetic field. Also, it is well known from relativistic electromagnetism that, $E^2 - c^2B^2$, is a Lorentz invariant that may be positive, zero or negative. Alternatively, for those people not familiar with relativity, it may be easier to think in terms of energies. We have

$$\frac{\varepsilon E^2/2}{B^2/2\mu} = \frac{E^2}{c^2B^2}. \tag{1.13}$$

We recover again the three previous cases, which now correspond to an electrically dominated system, a radiation field, and a magnetically dominated system, when this ratio is greater, equal, or lesser than one. We neglect the particular case in which we may have an electrostatic field and a magnetostatic field, both decoupled from each other. In the first case, denoted as Electroquasistatics, and in the third case, named Magnetoquasistatics, it is possible to simplify the Maxwell's equations. In both cases all the intrincacies due to electromagnetic wave phenomena and radiation may be disregarded, and as a consequence a great degree of simplification is achieved. Moreover, in the two latter cases it is possible to take the limit, $c \to \infty$, in a consistent way, thus obtaining two different Galilean limits to the electromagnetic field, with corresponding field transformations specific to each case.

1.2.1 Electroquasistatics: first Galilean limit of Maxwell's equations

In this case the electric energy is greater than the magnetic energy,

$$\frac{\varepsilon E^2/2}{B^2/2\mu} = \frac{E^2}{c^2B^2} > 1 \tag{1.14}$$

and we choose the electric field to nondimensionalize Maxwell's equations. Let us denote the typical magnitude of the electric field by E_0. A naturally derived scale for the charge density is obtained from Gauss's law as

$$q_0 = \varepsilon\frac{E_0}{l}, \tag{1.15}$$

where it is assumed that the length over which the electric field varies on the order of itself is also given by l.

To find the appropriate scale for the magnetic field we examine the fourth Maxwell equation

$$\nabla \times \mathbf{H} = \sigma \mathbf{E} + \varepsilon \frac{\partial \mathbf{E}}{\partial t}. \tag{1.16}$$

The ratio of both terms in the right hand side is

$$\frac{\varepsilon(\partial E/\partial t)}{\sigma E} \sim \frac{(\varepsilon/\sigma)}{\tau} = \frac{\tau_e}{\tau}, \tag{1.17}$$

with, $\tau_e = (\varepsilon/\sigma)$, the characteristic time for the relaxation of charge due to ohmic conduction, as indicated by the solution to the charge conservation equation. In effect, is

$$\frac{\partial q}{\partial t} + \nabla \cdot (\sigma \mathbf{E}) = 0. \tag{1.18}$$

Using Gauss's law this may be written as

$$\frac{\partial q}{\partial t} + \frac{\sigma}{\varepsilon} q = 0, \tag{1.19}$$

which has a solution,

$$q(t) = q(0) e^{-t/\tau_e}. \tag{1.20}$$

The charge decays exponentially to zero with the characteristic time τ_e.

We have two possible cases; either the ohmic current is larger (or equal) than the displacement current, or smaller (or equal).

The ohmic current is dominant

For the case in which the ohmic current determines the magnetic field, the scale for $\mathbf{H}$ comes from

$$\frac{H_0}{l} \nabla^* \times \mathbf{H}^* \sim \sigma E_0 \mathbf{E}^*, \tag{1.21}$$

with asterisks denoting dimensionless quantities of order one. The typical scale for the magnetic field is $B_0 = \mu H_0 = \mu \sigma E_0 l$, the Maxwell's equations may be written in nondimensional form, suppressing asterisks, as

$$\nabla \cdot \mathbf{E} = q, \qquad \nabla \times \mathbf{E} = -\left(\frac{\tau_m}{\tau}\right) \frac{\partial \mathbf{B}}{\partial t}, \tag{1.22}$$

$$\nabla \cdot \mathbf{B} = 0, \qquad \nabla \times \mathbf{H} = \mathbf{E} + \left(\frac{\tau_e}{\tau}\right) \frac{\partial \mathbf{E}}{\partial t}, \tag{1.23}$$

together with the charge conservation law

$$\frac{\partial q}{\partial t} + \left(\frac{\tau}{\tau_e}\right) q = 0, \tag{1.24}$$

with $\tau_m = \sigma\mu l^2$ a characteristic time for the diffusion of the magnetic field, as can be seen from the following argument. Combining Faraday's law with the fourth Maxwell equation, we arrive to the equation

$$\left(\nabla^2 - \sigma\mu\frac{\partial}{\partial t} - \frac{1}{c^2}\frac{\partial^2}{\partial t^2}\right)\mathbf{H} = 0. \tag{1.25}$$

In dimensionless form this is

$$\left(\nabla^2 - \frac{\tau_m}{\tau}\frac{\partial}{\partial t} - \frac{\tau_{em}^2}{\tau^2}\frac{\partial^2}{\partial t^2}\right)\mathbf{H} = 0, \tag{1.26}$$

where $\tau_{em} = (l/c)$ is the transit time of electromagnetic waves. Usually, this time is much smaller than the characteristic time τ of the system, and the equation reduces to

$$\left(\nabla^2 - \frac{\tau_m}{\tau}\frac{\partial}{\partial t}\right)\mathbf{H} = 0. \tag{1.27}$$

This is a diffusion equation with both terms of similar order for $\tau = \tau_m$, the typical time of magnetic diffusion.

An important relation between the electromagnetic times is

$$\tau_{em}^2 = \tau_e\,\tau_m. \tag{1.28}$$

Also from the nondimensional fourth Maxwell equation, since the displacement current is smaller or equal to the ohmic current, we must have that $\tau_e \leq \tau$. On the other hand, using the scale for the magnetic field, and equation (1.14)

$$\frac{\varepsilon E^2/2}{B^2/2\mu} \sim \frac{\varepsilon E_0^2/2}{(\mu\sigma l E_0)^2/2\mu} = \frac{\tau_e}{\tau_m} > 1. \tag{1.29}$$

Therefore $\tau_m < \tau_e$ and the characteristic times are ordered in the way

$$\tau_m < \tau_{em} < \tau_e \leq \tau. \tag{1.30}$$

If we now take the limit $c \to \infty$ then $\tau_{em} \to 0$. As a consequence, $(\tau_m/\tau) = 0$ in this limit, and the set of Maxwell's equations is simplified to

$$\nabla\cdot\mathbf{E} = q, \qquad \nabla\times\mathbf{E} = 0, \tag{1.31}$$

$$\nabla\cdot\mathbf{B} = 0, \qquad \nabla\times\mathbf{H} = \mathbf{E} + \left(\frac{\tau_e}{\tau}\right)\frac{\partial\mathbf{E}}{\partial t}. \tag{1.32}$$

The most important result is that the electric field is an irrotational field, and may then be derived from the gradient of a scalar function, the electric potential.

The displacement current is dominant

Consider now the case in which the displacement current dominates over the ohmic current. The appropriate scale for the magnetic field is derived from

$$\frac{H_0}{l}\nabla^* \times \mathbf{H}^* \sim E_0\frac{\varepsilon}{\tau}\frac{\partial \mathbf{E}^*}{\partial t^*}, \tag{1.33}$$

which implies that $cB_0 = \mu H_0 = E_0(u/c)$.

The dimensionless Maxwell's equations are

$$\nabla \cdot \mathbf{E} = q, \qquad \nabla \times \mathbf{E} = -\left(\frac{\tau_{em}}{\tau}\right)^2 \frac{\partial \mathbf{B}}{\partial t}, \tag{1.34}$$

$$\nabla \cdot \mathbf{B} = 0, \qquad \nabla \times \mathbf{H} = \left(\frac{\tau}{\tau_e}\right)\mathbf{E} + \frac{\partial \mathbf{E}}{\partial t}. \tag{1.35}$$

From the last equation we must have $\tau \leq \tau_e$. Also, from the ratio of electric to magnetic energies, eq. (1.14), we have

$$\frac{\varepsilon E^2/2}{B^2/2\mu} \sim \frac{\varepsilon E_0^2/2}{E_0^2(u/c^2)^2/2\mu} \sim \left(\frac{\tau}{\tau_{em}}\right)^2 > 1. \tag{1.36}$$

Therefore, $\tau_{em} < \tau$ and the characteristic times are ordered in the way

$$\tau_m < \tau_{em} < \tau \leq \tau_e. \tag{1.37}$$

If we take the limit $c \to \infty$, then $\tau_{em} \to 0$, and in this limit, the set of Maxwell's equations is simplified to

$$\nabla \cdot \mathbf{E} = q, \qquad \nabla \times \mathbf{E} = 0, \tag{1.38}$$

$$\nabla \cdot \mathbf{B} = 0, \qquad \nabla \times \mathbf{H} = \left(\frac{\tau}{\tau_\sigma}\right)\mathbf{E} + \frac{\partial \mathbf{E}}{\partial t}. \tag{1.39}$$

Again the most important fact is, that in this case also, the electric field is irrotational, and may be derived from the gradient of a scalar function, the electric potential.

To summarize, if the electric energy is strictly larger than the magnetic energy we are in the realm of Electroquasistatics, and the Maxwell equations are reduced to

$$\nabla \cdot \mathbf{D} = q, \qquad \nabla \times \mathbf{E} = 0, \tag{1.40}$$

$$\nabla \cdot \mathbf{B} = 0, \qquad \nabla \times \mathbf{H} = \mathbf{J} + \frac{\partial \mathbf{D}}{\partial t}, \tag{1.41}$$

together with the charge conservation law

$$\frac{\partial q}{\partial t} + \nabla \cdot \mathbf{J} = 0. \tag{1.42}$$

1.2.2 Magnetoquasistatics: second Galilean limit of Maxwell's equations

Let us consider briefly the case where the magnetic energy is strictly larger than the electric energy. Now the scale for the magnetic field is H_0 and to derive an appropriate scale for the electric field we again have two possibilities. Either the source for the electric field is the ohmic term and the scale for E_0 is given by $E_0 = (1/\sigma)(H_0/l)$ (from the fourth Maxwell equation), or else the origin of the electric field is the time variation of the magnetic field through Faraday's law. In the latter case is

$$\frac{E_0}{l}\nabla^* \times \mathbf{E}^* = -\frac{B_0}{\tau}\frac{\partial \mathbf{B}^*}{\partial t^*}, \tag{1.43}$$

and consequently $E_0 = B_0 l/\tau = \mu H_0 l/\tau$.

The ohmic current is the source of the electric field

With the ohmic scale $E_0 = (1/\sigma)(H_0/l)$, the set of dimensionless Maxwell's equations is

$$\nabla \cdot \mathbf{E} = q, \qquad \nabla \times \mathbf{E} = -\left(\frac{\tau_m}{\tau}\right)\frac{\partial \mathbf{B}}{\partial t}, \tag{1.44}$$

$$\nabla \cdot \mathbf{B} = 0, \qquad \nabla \times \mathbf{H} = \mathbf{E} + \left(\frac{\tau_e}{\tau}\right)\frac{\partial \mathbf{E}}{\partial t}, \tag{1.45}$$

where $\tau_e \leq \tau$, since the ohmic current is the dominant term. This set of equations is formally identical to the equations obtained in Electroquasistatics for the ohmic conduction dominant case. However, the ordering of times is different, since now

$$\frac{\varepsilon E^2/2}{B^2/2\mu} \sim \frac{\varepsilon(\mu B_0/\sigma l)^2/2}{B_0^2/2\mu} \sim \left(\frac{\tau_e}{\tau_m}\right) < 1. \tag{1.46}$$

Therefore

$$\tau_e \leq \tau, \qquad \tau_e < \tau_{em} < \tau_m. \tag{1.47}$$

Usually $\tau > \tau_{em}$, and we have any of the two orderings

$$\tau_e < \tau_{em} < \tau_m < \tau, \qquad \tau_e < \tau_{em} < \tau < \tau_m. \tag{1.48}$$

In both cases, if we take the limit $c \to \infty$, the Maxwell's equations are reduced to

$$\nabla \cdot \mathbf{E} = q, \qquad \nabla \times \mathbf{E} = -\left(\frac{\tau_m}{\tau}\right)\frac{\partial \mathbf{B}}{\partial t}, \tag{1.49}$$

$$\nabla \cdot \mathbf{B} = 0, \qquad \nabla \times \mathbf{H} = \mathbf{E}. \tag{1.50}$$

The important fact is that now the displacement current no longer exists in this limit, and consequently radiation is absent.

The electric field is induced by the time-varying magnetic field

For this second case, in which the origin of the electric field is the time variation of the magnetic field, the appropriate scale for the electric field is, from Faraday's law, $E_0 = B_0 l/\tau = \mu H_0 l/\tau$. The dimensionless Maxwell's equations are

$$\nabla \cdot \mathbf{E} = q, \qquad \nabla \times \mathbf{E} = -\frac{\partial \mathbf{B}}{\partial t}, \tag{1.51}$$

$$\nabla \cdot \mathbf{B} = 0, \qquad \nabla \times \mathbf{H} = \left(\frac{\tau_m}{\tau}\right)\mathbf{E} + \left(\frac{\tau_{em}}{\tau}\right)^2 \frac{\partial \mathbf{E}}{\partial t}. \tag{1.52}$$

In this case

$$\frac{\varepsilon E^2/2}{B^2/2\mu} \sim \frac{\varepsilon (B_0 l/\tau)^2/2}{B_0^2/2\mu} \sim \left(\frac{\tau_{em}}{\tau}\right)^2 < 1. \tag{1.53}$$

Therefore $\tau_{em} < \tau$, and the two orderings given in equation (1.48) are possible.

Taking the limit $c \to \infty$ the reduced set of Maxwell's equations is

$$\nabla \cdot \mathbf{E} = q, \qquad \nabla \times \mathbf{E} = -\frac{\partial \mathbf{B}}{\partial t}, \tag{1.54}$$

$$\nabla \cdot \mathbf{B} = 0, \qquad \nabla \times \mathbf{H} = \left(\frac{\tau_m}{\tau}\right)\mathbf{E}, \tag{1.55}$$

and again the displacement current does not contribute in this limit. Maxwell's equations are reduced to

$$\nabla \cdot \mathbf{D} = q, \qquad \nabla \times \mathbf{E} = -\frac{\partial \mathbf{B}}{\partial t}, \tag{1.56}$$

$$\nabla \cdot \mathbf{B} = 0, \qquad \nabla \times \mathbf{H} = \mathbf{J}, \tag{1.57}$$

plus the charge conservation equation.

The reader may easily show that it is not possible to have $B_0 > cE_0$, and simultaneously, to have the displacement current as the source for the magnetic field.

In conclusion, Maxwell's equations may be simplified when applied to situations where the ratio between the typical time scale of the system and the transit time of an electromagnetic wave through the system, that is τ/τ_{em}, is much larger than one. Depending on whether the electric energy is larger or smaller than the magnetic energy this nondimensional ratio affects the term $\partial \mathbf{B}/\partial t$ or else the displacement current $\partial \mathbf{D}/\partial t$. In the first case Faraday law reduces to $\nabla \times \mathbf{E} = 0$, and we are in the realm of Electrohydrodynamics (EHD). In the second case the displacement current is negligible compared to the conduction current and we are in the realm of Magnetohydrodynamics (MHD). In both cases, electromagnetic wave phenomena and radiation may be disregarded.

Given a physical system, previous experience or an order of magnitude analysis as the one shown above is of help to decide between both limits. To illustrate this point, consider a liquid in which an electric field $\mathbf{E}$ is present. Suppose further that the liquid obeys Ohm's law. If the magnetic field is determined by $\mathbf{J}$, then $B \sim \sigma\mu E l$. For typical systems in the laboratory $\varepsilon \sim 10^{-11}$ farad/m, $\mu \sim 10^{-6}$ henry/m, $l \sim 10^{-2}$ m, and we obtain $E/cB \sim 1/\sigma*$, with $\sigma*$ a dimensionless number equal in numerical value to the conductivity of the fluid in S.I. units (International System of Units). It is then clear that in dielectric liquids characterized by very low values of the conductivity, Maxwell's equations reduce to the electroquasistatic limit thus falling in the realm of EHD. On the contrary in highly conductive liquids, as mercury or molten metals, magnetic effects will be dominant which is the domain of Magnetohydrodynamics (MHD).

1.3 Electroquasistatics in moving fluids

In this section we elaborate the case of fluids subjected to electroquasistatic fields a little further. Consider first that the fluid is subjected to a time varying electric field of frequency ω. As discussed in the previous section, a necessary condition to reduce Maxwell's equations to the set of electroquasistatics equations, is that $\tau_{em} \ll \tau$ is true where τ is any characteristic time associated with electrical or mechanical processes taking place in the fluid. In particular, for a time varying electric field we may take as τ the period, T, of the electric field, and therefore we must have

$$\frac{l}{cT} \sim \frac{l\omega}{c} \ll 1, \tag{1.58}$$

The condition $\omega \ll (c/l)$ ensures that the magnetic field generated by the displacement current, which is of order $B_0 \simeq \mu\varepsilon l\omega E_0$, satisfies automatically the condition $cB_0 \ll E_0$.

The second condition that must be met by the physical system in order to be quasielectrostatic, is that the magnetic field due to the current density in the system, regardless of its origin, i. e. injection, dissociation or current due to particles, must also satisfy the condition $cB_0 \ll E_0$. Since from the fourth Maxwell's equation the scale for B_0 due to the current density is $B_0 = \mu l J_0$, we have

$$\frac{cB_0}{E_0} = \frac{cl\mu J_0}{E_0}, \tag{1.59}$$

and consequently

$$J_0 \ll E_0\sqrt{\frac{\varepsilon}{\mu}}\frac{1}{l}. \tag{1.60}$$

For typical systems in the laboratory $l \sim 10^{-2}$m, $\mu \simeq \mu_0 \simeq 10^{-6}$Hm^{-1} and $\varepsilon \sim 10^{-11}$Fm^{-1}, and we must have $J_0 \ll 3E_0$. In insulating liquids this condition is amply satisfied in practical situations as J_0 is in the range of microamperes per square centimetre, and E_0 is several kV per centimeter.

To conclude, in Electrohydrodynamics, Maxwell equations reduce to

$$\nabla \cdot \mathbf{D} = q, \qquad \nabla \times \mathbf{E} = 0, \qquad \frac{\partial q}{\partial t} + \nabla \cdot \mathbf{J} = 0, \tag{1.61}$$

with the magnetic field determined by the equations

$$\nabla \cdot \mathbf{B} = 0, \qquad \nabla \times \mathbf{H} = \mathbf{J} + \frac{\partial \mathbf{D}}{\partial t}. \tag{1.62}$$

The solution for $\mathbf{B}$ can be found once the electric field is known, although in EHD the explicit solution for the magnetic field is rarely needed.

1.3.1 Transformation between inertial frames

Let us now consider the transformation properties of quasielectrostatic fields. In this section primed quantities refers to the inertial system where the liquid is at rest. A nonrelativistic treatment is enough for the vast majority of problems, since fluid velocities are quite small compared to the velocity of light c. Consider a fluid moving as a whole with uniform velocity $\mathbf{u}$. Then, for a liquid moving with velocity $\mathbf{u}$, invariance under the Galilean transformation

$$\mathbf{x}' = \mathbf{x} - \mathbf{u}t, \qquad t' = t, \tag{1.63}$$

implies that for any vector function, $\mathbf{C}'(\mathbf{x}', t') = \mathbf{C}'(\mathbf{x}'(\mathbf{x}, t), t) = \mathbf{C}'(\mathbf{x}, t)$,

$$\nabla' = \nabla, \quad \frac{\partial \mathbf{C}'}{\partial t'} = \mathbf{u} \cdot \nabla \mathbf{C}' + \frac{\partial \mathbf{C}'}{\partial t} = (\nabla \cdot \mathbf{C}')\mathbf{u} - \nabla \times (\mathbf{u} \times \mathbf{C}') + \frac{\partial \mathbf{C}'}{\partial t}. \tag{1.64}$$

The total electric charge, dQ contained in the element of volume, dV, is a scalar, i. e. it takes the same value in all inertial frames of reference. Since $q = dQ/dV$ and dV is also a scalar, the charge density is a Galilean invariant $q = q'$. The Galilean covariance of the electrical equations (1.61) implies that the equations for the primed quantities in the primed system are of the same form. Using the equations (1.64) in the primed system and equating the resulting equations to those of the unprimed system we obtain

$$\mathbf{E} = \mathbf{E}', \qquad \mathbf{D} = \mathbf{D}', \qquad \mathbf{J} = \mathbf{J}' + q'\mathbf{u}. \tag{1.65}$$

The polarization density, $\mathbf{P}$, defined by $\mathbf{D} = \varepsilon_0 \mathbf{E} + \mathbf{P}$, transforms according to $\mathbf{P} = \mathbf{P}'$.

The equations for the magnetic induction field $\mathbf{B}$ and the magnetic field intensity $\mathbf{H}$ are

$$\nabla \cdot \mathbf{B} = 0, \qquad \nabla \times \mathbf{H} = \mathbf{J} + \frac{\partial \mathbf{D}}{\partial t}. \tag{1.66}$$

Using equations (1.64) in the primed system, the following transformation properties are derived

$$\mathbf{B} = \mathbf{B}', \qquad \mathbf{H} = \mathbf{H}' + \mathbf{u} \times \mathbf{D}', \tag{1.67}$$

and for the magnetization density $\mathbf{M}$ we have $\mathbf{M} = \mathbf{M}' - \mathbf{u} \times \mathbf{P}'$.

Let us consider now a fluid in motion, not necessarily uniform. An inertial frame for which the fluid is instantaneously at rest at some point P, of vector position $\mathbf{x}$, at time t, is called the instantaneous rest frame of the fluid at that point. It is postulated that the electrical quantities $\mathbf{E}, \mathbf{D}, q, \mathbf{J}, \mathbf{P}$ in the laboratory are related to the quantities defined in the instantaneous frame of reference, for which the fluid is at rest, through the transformations given above. Thus the electrical equations reduce to the quasi-electrostatic equations, i.e. Gauss's law, the irrotational nature of the electric field and charge conservation equation,

$$\nabla \cdot \boldsymbol{D} = q, \qquad \boldsymbol{E} = -\nabla \phi, \qquad \frac{\partial q}{\partial t} + \nabla \cdot \boldsymbol{J} = 0, \tag{1.68}$$

where ϕ is the electric potential.

We will assume that the electrical displacement is given by $\boldsymbol{D} = \varepsilon \boldsymbol{E}$ with ε the dielectric constant. This is not a restriction in Electrohydrodynamics as most fluids satisfy this relation. The current density $\mathbf{J}$ is related to $\mathbf{J}'$, the current density for a fluid at rest, by $\mathbf{J} = \mathbf{J}' + q\mathbf{u}$.

1.4 Boundary and jump conditions

Consider the simple physical system, depicted schematically in fig. 1.1, in which we have two electrodes immersed in two nonmagnetic fluids of dielectric constants ε_1 and ε_2, separated by a moving interface.

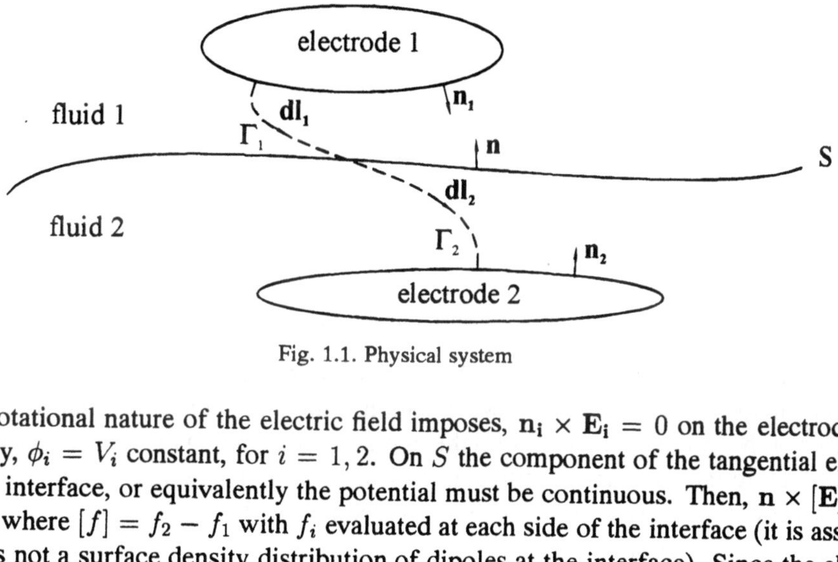

Fig. 1.1. Physical system

The irrotational nature of the electric field imposes, $\mathbf{n_i} \times \mathbf{E_i} = 0$ on the electrodes, or equivalently, $\phi_i = V_i$ constant, for $i = 1, 2$. On S the component of the tangential electric field to the interface, or equivalently the potential must be continuous. Then, $\mathbf{n} \times [\mathbf{E}] = 0$, or $[\phi] = 0$, where $[f] = f_2 - f_1$ with f_i evaluated at each side of the interface (it is assumed that there is not a surface density distribution of dipoles at the interface). Since the electric field is conservative, we have

$$V = \phi_1 - \phi_2 = \int_{\Gamma_1} \mathbf{E}_1 \cdot dl_1 + \int_{\Gamma_2} \mathbf{E}_2 \cdot dl_2, \tag{1.69}$$

along any path connecting the two electrodes. Gauss's law implies that, at each electrode, there is a surface charge density given by, $q_{si} = \mathbf{n}_i \cdot (\varepsilon_i \mathbf{E}_i) \quad i = 1, 2$, and at the interface $q_s = \mathbf{n} \cdot [\varepsilon \mathbf{E}]$.

The charge conservation equation imposes the condition that there can be no net generation of charge at the metal-liquid boundaries, or at the interface, S. However, if we consider only the volume occupied by the fluids, and the surface electrodes as the boundaries to it, then we need to specify the rate of charge injection (ejection) into (out) the fluids from (to) the electrodes. These injection and ejection laws have to be defined for each type of charge carrier present in the fluids. Examples of these laws will be given in chapter 2, 6 and 7.

The exact expression for the jump condition relative to the charge conservation equation on the interface is a subtle point that has led to some confusion. Here we establish this formula in a simple way. The surface, S, that may be moving and deforming, will be defined geometrically by an implicit function $F(\mathbf{r}, t) = 0$ and parametrized in such a way that its position at time $t + dt$ is defined unambiguosly by $F(\mathbf{r} + v_n\mathbf{n}\,dt, t + dt) = 0$, where $\mathbf{n} = \nabla F/|\nabla F|$ is the normal to the surface and v_n the normal velocity. This parametrization excludes such cases as cavitation at the interface. The actual fluid particles that define the surface may have tangential velocities that are not included in this geometrical definition.

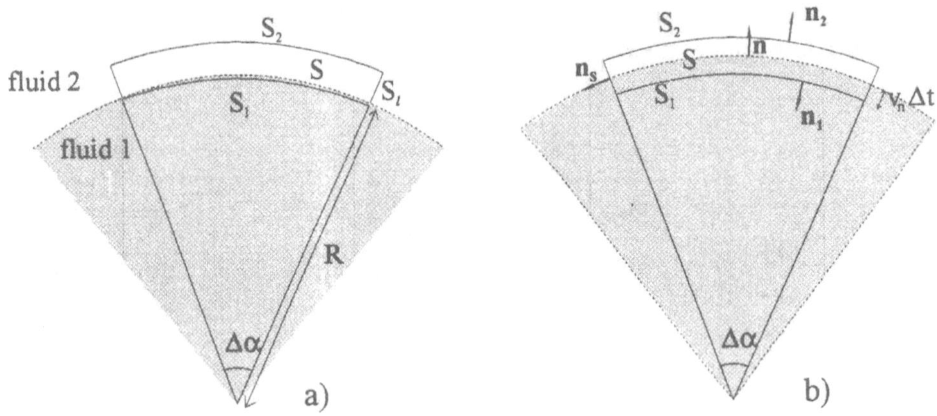

Fig. 1.2. A moving surface element: a) at time $t = 0$, and b) at time $t = \Delta t$.

For simplicity, consider a cylindrical pillbox of unit length in the plane perpendicular to the paper, enclosing the interface between region 1 and 2 in such a way that when $t = 0$ the lower face of the pillbox coincides with the interface, and contains all the surface charge (see fig. 1.2). Taking into account the singular distributions of charge and current density on the interface, the charge conservation law is expressed for the fixed pillbox as

$$\frac{\partial}{\partial t}\int_S q_s dS + \frac{\partial}{\partial t}\int_V q_s dV + \int_{C_s} \mathbf{K}_s \cdot \mathbf{n}_s ds + \int_{S_1 U S_2 U S_l} \mathbf{J} \cdot \mathbf{n}(dS \cdot 1) = 0, \qquad (1.70)$$

with q_s the surface charge density, q the charge density in the bulk, $\mathbf{K}_s$ the surface current density tangential to the interface, as measured by an inertial observer in the laboratory (hereafter the subindex "s" will denote a quantity proper of, or tangential to the surface, while the subindex "n" will mean normal to the surface). The current in the bulk of the fluids, $\mathbf{J}_i$ ($i = 1, 2$), may be expressed as $\mathbf{J}_i = q_i\mathbf{u}_i + \mathbf{J}'_i$, with $\mathbf{u}_i$ the velocity of fluid i, and $\mathbf{J}'_i$ the current density measured by a galilean observer in the instantaneous rest frame of the element of volume. Evaluation of these integrals is straightforward for a small surface element, giving for the first term in (1.70)

$$\begin{aligned}\frac{\partial}{\partial t}\int_S q_s dS &\simeq \frac{1}{\Delta t}[q_s(\mathbf{x}_s + v_n\Delta t, t + \Delta t)(R + v_n\Delta t)\Delta\alpha - q_s(\mathbf{x}_s, t)R\Delta\alpha] \\ &\simeq \left(\frac{\partial q_s}{\partial t} + v_n\mathbf{n} \cdot \nabla q_s + v_n\frac{1}{R}q_s\right)\Delta S. \qquad (1.71)\end{aligned}$$

R is the radius of curvature of the surface. For a general surface, the three-dimensional divergence of the normal vector is associated to the intrinsical mean curvature, K, through

$$\nabla \cdot \mathbf{n} = \left(\frac{1}{R_1} + \frac{1}{R_2}\right) = 2K, \tag{1.72}$$

and the last term of the equation (1.71) may be written as $v_n(\nabla \cdot \mathbf{n})q_s = 2Kv_nq_s$.

For the second term of (1.70) we have

$$\frac{\partial}{\partial t}\int_V q dV \simeq \frac{1}{\Delta t}\left[q_1(R\Delta\alpha)(v_n\Delta t) - q_2(R\Delta\alpha)(v_n\Delta t)\right] \simeq -[q]v_n\Delta S. \tag{1.73}$$

The third term gives

$$\int_{C_s} \mathbf{K}_s \cdot \mathbf{n}_s ds \simeq \nabla_s \cdot \mathbf{K}_s \Delta S. \tag{1.74}$$

The operator ∇_s must be understood as the ∇ operator minus its projection onto the normal direction, $\nabla_s = \nabla - \mathbf{n}(\mathbf{n} \cdot \nabla)$.

The result for the fourth term of (1.70) is

$$\int_{S_1 U S_2 U S_l} \mathbf{J} \cdot \mathbf{n}(dS \cdot 1) \simeq (-[q]v_n + \mathbf{n} \cdot [\mathbf{J}'])\Delta S, \tag{1.75}$$

where the usual condition $S_l \ll S_i$ has been taken into account.

Collecting all terms, and letting the limits Δt and ΔS tend to zero we finally obtain

$$\frac{\partial q_s}{\partial t'} + 2K\, v_n\, q_s + \nabla_s \cdot \mathbf{K}_s + \mathbf{n} \cdot [\mathbf{J}'] = 0, \tag{1.76}$$

where the primed time derivâtive means the derivative when we follow the surface along a direction normal to itself, that is

$$\frac{\partial}{\partial t'} = \frac{\partial}{\partial t} + v_n \mathbf{n} \cdot \nabla. \tag{1.77}$$

1.5 Electrical energy

The work required to bring a charge δQ against an electric field, $\mathbf{E}$, in a quasistatic process from infinity to the point $\mathbf{x}$ is given by

$$\delta W_e = -\delta Q \int_{\infty}^{\mathbf{x}} \mathbf{E}(\mathbf{x}') \cdot d\mathbf{x}' = \delta Q \int_{\infty}^{\mathbf{x}} \nabla\phi(\mathbf{x}') \cdot d\mathbf{x}' = \delta Q \phi(\mathbf{x}) \tag{1.78}$$

Therefore, the work required to build a given distribution of charge responsible for the existence of an electric field in a material medium is

$$\begin{aligned} W_e &= \int\int_0^q \delta q' \phi'(\mathbf{x}) dV = \int\int_0^{\mathbf{D}} (\nabla \cdot \delta\mathbf{D}')\phi'(\mathbf{x}) dV \\ &= \int\int_0^{\mathbf{D}} \nabla \cdot (\delta\mathbf{D}'\phi'(\mathbf{x})) dV - \int\int_0^{\mathbf{D}} \nabla\phi' \cdot \delta\mathbf{D}' dV \\ &= \int\int_0^{\mathbf{D}} \mathbf{E}' \cdot \delta\mathbf{D}' dV. \end{aligned} \tag{1.79}$$

The change in electrical energy is equal to the electrical work done to build the electric field, and the computation of this energy depends on the process. It is shown below that for a quasistatic isothermal process at constant mass density, this work may be easily computed, and depends only on the values of the electric field in the initial and final situations. It is postulated that this electrical energy is a state function of the system in the thermodynamic sense. In general, if this process takes place in finite time, the polarization may or may not follow the electric field. In the latter case part of the work is dissipated in the form of heat (see chapter 4), and it makes no sense to identify the work done with the increase of any thermodynamic state function.

Since, as shown in the following chapter, $\varepsilon = \varepsilon(\rho, T)$ with ρ the mass density and T the absolute temperature, we may find the electrical work needed to create an electric field in a quasistatic isothermal process at constant mass density. It is given by

$$W_e = \int \int_0^{\mathbf{D}} \frac{\mathbf{D}'}{\varepsilon} \cdot \delta \mathbf{D}' dV == \int \frac{1}{\varepsilon} \int_0^{\mathbf{D}} \mathbf{D}' \cdot \delta \mathbf{D}' dV = \int \frac{1}{2} \frac{\mathbf{D}^2}{\varepsilon} dV. \tag{1.80}$$

To maintain the fixed temperature, the dielectric material must be in contact with a heat bath. Due to the presence of heat flows we can not equate the work to the increase of internal energy of the dielectric. However, this work is equal to the change of free energy of the system, a thermodynamic potential, defined by

$$\mathcal{F} = \mathcal{U} - T\mathcal{S} = \int F dV - T \int S dV, \tag{1.81}$$

with $\mathcal{F}$, $\mathcal{U}$, and $\mathcal{S}$ the free energy, internal energy and entropy, and F, U, and S the corresponding quantities per unit volume. Denoting with a subscript o the value of these quantities in the absence of the electric field, we have

$$\mathcal{F} = \mathcal{F}_o + \int \frac{1}{2} \frac{\mathbf{D}^2}{\varepsilon} dV. \tag{1.82}$$

Given that

$$d\mathcal{F} = -\mathcal{S} dT + \int \mathbf{E} \cdot d\mathbf{D} dV, \tag{1.83}$$

we have

$$\mathcal{S} = -\left(\frac{\partial \mathcal{F}}{\partial T}\right)_{\mathbf{D}} = \mathcal{S}_o - \int \frac{1}{2} \mathbf{D}^2 \frac{\partial}{\partial T} \left(\frac{1}{\varepsilon}\right) dV = \mathcal{S}_o + \int \frac{1}{2} \mathbf{E}^2 \frac{\partial \varepsilon}{\partial T} dV. \tag{1.84}$$

The internal energy is

$$\mathcal{U} = \mathcal{F} + T\mathcal{S} = \mathcal{U}_o + \int \frac{1}{2} \mathbf{E}^2 \frac{\partial}{\partial T} (\varepsilon T) dV. \tag{1.85}$$

Therefore, the contribution of the electric field to the internal energy is

$$\mathcal{U}_e = \int \frac{1}{2} \mathbf{E}^2 \frac{\partial \varepsilon T}{\partial T} dV = \int \frac{1}{2} \mathbf{E}^2 \left[\varepsilon + \left(\frac{\partial \varepsilon}{\partial T}\right)_\rho\right] dV. \tag{1.86}$$

It is important to note that for those materials for which $\varepsilon = \varepsilon(\rho)$, that is for all fluids with molecules without permanent dipoles as we will see in the next chapter, the internal electrical energy reduces to

$$\mathcal{U}_e = \int \frac{1}{2}\varepsilon \mathbf{E}^2 dV. \tag{1.87}$$

For polar fluids, i. e. fluids with molecules endowed with permanent dipoles, a change of the electric field will give rise to an absorption of heat per unit volume given by

$$\delta Q = T\delta S_e = T\delta \left[\frac{1}{2}\mathbf{E}^2 \left(\frac{\partial \varepsilon}{\partial T}\right)_\rho\right]. \tag{1.88}$$

For example, if we create an electric field in a fluid in contact with a thermal bath there will be a heat absorbed by the body equal to

$$Q = T \left[\frac{1}{2}\mathbf{E}^2 \left(\frac{\partial \varepsilon}{\partial T}\right)_\rho\right]. \tag{1.89}$$

As shown in the next chapter is, $(\partial\varepsilon)/\partial T)_\rho < 0$, and consequently the heat is given off to the thermal bath. On the contrary, if the field is switched off the fluid absorbs energy from the thermal bath. This may be easily understood by taking into account, that for the latter situation where the dipoles have a preferred orientation, if the electric field is switched off, is randomised through collisions with the molecules of the fluid. This implies an increase of disorder and therefore an increase of entropy in the fluid.

In an adiabatic process the entropy must be kept constant. Hence, we must have

$$\rho c_p \ln T_i + \left[\frac{1}{2}\mathbf{E}^2 \left(\frac{\partial \varepsilon}{\partial T}\right)_\rho\right]_i = \rho c_p \ln T_f + \left[\frac{1}{2}\mathbf{E}^2 \left(\frac{\partial \varepsilon}{\partial T}\right)_\rho\right]_f. \tag{1.90}$$

where $c_p = T(\partial S_o/\partial T)_p$ is the specific heat at constant pressure, and i and f indicate the initial and final states. The adiabatic temperature change when we switch off an electric field in a polar liquid is then

$$\Delta T = T_i - T_f \sim \frac{-(1/2)E^2(\partial\varepsilon/\partial T)_\rho}{\rho c_p}T. \tag{1.91}$$

According to Onsager's formula for the permittivity of polar liquids (see next chapter) it is

$$\frac{(\varepsilon_r - n^2)(2\varepsilon_r + n^2)}{\varepsilon_r} = (n^2+2)^2 \frac{\rho p_o^2}{9k_B T}, \tag{1.92}$$

with ε_r the relative permittivity, n the index of refraction, p_o the permanent dipole of the molecule, ρ mass density, M the molecular weight, and k_B the Boltzmann constant. Therefore

$$e_1 = \frac{1}{\varepsilon}\left(\frac{\partial\varepsilon}{\partial T}\right)_\rho = -\frac{(2+n^2)(\varepsilon_r + 2n^2)}{3n^2}\frac{\rho p_0^2}{27 M k_B \varepsilon_0 T^2}. \tag{1.93}$$

Taking $\varepsilon_r = 40, n = 2, M = 40 \times 10^{-27}$ kg, $\rho = 10^3$kgm^{-3}, and $p_o = 2 \times 10^{-30}$ Cm, values typical of polar liquids, we obtain at ambient temperature $e_1 \simeq -2.5 \times 10^{-3}$. Because for liquids $c_p \sim 10^3$JkgK^{-1}, we see that switching off an initial field of the order of 10^6Vm^{-1} in a polar fluid, the variation in ΔT is of the order of 10^{-4}K. For polar gases the effect is smaller.

1.5.1 Electrical energy conservation equation

The conservation equation for the electrical energy in Electrohydrodynamics is a particular case of the Poynting theorem of electromagnetism. Multiplying the fourth Maxwell equation by $\mathbf{E}$ gives:

$$\mathbf{E} \cdot (\nabla \times \mathbf{H}) = \mathbf{E} \cdot \mathbf{J} + \mathbf{E} \cdot \frac{\partial \mathbf{D}}{\partial t}. \tag{1.94}$$

The identity

$$\nabla \cdot (\mathbf{E} \times \mathbf{H}) = \mathbf{H} \cdot (\nabla \times \mathbf{E}) - \mathbf{E} \cdot (\nabla \times \mathbf{H}), \tag{1.95}$$

reduces in EHD to

$$\nabla \cdot (\mathbf{E} \times \mathbf{H}) = -\mathbf{E} \cdot (\nabla \times \mathbf{H}), \tag{1.96}$$

as the electric field is irrotational. Using this relationship in equation (1.94) gives

$$\mathbf{E} \cdot \frac{\partial \mathbf{D}}{\partial t} + \nabla \cdot (\mathbf{E} \times \mathbf{H}) = -\mathbf{E} \cdot \mathbf{J}, \tag{1.97}$$

which may be written in the form

$$\frac{\delta W_e}{\delta t} + \nabla \cdot (\mathbf{E} \times \mathbf{H}) = -\mathbf{E} \cdot \mathbf{J}, \tag{1.98}$$

with

$$W_e = \int_0^{\mathbf{D}} \mathbf{E} \cdot \delta \mathbf{D}, \tag{1.99}$$

where the integral is over unit volume. For an arbitrary volume V, and recalling that for an isothermal process at constant mass density W_e can be integrated, we obtain

$$\frac{\partial \mathcal{F}_e}{\partial t} + \int_V (\mathbf{E} \times \mathbf{H}) \cdot \mathbf{n} dS = -\int_S \mathbf{E} \cdot \mathbf{J} dV, \tag{1.100}$$

where

$$\mathcal{F}_e = \int_V \frac{1}{2} \varepsilon E^2 dV, \tag{1.101}$$

is the electrical free energy of the system. This is the more common form of the conservation of electrical energy in EHD.

1.5.2 Uniqueness theorem for lossy media

Uniqueness theorems are very important in all branches of physics and mathematics, but unfortunately in most EHD problems we have to face the fact that there may be multiple solutions, and the one chosen by the system usually depends on its history and stability properties. However, in the case of lossy media, and for a given initial electric field, and given values of the tangential component of the electric field over the boundaries for all $t \geq 0$, the solution is unique.

Lossy media are defined as those for which

$$\int_V \mathbf{E} \cdot \mathbf{J} \, dV > 0, \tag{1.102}$$

holds for all physical admissible solutions $\mathbf{J}$ and $\mathbf{E}$. A very important class of lossy media are fluids in the ohmic regime.

First, suppose that there are two different solutions, $\mathbf{E}_1$, and, $\mathbf{E}_2$ satisfying the same initial and boundary conditions. The field $\mathbf{E}(\mathbf{x}, t) = \mathbf{E}_1(\mathbf{x}, t) - \mathbf{E}_2(\mathbf{x}, t)$ is also solution of the electrical equations satisfying homogeneous initial and boundary conditions, i.e., $\mathbf{E}(\mathbf{x}, 0) = 0$ in V, and $\mathbf{n} \times \mathbf{E}(\mathbf{x}_s, t)$ in S for $t \geq 0$. The latter condition implies that the flux of electromagnetic energy through S is zero for all $t \geq 0$, and the former that the initial energy of the electric field is zero at $t = 0$.

Equation (1.100) implies that, for these lossy media,

$$\frac{\partial \mathcal{F}_e}{\partial t} \leq 0, \qquad t \geq 0, \tag{1.103}$$

and since the energy is definite positive, and zero for $t = 0$, is zero for all times greater than zero. Therefore $\mathbf{E}$ must be zero, and consequently $\mathbf{E}_1 = \mathbf{E}_2$, i.e. the solution is unique.

REFERENCES

1. Landau L. M. and Lifshitz E. M.: Théorie des Champs, MIR, Moscow 1989.

2. Melcher J. R.: Continuum Electromechanics, The MIT Press, Cambridge, Massachusetts, 1981.

3. Landau L. M. and Lifshitz E. M.: Electrodynamique des Milieux Continus, MIR, Moscow 1989.

4. Panofsky W. K. H. and Phillips M.: Classical Electricity and Magnetism, Addison-Wesley, Reading, USA, 1962.

5. Van Vleck J. H.: The Theory of Electric and Magnetic Susceptibilities. Oxford University Press, London, 1932.

Chapter 2

CONDUCTION AND POLARIZATION

In the first part of this chapter we discuss the transport and generation of free charge carriers in a fluid subjected to an electric field. After formulating the charge conservation equation for each chemical species relevant to the process of electrical conduction, we discuss the concept of mobility. Then, the generation of ions is considered both in the bulk of the fluid, and at fluid-electrode interfaces, with emphasis on the physical scales responsible for the different processes. Simple injections laws will be formulated and discussed. More complicated injection laws will be considered by A. Zhakin in chapters 6 and 7.

In the second part of the chapter we examine the behavior of bounded or blocked charge carriers. We begin with a discussion of the reaction of individual atoms and molecules to the presence of an electric field. Then, the behavior of these same molecules as part of the fluid is examined . Although the models that we are going to discuss are simplifications of the reality, they give important insights on the fundamental variables that affect the behavior of the permittivity. A basic understanding of the equation of state for the permittivity is essential to answer several important questions in electrohydrodynamics. We end this chapter with a discussion of interfacial polarization, that is the predominant mechanism responsible for dielectric heating in EHD of nonhomogeneous fluids subject to a.c. fields of low frequency (well below the microwave frequencies).

This chapter focuses on the basic ideas of conduction and polarization, and obviously does not cover the complexity of these processes in real fluids. Conduction phenomena are further discussed by Zhakin and Watson in chapters 5, 6, 7, and 10 respectively. For gases we refer the reader to [1], and for liquids to [2] where additional information may be found, particularly in chapters 1 and 2 written by Barnikas and Schmidt respectively.

2.1 Electrical conduction

When a d.c. electrical potential difference is applied across a fluid by means of two electrodes, an electrical current flows between them. This current is composed of two terms, one is due to transport of free charge carriers present in the liquid, and the other is the displacement current, which is given by $\partial \mathbf{D}/\partial t$. The first term is called the conduction current and depends on the type and number of charged species, its velocity and the chemical reactions that they may undergo. In what follows the current density **J** will refer only to the conduction current.

2.1.1 The charge conservation equation

Let us denote the number density of species of type k by n_k, the velocity by $\mathbf{u}_k$ and the sources and sinks by P_k. The charge, which may be zero or an integer (positive or negative) of the absolute value of the electron charge $e = 1.6 \times e^{-19}$C, we will denote by e_k. For any fixed volume V we must have that the rate of change of n_k within V is equal to the flow of species, n_k, through the surface limiting the volume V plus the production of n_k due to chemical reactions inside V

$$\int_V \frac{\partial n_k}{\partial t} dV + \int_S n_k \mathbf{u}_k \cdot \mathbf{n} dS = \int_V P_k dV. \tag{2.1}$$

Applying Gauss's theorem it follows that,

$$\int_V \left[\frac{\partial n_k}{\partial t} + \nabla \cdot (n_k \mathbf{u}_k) \right] dV = \int_V P_k dV. \tag{2.2}$$

Since this is true for an arbitrary volume

$$\frac{\partial n_k}{\partial t} + \nabla \cdot (n_k \mathbf{u}_k) = P_k. \tag{2.3}$$

By mass conservation,

$$\sum_1^N P_k = 0, \tag{2.4}$$

where the sum extends to all species either charged or neutral.

Representing by d/dt the material (substantive) derivative or rate of change of any physical quantity, A, following a fluid element, which is given by (see chapter 3)

$$\frac{dA}{dt} = \frac{\partial A}{\partial t} + (\mathbf{v} \cdot \nabla) A, \tag{2.5}$$

we also have

$$\frac{dn_k}{dt} + n_k \nabla \cdot \mathbf{u}_k = P_k. \tag{2.6}$$

We define in the usual way, the charge density q and the current density **J** in terms of the number density n_i and velocities $\mathbf{u}_i$ of the M charged species ($M \leq N$) present in our physical system,

$$q = \sum_1^M e_i n_i, \qquad \mathbf{J} = \sum_1^M e_i n_i \mathbf{u}_i. \tag{2.7}$$

Then, multiplying equation (2.3) by e_k, remembering that this is zero for neutral species, and summing we have

$$\sum_{1}^{M} e_i P_i = 0, \tag{2.8}$$

because there can be no net generation of charge, and

$$\frac{\partial q}{\partial t} + \nabla \cdot \mathbf{J} = 0, \tag{2.9}$$

which is the charge conservation equation. To proceed further we need to know the type, the number density n_i, and the velocity of each species of charge carriers.

2.1.2 Transport of charge carriers

Charge carriers are typically electrons and ions. The behavior of electrons in gases or liquids is quite complicated. Its generation may be due to ionizing collisions with the neutral molecules and ions, injection from the cathode, photoionization, secondary emission at the cathode, etc. Its velocity is also strongly dependent on the local electric field, and its computation is a formidable problem [1],[2]. Fortunately, there exist many situations where the electrons are quickly trapped by electronegative impurities or molecules of the fluid. For these situations we may consider that the positive and negative ions are the only charge carriers present in the fluid. It is then possible to approximate the behavior of these ions as spheres moving in a continuum, following the laws of an ideal gas of particles in the volume occupied by the fluid.

Consider a ion of charge e_i, positive or negative, in a fluid at rest. There are three forces acting upon the ion. The Coulomb force, $e_i\mathbf{E}$, due to the presence of the electric field $\mathbf{E}$. The friction force due to the surrounding fluid, which, in our approximation, is given by the Stokes law, $6\pi\eta R_h \mathbf{u}_i$, where η is the dynamic viscosity of the fluid, R_h the equivalent hydrodynamic radius of the ion, and $\mathbf{u}_i$ the relative velocity of the ion with respect to the liquid. The ion, together with the cloud of molecules, possibly adhered to it by dipolar forces (*solvated ion*), defines the hydrodynamic radius R_h. Finally, we have the force acting upon the ion due to the partial pressure of other ions, which according to the ideal gas law is given by $p = n_i kT$ with n_i the number density of ions of the type i per unit volume. Therefore, Newton's law becomes

$$m\frac{d\mathbf{u}_i}{dt} = e_i\mathbf{E} - 6\pi\eta R_h \mathbf{u}_i - \frac{1}{n_i}\nabla(n_i k_B T). \tag{2.10}$$

The inertia term compared to the friction term gives us the time needed for the ion to reach its limit velocity in the fluid. From $m/\tau_{iv} \sim 6\pi\eta R_h$ we have

$$\tau_{iv} = \frac{m}{6\pi\eta R_h}. \tag{2.11}$$

Even for fluids with low viscosity, such as water, for which $\eta = 10^{-3}\mathrm{kg\,m^{-1}s^{-1}}$ we have for an ion such as OH^-, with an ionic radius of the order of 2×10^{-10}m, that it is $\tau_{iv} \sim 10^{-14}$s.

Hence, we may neglect the inertia term. Consequently, the velocity of the ion in a fluid at constant temperature is given by

$$\mathbf{u}_i = \mu_i \mathbf{E} - \frac{k_B T}{e_i \mu_i} \frac{1}{n_i} \nabla n_i, \tag{2.12}$$

where we have introduced the mobility μ_i given by

$$\mu_i = \frac{e_i}{6\pi\eta R_H}. \tag{2.13}$$

Notice that for negative ions μ_i is negative. This constant is also obtained equating the driving force, $e_i\mathbf{E}$, to the resistive force, $6\pi R_h \eta \mathbf{u}_i$. Taking $R_h \sim 10^{-9}$ m, it is $\mu_i \eta \sim 10^{-11}$ C/m. This relation is known as *Walden's rule*. The product $\mu_i\eta$ is not rigorously constant for a given ion in different solvents, because solvation is dependent upon the permittivity of the liquid. Nevertheless, it is correct in order of magnitude. Thus, for a viscosity $\eta \sim 10^{-2}$ kg/m s, we have for the absolute value of mobility $\mu_i \sim 10^{-8}$m^2/Vs, and for an electric field $E \sim 1$ kV/cm we find $u \sim 10^{-3}$ m/s. Since the Reynolds number for the flow around the ion is $Re = \rho\eta u / R_h \sim 10^{-7} \ll 1$, the use of the Stokes formula is justified.

The second term in (2.12), which is proportional to the gradient of the ion concentration, has the classical form of a diffusion term. We may then put equation (2.12) in the more recognizable way

$$\mathbf{u}_i = \mu_i \mathbf{E} - D_i \frac{1}{n} \nabla n \tag{2.14}$$

with D_i the coefficient of molecular diffusion. Both forms are equivalent. In effect, for a liquid at rest in thermodynamic equilibrium, $\mathbf{u}_i = 0$. Also, according to Boltzmann's law we have $n_i \propto \exp -e_i\phi / k_B T$, and from (2.12), taking into account that $\mathbf{E} = -\nabla\phi$, we obtain

$$\frac{D_i}{\mu_i} = \frac{k_B T}{e_i}, \tag{2.15}$$

with k_B the Boltzmann constant. This is known as Einstein's relation. For an ion of charge e_i, $D_i/\mu_i = 1/40$ V, at ambient temperature, and unless the electric field is of order of 0.025 V/m, diffusion may be safely neglected in the bulk of liquids.

The mobility of ions in gases and liquids is quite different due to the different structure of liquid and gases. In gases the viscosity arises due to momentum transfer across a plane by the molecules above and below that plane that are within a distance of the mean free path, which is much larger than the size of a molecule. In a liquid, on the contrary, the molecules are so closely packed that only the molecules adjacent to the plane produces this transfer of momentum. In gases, if we increase the temperature, at constant pressure, we increase the momentum transfer since each molecule has an average velocity that goes as $(k_B T)^{1/2}$. Therefore, the viscosity is an increasing function of the temperature, and the mobility is a decreasing function of T. If we increase the pressure, at constant temperature, the viscosity should be approximately the same, since although the number of molecules participating in the momentum transfer increases, simultaneously the mean free path decreases in the same proportion. In liquids it is experimentally found that the viscosity is always a decreasing function of the temperature. Therefore, in liquids the mobility increases as the temperature increases. Because of the incompressibility of liquids the mobility is, to a first approximation, independent of pressure.

Until now we have considered the fluid at rest. For a fluid in motion the ions are advected with the flow velocity $\mathbf{u}$. Then, each charge carrier moves in the liquid with a velocity, which is the sum of three terms

$$\mathbf{u}_i = \mu_i \mathbf{E} - D_i \frac{\nabla n_i}{n_i} + \mathbf{u}. \tag{2.16}$$

The first term gives the contribution due to the drift of ions that move relative to the liquid with velocity $\mu_i \boldsymbol{E}$ with μ_i its ionic mobility. The second term accounts for molecular diffusion, with D_i the molecular charge diffusion coefficient, and n_i the ion number density. The third term is due to convection of charge carriers by the liquid velocity field.

For the case where we have only charge carriers of the type i, the current density is given by

$$\mathbf{J} = \mu_i q_i \mathbf{E} - D_i \nabla n_i + q_i \mathbf{u} = \mathbf{J}' + q_i \mathbf{u}. \tag{2.17}$$

with $q_i = e_i n_i$ the charge density. In general

$$\mathbf{J} = \sum e_i (\mu_i n_i \mathbf{E} - D_i \nabla n_i) + q\mathbf{u} = \mathbf{J}' + q\mathbf{u}. \tag{2.18}$$

with $q_i = \sum e_i n_i$.

In most circumstances these charge carriers originate from dissociation of impurities in the bulk, and/or are injected into the liquid through the electrodes. In the following we make a quick review of these two processes.

2.1.3 Dissociation

In gases ions appear in pairs as the result of natural processes such as natural radiation, mainly due to cosmic rays. When the electric field is very high, ions may be produced massively by ionizing collisions. The nature of the gas determines the creation of ions in the bulk through processes at the atomic level [1].

In liquids the creation of ions by dissociation is governed by physical and chemical processes. At low values of the electric field, the origin of the conductivity is due to dissociation of impurities contained in the liquid, or of the liquid itself. The contribution of impurities, or electrolytes, to the liquid resistivity is different depending on their nature. We discuss these processes below.

Ionophore electrolytes are ionic salts and, thus, ions exist even before the impurity is introduced in the liquid. The action of the solvent is to break the ionic lattice by means of ion–dipole interactions and to solvate the free ions. We have

$$(\mathrm{A^+B^-}) \underset{\alpha_\mathrm{r}}{\overset{\alpha_\mathrm{d}}{\rightleftarrows}} \mathrm{A^+ + B^-}, \tag{2.19}$$

where (A^+B^-) stands for the concentration of ionic pairs, A^+ and B^- for the concentration of free ions, and α_d, α_r are the dissociation and recombination coefficients.

Ionogen electrolytes are formed by covalent bonds and they are unable to produce free ions by themselves. Nevertheless, these impurities may react with the solvent or with other substances in the dissolution to generate an ionophore electrolyte. Therefore, the dissociation of ionogen impurities includes a first step of chemical character that will depend on the specific nature of reactants:

$$C + D \underset{\alpha_2}{\overset{\alpha_1}{\rightleftarrows}} (A^+B^-) \underset{\alpha_r}{\overset{\alpha_d}{\rightleftarrows}} A^+ + B^-, \tag{2.20}$$

where C and D stand for the impurity molecule and/or another impurity species or solvent, and α_1, α_2 are kinetic rate constants.

The equilibrium between the ionic pairs and free ions can be described by means of models that include only physical interactions. At any moment,

$$-\frac{dc}{dt} = \frac{dn_+}{dt} = \frac{dn_-}{dt} = \alpha_d c - \alpha_r n_+ n_-, \tag{2.21}$$

where c is the concentration of ionic pairs, n_+ and n_- the concentrations of cations and anions. In the thermodynamic equilibrium

$$\alpha_d c_0 = \alpha_r n_+ n_- = \alpha_r n_0^2, \tag{2.22}$$

holds since the solution must be electroneutral. The subscript "0" indicates values at thermodynamic equlibrium.

The length scales appearing in the problem are: a, the size of the ions; $n^{-1/3}$, the average distance of ions in the solution; $l_B = e^2/4\pi\varepsilon k_B T$, called the Bjerrum distance, is the distance at which the Coulombian interaction electrostatic energy between two ions of opposite polarity equals their thermal energies; $l_O = (e/4\pi\varepsilon E)^{1/2}$, the Onsager distance, at which the force due the the external electric E equals the attractive force between two ions of opposite polarity; $l_E = k_B T/2eE$, the distance at which the electric potential difference between two opposite ions due to an external field, E, equals their thermal energies; and $\lambda_D = (\varepsilon kT/ne^2)^{1/2}$, called the Debye length, is the radius of the ionic atmosphere. Due to the Coulombian interaction each ion tends to be surrounded by ions of opposite polarity. For distances greater than λ_D the electrostatic potential due to the ion is screened and it is no longer equal to the Coulombian potential.

For those situations for which $n^{-1/3} \ll \lambda_D$, we also have $l_B \ll n^{-1/3}$ and the interactions between the two ions of an ionic pair are Coulombian.

Debye [3] obtained the recombination constant evaluating the collision frequency of an ion subjected to Brownian motion and the Coulomb interaction with the rest of ions (see chapter 5). He found

$$\alpha_r = \frac{e\,(\mu_+ + \mu_-)}{\varepsilon}(1 - \exp(-l_B/a))^{-1}, \tag{2.23}$$

where e is the absolute electronic charge, ε the permittivity of the liquid, $\mu_+ = \mu_i$ for $e_i > 0$, and $\mu_- = -\mu_i$ for $e_i < 0$, are the mobilities of cations and anions, a the distance between ions in an ionic pair and l_B the Bjerrum distance. Then, for $l_B \gg a$ the recombination constant reduces to the Langevin's constant for gases [4],

$$\alpha_r = \frac{e(\mu_+ + \mu_-)}{\varepsilon}. \tag{2.24}$$

Now, if $l_B/a \gg 1$, the majority of ions are associated in ionic pairs, while if $l_B/a \leq 1$ ions stand free. Liquids for which is $l_B \gg a$ ($l_B \leq a$) are called nonpolar liquids (polar liquids). At ambient temperatures the transition takes place around $\varepsilon \simeq 2$. In polar liquids, all the impurities will be dissociated in the form of free ions, and therefore the density of ions is determined from the first step of equation (2.20). In nonpolar liquids, there will exist a dynamic equilibrium between free ions and ionic pairs, i.e., cations and anions associated by Coulomb forces. The dissociation will be favored if l_B decreases.

The dissociation constant may be obtained from the ionic association constant at thermodynamic equilibrium, α_A, defined through $c_0 = \alpha_A n_0^2$, and given by

$$\alpha_A = \frac{c_0}{n_0^2} = \frac{\alpha_r}{\alpha_d}. \tag{2.25}$$

Following Fuoss [5] we write

$$\alpha_A = \frac{4}{3}\pi a^3 \exp(l_B/a), \tag{2.26}$$

and consequently

$$\alpha_d = \frac{3e(\mu_+ + \mu_-)}{4\pi\varepsilon a^3} \exp(-l_B/a). \tag{2.27}$$

The rate of generation of free ions is modified by the presence of an external electric field, E. The exact evaluation of this effect is due to Onsager [6] (see chapter 5 in this same volume), who completed the dissociation constant in the following way,

$$\alpha_d(E) = \alpha_d^0 F(b) = \alpha_d^0 \frac{I_1(2b)}{b}, \tag{2.28}$$

with $b = l_B/l_O$, α_d^0 the rate of dissociation at zero field and I_1 the modified Bessel function of first kind and order one. In nonpolar liquids and at high fields ($E \geq 10^6$ V/m) this effect may increase the conductivity of a solution by several orders of magnitude. On the other hand, the recombination constant is almost unaffected by the presence of the electric field.

In the limit $b \ll 1$ and $\lambda_D \gg l_B \gg a$, the Thomson model [7] is applicable. In this model we have two distinct regimes, depending whether the ions recombine or not in the volume.

Ohmic regime: the electric field does not appreciably modify the thermodynamic equilibrium given by equation (2.22), and we have $n^+ = n^- = n_0 = (\alpha_d c/\alpha_r)^{1/2}$. Hence, the current density is given by

$$\mathbf{J} = e(\mu_+ + \mu_-)\sqrt{\frac{\alpha_d c}{\alpha_r}}\mathbf{E} = \sigma\mathbf{E}, \tag{2.29}$$

where μ_+ and μ_- are the mobilities of positive and negative ions, respectively. The current is proportional to the applied field, with σ the conductivity, given by $\sigma = e(\mu_+ + \mu_-)(\alpha_d c/\alpha_r)^{1/2}$.

Saturation regime: in this limit the ions drift so fast that the probability of recombining with each other is negligible. Hence, the current density reaches the limiting value $J_s = e\alpha_d c_0 d$, independent of the electric field. Here d is the distance between the electrodes.

Field-enhanced dissociation: for intense electric fields we arrive at the point where b is no longer negligible and we have enhanced dissociation by the electric field. We have two possibilities; either we are in the ohmic regime and $J = \sigma(E)E$ with $\sigma(E) = \sigma_0(F(b))^{1/2}$ and σ_0 the conductivity at zero electric field, or else we have already attained the saturation regime and this gives $J_s = J_s^0 F(b)$. In both cases the current density may increase by several orders of magnitude.

2.1.4 Injection

The metal/fluid interface is a region of special interest. First, electron transfer reactions at the electrodes, make possible the neutralization of incoming ions from the bulk, thus allowing the continuous flux of current. Second, ions with the same polarity as the electrode are created and injected into the fluid, a process known as *ion injection.* These injections of ions can take place at one or both of the electrodes (unipolar or ambipolar). We may inject ions by many different mechanisms, corrosion chemical reactions, impact ionization, ionizing radiations, etc. Here, it will be assumed that the first step is always an electron transfer from the metal electrode to a molecule of the fluid, or viceversa. This electronic transfer may be enhanced by controlled local irradiation with ultraviolet rays, x-rays, etc. But we will be interested in those process linked to the contact metal-fluid.

Emission ef electrons from a metal into the vacuum is well known. One mechanism is thermoionic emission enhanced by lowering the potential barrier with an electric field and correcting for the image force. This is called Schottky injection; we have for the emitted current density

$$J = J_o \exp\left(\frac{BE^{1/2}}{T}\right), \tag{2.30}$$

with J_o the extrapolated value of the current for zero electric field. The constant B takes the value $0.441\ \mathrm{Km^{1/2}V^{-1/2}}$. In this process the electrons with enough high energy overcome the potential barrier energy between the metal and the vacuum. However, for very intense electric fields the width of the potential barrier is so diminished that the electrons can tunnel through the barrier. This process, first discussed by Fowler and Nordheim, is called field emission. The emitted current density is given by

$$J = \frac{1.54 \times 10^{-6}}{W} E^2 \exp\left(-\frac{6.83W^{3/2}}{E}\right), \tag{2.31}$$

with W the height of the potential barrier above the Fermi level of the metal, measured in electron-volts. This injection law begins to dominate above $E \sim 10^8 \mathrm{Vm}^{-1}$. See [8] for a good review of these field emission processes.

In gases, the Schottky emission or field emission are the dominant mechanisms. However, there is an upper limit to the current that may be injected because of the space-charge limitation of the electric field. This may be understood in the following way: above a certain threshold of the electric field, the electrons in the bulk of the gas have enough energy to ionize the molecules of the gas. As a result, a large number of positive and negative carriers are created. In uniform fields these ionization processes will cause breakdown of the gas, but in nonuniform fields, such as those produced by a wire or a needle in front of a coaxial or of a plane electrode, the region of ionization may be limited to the neighbourhood of the part of the electrode with the smallest radius of curvature. Ions of opposite polarity to that of the electrode are collected, while ions of the same polarity drift toward the other electrode, giving rise to an important space charge that decreases the value of the electric field. This may be enough to lower the value of the electric field below the critical value, E_c, defining the onset of ionization (the corona onset). Once the space charge has drifted to the other electrode the electric field increases again and the emission begins. In this way we may have an oscillating current (Trichel pulses). Under stationary conditions an equilibrium is reached such that the value of the electric field at the emitting electrode is approximately constant (Peek's law) and equal to E_c.

In liquids Halper and Gomer, [9], [10], have shown that field emission occurs at very intense values of the electric field. At such high values the electrical double layer is destroyed and tunneling of electrons from the metal to the molecule (field emission) or from the molecule to the metal (field ionization) takes place. However, in most situations of practical interest in EHD, the electric fields are not intense enough, and the electrical double layer mediates the electronic transfer at the interface. In what follows, we will disregard field emission and field ionization as injection mechanisms in liquids.

We only consider simple mechanisms of injection, mediated by the electrical double layer. We distinguish two cases: nonpolar and polar liquids.

Injection by electrochemical reactions in liquids

Prior to the application of the electric field, neutral molecules, ions and ionic pairs attach to the metal surface by chemical and/or electrostatic image forces. Electrons are transferred from the metal to the molecules, ions and ionic pairs or vice versa, giving rise to an electrical double layer. There is a first layer of absorbed solvent molecules and to a lesser extend also molecules, ions and ionic pairs originated from the dissolved impurities. This first layer define a plane called the inner Helmholtz plane. Here the absorption can be due to chemical bonds and it may happen that the absorbed ions have the same charge as the electrode. A second layer, called the outer Helmholtz plane, is formed with solvated ions of polarity opposed to the electrode. Both layers define the compact double charged sublayer, also called the Stern layer, and its width, l_H, is of the order of the solvated ionic radius. The potential drop across the compact sublayer is of the order of 0.1 V, and the electric fields are extremely intense ($\sim 10^8$ to 10^9 V/m). Another region is defined by the distance $l_B/4$ where the ions are subjected to their image force. This distance is obtained by equating the interaction energy of ion-metal, $-e^2/16\pi\varepsilon x$, with the ion thermal energy. In addition, the electrode may be considered as a charged sphere with a radius $R \to \infty$, and its ionic atmosphere is characterized by the Debye length λ_D. The latter is called the diffuse layer. In what follows, we will consider that $l_B \ll \lambda_D$ holds. Finally, in the presence of an external imposed field, there is a potential barrier given by $\phi(x) = -e/16\pi\varepsilon x - xE$. The distance to the electrode where $\phi(x)$ is a maximum is $l_O/2$.

Again we have to distinguish two cases. Polar liquids for which $l_B \leq l_H$ and therefore the ions, once produced, are immediately injected into the liquid, and nonpolar liquids for which to inject the ions in the bulk, they need to overcome the potential barrier created by the image force and the external field. For simplicity, we will consider the case of pure of pure injection, i.e. if there is dissociation in the bulk, the density of dissociated ions in the vicinity of the injecting electrode have a negligible influence on the local electric field. If this is not the case, more general models need to be considered (chapter 6, section 6.4.2).

Injection in polar liquids

Suppose that ions are injected following the reaction [11],

$$\mathrm{M} + \mathrm{A} \underset{\alpha_{\mathrm{rd}}}{\overset{\alpha_{\mathrm{ox}}}{\rightleftharpoons}} \mathrm{M(e)} + \mathrm{A}^+, \tag{2.32}$$

with M denoting the metal, and A the electron donor molecule in the liquid phase. Let us denote with a superscript o the values at thermodynamic equilibrium of the species concentration per unit area, n_A^o at $x = 0$ and n_{A+}^o at $x = l_H$, and the oxidation and reduction constants α_{ox}^o, α_{rd}^o. The dynamical equilibrium represented by equation (2.32) requires that:

$$J_{\mathrm{o}} = e\alpha_{\mathrm{ox}}^o n_A^o = J_{\mathrm{r}} = e\alpha_{\mathrm{rd}}^o n_{A+}^o = J_{\mathrm{s}}. \tag{2.33}$$

In the presence of an external field, transfer of electrons to the metal is favored. The new oxidation and reaction constants take the values

$$\alpha_{\mathrm{ox}} = \alpha_{\mathrm{ox}}^o \exp\left[\frac{eEl_H}{k_BT}\right], \qquad \alpha_{\mathrm{rd}} = \alpha_{\mathrm{rd}}^o \exp\left[-\frac{eEl_H}{k_BT}\right]. \tag{2.34}$$

Some of the ions created at $x = l_H$ are injected into the bulk of the liquid. Denoting the number density of the injected ions by $n^{\rm i}_{A+}$, we have

$$e\alpha^o_{\rm ox}\exp\left[\frac{eEl_H}{k_BT}\right]n_A = e\alpha^o_{\rm rd}\exp\left[-\frac{eEl_H}{k_BT}\right]n_{A+} + en^{\rm i}_{A+}\mu_+E. \tag{2.35}$$

If the concentration of neutral molecules at the surface electrode is constant, i.e. $n^o_A = n_A$, it follows that $n_{A+} = Cn^o_A$, $n^{\rm i}_{A+} = Cn^o_A$. Denoting by $q_{\rm i}$ and $J_{\rm i}$ the injected charged density and the injected current density, given by $q_{\rm i} = en^{\rm i}_{A+}/l_H$ and $J_{\rm i} = q_{\rm i}\mu_+E$ respectively, we obtain

$$\frac{J_{\rm i}}{J_{\rm s}} = \frac{\exp(\vartheta E/E_{\rm s})}{E/E_{\rm s} + \exp(-\vartheta E/E_{\rm s})}\frac{E}{E_{\rm s}}, \tag{2.36}$$

with $E_{\rm s} = \alpha^o_{\rm rd}l_H/\mu_+$ and $\vartheta = el_HE_{\rm s}/k_BT$.

There are two distinct limits depending on the ratio $\vartheta E/E_{\rm s}$. If this ratio is much less than one, the electric field does not modify the voltage drop across the double layer, and we obtain for the current density

$$\frac{J_{\rm i}}{J_{\rm s}} = \frac{\vartheta E/E_{\rm s}}{1 + E/E_{\rm s}}. \tag{2.37}$$

If, in addition, $E \ll E_s$, $J_{\rm i}$ is proportional to the electric field. Though the current density is proportional to the electric field, as in the ohmic regime case, transient currents are different for both regimes due to the different spatial distributions of the charge carriers.

The injected charge density is given by

$$q_{\rm i} = \frac{J_{\rm s}/E_{\rm s}}{\mu_+(1 + E/E_{\rm s})}. \tag{2.38}$$

This charge density is constant, i.e. *autonomous injection*, for values of $E \ll E_s$ only. Since $E_s \sim 10^5$V/m typically, and injection needs high fields, autonomous injection is the exception, rather than the rule, in polar liquids.

The other limit is characterized by $\vartheta E/E_{\rm s} \gg 1$, in which case we have

$$J_{\rm i} = J_{\rm s}\exp(\vartheta E/E_{\rm s}), \qquad q_{\rm i} = (J_{\rm s}/E_{\rm s}\mu_+)\exp(\vartheta E/E_{\rm s}), \tag{2.39}$$

and both the current density and the injected charge increase exponentially.

Injection in nonpolar liquids

In nonpolar liquids ion injection may be described by a two-step process: first the ions are created at the electrodes, and second followed the ions escape out of the image force region [12]. This reaction may be written as follows,

$$\mathrm{M} + (\mathrm{A}^+\mathrm{B}^-) \underset{\alpha_{\rm rd}}{\overset{\alpha_{\rm ox}}{\rightleftarrows}} \mathrm{M(e)} + (\mathrm{A}^+\mathrm{B}) \overset{\alpha_{\rm ext}}{\longrightarrow} \mathrm{A}^+_{\rm free}, \tag{2.40}$$

where M represents the metal that captures an electron (e) from the ionic pair. In the kinetics of this reaction, the oxidation and reduction constants of the ionic pair are $\alpha_{\rm ox}$ and $\alpha_{\rm rd}$, and the extraction constant of the ion from the image force is $\alpha_{\rm ext}$.

If the density of ionic pairs is constant, the injection process is governed by

$$\left.\begin{array}{r}\dfrac{dn_+^{\rm free}}{dt} = \alpha_{\rm ext} n_+ \\ (\alpha_{\rm rd} + \alpha_{\rm ext}) n_+ = \alpha_{\rm ox} n\end{array}\right\} \Rightarrow \quad \frac{dn_+^{\rm free}}{dt} = \alpha_{\rm ext}\frac{\alpha_{\rm ox}}{\alpha_{\rm rd}+\alpha_{\rm ext}} n, \tag{2.41}$$

The injected current density will then be

$$J_{\rm i} = e\alpha_{\rm ext} n_+ = \alpha_{\rm ext}\frac{\alpha_{\rm ox}}{\alpha_{\rm rd}+\alpha_{\rm ext}} en, \tag{2.42}$$

where n and n_+ are now the surface density of ionic pairs and ions formed by electron transfer at the electrode.

The rate of extraction of ions may be determined by studying the transport mechanisms participating at the interface: ionic migration and diffusion. It is

$$J_{\rm i} = -\mu_+ q(x)\frac{d\phi(x)}{dx} - D_+\frac{dq(x)}{dx}, \tag{2.43}$$

where $q(x)$ is the space charge density at a distance x of the electrode, D_+ the diffusion coefficient and $\phi(x)$ the electrical potential given by

$$\phi(x) = \frac{-e}{16\pi\varepsilon x} - xE. \tag{2.44}$$

Equation (2.43) is easily integrated after multiplying by the factor $exp(e\phi/k_BT)$. Using Einstein's relation, $D_+/\mu_+ = k_BT/e$, we may write for the case where the electric field is not too intense, i.e. $l_O = \sqrt{e/16\pi\varepsilon E} \gg l_H$ and therefore the lower limit in the integral may be taken as zero,

$$J_{\rm i} = q(l_H)\frac{\exp\left(e\phi(l_H)/k_BT\right)}{b\,K_1(b)}\mu_+ E, \tag{2.45}$$

where $b = l_B/l_O$ and K_1 the modified Bessel function of the second kind and order one. Identifying the space charge density at the distance l_H, $q(l_H)$, with en_+/l_H gives [13]

$$\alpha_{\rm ext} = \frac{\mu_+ E}{l_H}\frac{\exp\left(e\phi(l_H)/k_BT\right)}{b\,K_1(b)}, \tag{2.46}$$

Finally, the expression for the injected current may finally be written as

$$J_{\rm i} = \alpha_{\rm ext}\frac{\alpha_{\rm ox}^o \exp(eEl_H/2k_BT)}{\alpha_{\rm ext} + \alpha_{\rm rd}^o \exp(-eEl_H/2k_BT)} en. \tag{2.47}$$

If the extraction kinetics is quite fast compared to the electronic transfer we recover the exponential law of injection obtained for the polar liquids. But in most cases, the stage of extraction is the slower process and the current density may be simplified to

$$J_{\rm i} = \alpha_{\rm ext}\frac{\alpha_{\rm ox}}{\alpha_{\rm rd}} en. \tag{2.48}$$

Since the injected current density in the bulk is given by

$$J_{\rm i} = q_{\rm i}\mu_+ E, \tag{2.49}$$

the injected charge density in the liquid at the electrode is

$$q_{\rm i} = \frac{en\,\alpha^o_{\rm ox}}{l_H\,\alpha^o_{\rm rd}}\frac{\exp(-e^2/16\pi\varepsilon l_H k_B T)}{b\,K_1(b)} = \frac{q^0_{\rm i}}{b\,K_1(b)}, \tag{2.50}$$

with $q_{\rm i}^0$ a constant that depends on the liquid and electrolyte nature. This expression has been successfully verified [12], [14], [15], in a large variety of organic solvents and salts up to fields of 2×10^7 V/m. Finally it is to be noticed that for values of $10^5 \leq E \leq 4.10^6$ V/m we may safely assumed that injection is *autonomous*.

2.2 Polarization

Polarization was introduced in chapter 1 in a formal way as the solution of equation $\rho_b = -\nabla \cdot \mathbf{P}$ with ρ_b the density of bounded charges. Here we define the polarization vector as an average of the molecular dipoles when the fluid is subject to an electric field. Since fluids are isotropic, the vector $\mathbf{P}$ will be in the direction of the applied field.

$$\mathbf{P}_m = \chi_e \mathbf{E}. \tag{2.51}$$

This relation defines χ_e called the electrical susceptibility. We consider first a single molecule subjected to an electric field, and then the same molecule as part of a fluid.

2.2.1 Polarizability

Because of the internal forces between the constituents atoms of a molecule the center of gravity of the positive charge distribution may or may not be coincident with the center of the negative charge distribution. In the first case we refer to the molecule as nonpolar, while in the second case the molecule, endowed with a permanent electric dipole, is called polar. In both cases, when the molecule is subjected to an electric field, the centers of gravity of the positive and negative charge distributions are displaced in opposite directions until a new balance of the electrical forces with the restoring internal force is achieved. As a result a new electric dipole appear. For values of the electrical fields much smaller than the internal molecular fields, this induced dipole is proportional to the electric field.

$$\mathbf{p}_m = \alpha \mathbf{E}, \tag{2.52}$$

with α called the polarizability of the molecule.

Two different mechanisms contribute to the induced dipole. The electronic polarization, which is the separation of the center of gravity of the electronic cloud from the nucleus, and the ionic one, which is the separation of two atoms or groups of atoms that were already partially or totally ionized previous to the application of the electric field.

Consider the electronic polarization. An exact computation of this dipole requires a quantum mechanical computation that it is feasible only for simple atoms. However, an elementary classical model in which the electron is modeled as a charge density cloud, with the nucleus displaced a small distance from its center, allows us to obtain the correct order of magnitude for the induced dipole

$$\mathbf{p}_m = 4\pi\varepsilon_0 a^3 \mathbf{E} = \alpha_e \mathbf{E} \tag{2.53}$$

with a the atomic radius, and $\alpha_e = 4\pi\varepsilon_0 a^3$ the electronic polarizability. For the hydrogen atom $a = 0.53 \times 10^{-10}$m the Bohr radius, and $\alpha_e = 1.6 \times 10^{-41}$Fm^2, in good agreement with the measured value. The electronic polarizability increases as the radius increases, and it is comprised between one tenth of this value (for the smaller positive ions) and one hundred (for the largest negative ions).

Consider now the dipole induced due to the ionic polarization, for a diatomic molecule composed of two ions of charge q separated a distance d_o in the absence of an external electric field. When we subject the molecule to a field $\mathbf{E}$, along the axis of the molecule, the new equilibrium is given by

$$qE = k(d - d_o) = k\delta d \tag{2.54}$$

with $k = m\omega_o^2$, the oscillator constant, m the reduced mass of the molecule, $1/m = 1/m_1 + 1/m_2$, and ω_o the angular vibration frequency of the diatomic molecule. For values of the electric field much lower than the interatomic electric field we may take the frequency ω_o as the natural frequency, which is typically in the short wave-length infra-red. We have

$$\delta d = qE/k, \tag{2.55}$$

and therefore the induced electric moment will be

$$p_m = q\delta d = \frac{q^2}{k}E = \frac{q^2}{m\omega_o^2}E = \alpha_i E. \tag{2.56}$$

The ionic polarizability is given by $\alpha_i = q^2/(m\omega_o^2)$. Taking into account that the typical vibrational frequencies are in the infrared, the values for α_i are of the same order or smaller than the electronic polarizability.

Consider now a molecule with permanent dipole, $\mathbf{p}_o$, in thermal equilibrium at temperature T. In the absence of an external electric field, and due to thermal motions the orientation of the molecule varies at random with a zero ensemble average for any given direction in space. If we subject the molecule to an electric field, the molecule will tend to align with the electric dipole parallel to the field. Thermal agitation opposes a preferential direction, but under conditions of thermal equilibrium, the probability for the molecule of having energy W is given by the Boltzmann factor $\exp(-W/k_B T)$ with k_B the Boltzman constant and T the absolute temperature. From elementary considerations the energy of the molecule in the electric field is $W = -\mathbf{p}_o \cdot \mathbf{E}$.

The average permanent dipole in the direction of the electric field $\mathbf{E} = E\mathbf{e}_z$ is, then, given by $<\mathbf{p}_o>=<p_{oz}>\mathbf{e}_z$ with

$$<p_{oz}>=p_0<\cos\theta>=p_o\frac{\int_0^\pi \cos\theta \exp(p_oE\cos\theta/k_BT)2\pi\sin\theta d\theta}{\int_0^\pi \exp(p_oE\cos\theta/k_BT)2\pi\sin\theta d\theta}. \qquad (2.57)$$

Defining $x = \cos\theta$, $y = p_oE/k_BT$ the integral is elementary and the result is

$$p_0<\cos\theta>=p_oL(y), \qquad (2.58)$$

with $L(y)$ the Langevin function given by

$$L(y) = \coth y - \frac{1}{y}. \qquad (2.59)$$

At ambient temperatures is

$$y \simeq \frac{10^{-30}}{300\times 1.38\,10^{-23}}E = 2.4\times 10^{-10}E, \qquad (2.60)$$

and even for very intense fields $E \sim 10^8\mathrm{Vm}^{-1}$ is $y \ll 1$. Hence, the average dipolar permanent dipole is given by

$$\mathbf{p}_o = \frac{p_o^2}{3k_BT}\mathbf{E} = \alpha_{\mathrm{or}}\mathbf{E}. \qquad (2.61)$$

Values of p_o are within the order of 1 to ten times 10^{-30}Cm and therefore at ambient temperatures the values of α_{or} are comprised between $10^{-40}\,\mathrm{Fm}^2$ and $10^{-38}\,\mathrm{Fm}^2$, to be compared with α_e or α_i, both of the same order and within the range $(10^{-41}, 10^{-39})\,\mathrm{Fm}^2$. For example, for isolated water molecules is $p_o = 6.13\times 10^{-30}\,\mathrm{Cm}$.

Summarizing the results, we have

$$<\mathbf{p}>=(\alpha_e+\alpha_i+\alpha_{\mathrm{or}})\mathbf{E} = \alpha\mathbf{E}, \qquad (2.62)$$

with

$$\alpha = \alpha_e + \alpha_i + \frac{p_o^2}{3k_BT}. \qquad (2.63)$$

2.2.2 Permittivity

The polarizability refers to the behavior of a single molecule, while the polarization vector, $\mathbf{P}$, is an average over many molecules. Both quantities are related by

$$\mathbf{P} = N\alpha\mathbf{E}_m, \qquad (2.64)$$

with N the number of molecules per unit volume, α the polarizability of the molecule, and $\mathbf{E}_m$ the average electric field at the molecular position.

From the classical relations

$$\mathbf{D} = \varepsilon_0 \mathbf{E} + \mathbf{P}, \qquad \mathbf{D} = \varepsilon \mathbf{E}, \tag{2.65}$$

we have, $\mathbf{P} = (\varepsilon - \varepsilon_0)\mathbf{E}$, and therefore

$$(\varepsilon - \varepsilon_0)\mathbf{E} = N\alpha \mathbf{E}_m, \tag{2.66}$$

with $\mathbf{E}$ the average electric field in the fluid.

The relation between the electric field averaged over the volume of a molecule and the macroscopic field, which is an average over a small volume from the macroscopic point of view, but still containing a large number of molecules, determines the relation between the permittivity and the molecular polarizability.

Gases at ordinary or low pressures

When the electric field due to the induced or permanent dipole of the molecules of the gas at any given position is negligible compared to the macroscopic field, we may safely take $\mathbf{E}_m = \mathbf{E}$. For this simple case

$$(\varepsilon - \varepsilon_0) = N\alpha = N\left(\alpha_e + \alpha_i + \frac{p_o^2}{3k_B T}\right). \tag{2.67}$$

For gases under standard conditions the number N of molecules per unit of volume is the Loschmidt number, $N \simeq 2.68 \times 10^{25}\mathrm{m}^{-3}$, and since α is of order $10^{-39}\,\mathrm{Fm}^2$ we have $\varepsilon \simeq \varepsilon_0$.

Gases at high pressures and liquids

In liquids, and gases at high pressures, is not any longer true that the contribution of the molecules at a given point is negligible, and we have to account for this interaction.

Lorentz model: Lorentz modeled the fluid as a continuum of permittivity ε and polarization $\mathbf{P}$ outside a sphere of radius R, and, inside the sphere, as a true collection of molecules. The contribution of the continuum to the field at the center of the cavity, $\mathbf{E}_c$, is due to the surface polarization charges $\sigma_P = -\mathbf{P}\cdot\mathbf{n}$. The calculation of $\mathbf{E}_c$ is an elementary problem of electrostatics [16]. Applying Coulomb's law, integrating the surface charge density over the surface of the sphere R, we have $\mathbf{E}_c = \mathbf{P}/3\varepsilon_0$, for the field at the center of the cavity.

The contribution due to the molecules, within the sphere of radius R, is quite difficult to compute. Fortunately, there are situations where this contribution is negligibly small. In particular, this holds true if the distribution of molecules is isotropic or random inside the sphere and the molecular dipoles are identical. Then

$$\mathbf{E}_m = \mathbf{E} + \frac{\mathbf{P}}{3\varepsilon_0} = \left(1 + \frac{\varepsilon - \varepsilon_0}{3\varepsilon_0}\right)\mathbf{E}. \tag{2.68}$$

Substituting this expression in equation (2.66) we end up with

$$\frac{\varepsilon - \varepsilon_0}{\varepsilon + 2\varepsilon_0} = \frac{N\alpha}{3\varepsilon_0} = \left(\frac{N_A \alpha}{3M\varepsilon_0}\right)\rho, \tag{2.69}$$

with $N_A = 6.022 \times 10^{23}$ the Avogadro number, M the molecular weight, and ρ the mass density. This relation, known as the Clausius-Mossoti relation, is valid for a wide class of nonpolar fluids and weakly polar liquids.

Onsager model: for polar fluids with large permanent dipoles, the Clausius-Mossoti relation is not accurate enough, and a refinement of the Lorentz model is needed. Onsager modeled the fluid as a continuum with permittivity ε, with the molecule in a cavity of its size, and computed the average molecular dipole field in the medium in three steps. In the first two steps he computed the electric field acting upon the molecule, and its instantaneous dipole. In step 1 the permanent dipole of the molecule is introduced in the cavity in the absence of the macroscopic electric field $\mathbf{E}$, and the resulting induced plus permanent dipole is computed. In step 2, first the molecule is absent and the electric field is turn on. Then, the molecule is introduced without its permanent dipole, and the new induced dipole is computed. In step 3 the final dipole is the sum of the dipoles obtained in steps 1 and 2, and the average of this final dipole is found following the Langevin model.

In the absence of the electric field, the permanent dipole of the molecule of magnitude $\mathbf{p}_0$ polarizes the continuum, thus producing a field at the position of the molecule. This new electric field, called the reaction field $\mathbf{E}_{\text{or}}$, induces in turn a dipole in the molecule proportional to its electronic and ionic polarizability. It is then the total dipole (in the absence of the macroscopic field $\mathbf{E}$), denoted also with the subscript "or",

$$\mathbf{p}_{\text{or}} = \mathbf{p}_o + \alpha_{ei}\mathbf{E}_{\text{or}}, \tag{2.70}$$

where we have denoted by $\alpha_{ei} = \alpha_e + \alpha_i$, the sum of electronic and ionic polarizabilities.

The electric field due to a dipole, $\mathbf{p}_{\text{or}}$, in the center of a cavity, is a classical problem in potential theory [16], and the result is

$$\mathbf{E}_{\text{or}} = f\mathbf{p}_{\text{or}}, \qquad f = \frac{2}{4\pi\varepsilon_0 a^3} \times \frac{\varepsilon - \varepsilon_0}{2\varepsilon + \varepsilon_0}. \tag{2.71}$$

It is therefore $\mathbf{p}_{\text{or}} = \mathbf{p}_o + \alpha_{ei} f \mathbf{p}_{\text{or}}$ and we have

$$\mathbf{p}_{\text{or}} = \frac{1}{1 - \alpha_{ei} f}\mathbf{p}_o, \qquad \mathbf{E}_{\text{or}} = \frac{f}{1 - \alpha_{ei} f}\mathbf{p}_o. \tag{2.72}$$

Notice that for a nonpolarizable molecule is $\alpha_{ei} = 0$ and it will be $p_{0r} = p_o$.

In step 2, the electric field $\mathbf{E}$ is now turned on in the absence of the molecule. The field in the spherical cavity is [16]

$$\mathbf{E}_c = g\mathbf{E}, \qquad g = \frac{3\varepsilon}{2\varepsilon + \varepsilon_0} \tag{2.73}$$

Let us now introduce the molecule without its permanent dipole $\mathbf{p}_o$. Initially, we have a field $\mathbf{E}_c$, that will induce a new dipole $\mathbf{p}_c = \alpha_{ei}\mathbf{E}_c$. This new dipole will create a reaction field, that in turn will modify the previous induced dipole. By the same arguments of step 1 we arrive at

$$\mathbf{p}_{\rm cr} = \alpha_{ei}(\mathbf{E}_c + \mathbf{E}_{\rm cr}) = \alpha_{ei}(\mathbf{E}_c + f\mathbf{p}_{\rm cr}), \tag{2.74}$$

hence

$$\mathbf{p}_{\rm cr} = \frac{g}{1-\alpha_{ei}f}\mathbf{E}, \tag{2.75}$$

and the electric field in the cavity is

$$\mathbf{E}_c + f\mathbf{p}_{\rm cr} = g\mathbf{E} + \frac{\alpha_{ei}g}{1-\alpha_{ei}f}\mathbf{E} = \frac{g(1+\alpha_{ei}) - \alpha_{ei}fg}{1-\alpha_{ei}f}\mathbf{E} \simeq \mathbf{E}_c = g\mathbf{E}, \tag{2.76}$$

as α_{ei} is negligible compared to 1. This is the end of step 2.

The total dipole of the molecule is $\mathbf{p}_m = \mathbf{p}_{\rm or} + \mathbf{p}_{\rm cr}$. Performing the average over the dipole, according to the Langevin model,

$$< p_{m_z} >=< p_{\rm or}\cos\theta + p_{\rm cr} >= \frac{\int_0^\pi (p_{\rm or}\cos\theta + p_{\rm cr})e^{-W/k_BT}2\pi\sin\theta d\theta}{\int_0^\pi e^{-W/k_BT}2\pi\sin\theta d\theta}, \tag{2.77}$$

with

$$W = -(\mathbf{p}_{\rm or} + \mathbf{p}_{\rm cr})\cdot(\mathbf{E}_c + \mathbf{E}_{0r}) = -gp_{\rm or}E\cos\theta + W_o, \tag{2.78}$$

the electrical energy of the dipole. The constant W_o factors out of the integrals, and defining $x = cos\theta$, $y = gp_{\rm or}E/k_BT$, the integrand may be expanded using a Taylor series in y, and keeping only the term linear in y, results

$$< \mathbf{p}_m >= \frac{g}{1-\alpha_{ei}}\left[\alpha_{ei} + \frac{p_o^2}{3(1-\alpha_{ei})fk_BT}\right]\mathbf{E}. \tag{2.79}$$

Then, from $\mathbf{P} = N < \mathbf{p}_m >$, we finally get

$$\varepsilon - \varepsilon_0 = \frac{Ng}{1-\alpha_{ei}}\left[\alpha_{ei} + \frac{p_o^2}{3(1-\alpha_{ei})fk_BT}\right]. \tag{2.80}$$

Since the cavity corresponds to the volume of a molecule, $N = 1/[(4/3)\pi a^3]$, and $f = 2N(\varepsilon - \varepsilon_0)/9\varepsilon_0$. Equation (2.80) is known as the Onsager relation [17]. Onsager's model and method of calculation yields the Clausius-Mossoti relation for nonpolar liquids.

The Onsager's relation needs to be corrected for those liquids with strong correlation between the molecules in some specific directions. This is the case of water, where the molecules form chains linked with hydrogen bonds. Kirkwood and Frölich [18] found that it was necessary to correct for this anisotropy with a factor

$$h = 1 + < \sum_j \cos\theta_{ij} >, \tag{2.81}$$

with the sum extended to all molecules contained in a small volume around molecule i. This allowed them to explain the anomalous high value of the water permittivity.

Permittivity as function of frequency

Some comments are due about the behavior of the permittivity as a function of the frequency of an applied alternating electric field. According to the mechanisms discussed above, we see that the response of the molecule to the orientational polarizability ceases at frequencies above the rotational frequency spectra of the molecule, which is found in the microwave region. The response to the ionic polarization ceases above the vibrational frequencies, typically found in the short wavelength of the infrared region, and the electronic polarizability ceases above the ultraviolet range. In the optical range both the orientational and ionic polarizabilities are zero, and the permittivity at theses frequencies is denoted by ε_∞. Taking into account the definition of index of refraction as $n^2 = \varepsilon_\infty/\varepsilon_0$ the Clausius-Mossoti relation is now

$$\frac{n^2-1}{n^2+2} = \frac{N\alpha_e}{3\varepsilon_0}. \tag{2.82}$$

Substracting this equation, known as Lorentz-Lorentz equation, from equation (2.69) we obtain

$$\frac{\varepsilon_r - n^2}{\varepsilon_r + 2n^2} = (n^2+2)\frac{Np_0^2}{27k_BT}, \tag{2.83}$$

with ε_r the relative permittivity at zero frequency.

For polar liquids we substract equation (2.82) from the Onsager's relation equation (2.80), and we obtain

$$\frac{(\varepsilon_r - n^2)(2\varepsilon_r + n^2)}{\varepsilon_r} = (n^2+2)^2\frac{Np_0^2}{27k_BT}, \tag{2.84}$$

known as Onsager's formula for the permittivity. For those liquids with strong correlation between the molecules in some specific directions, the right hand side of this formula has to be multiplied by the factor defined by equation (2.81), according to the correction introduced by Kirkwood and Frölich. Since in EHD the alternating fields have a frequency that usually does not go beyond the MHz's range, all three mechanisms act instantaneously and the permittivity may be taken as the static permittivity.

2.2.3 Interfacial polarization: Maxwell-Wagner effect

Until now we have considered a fluid with a uniform composition. However if we have interfaces which either solid-liquid (particles inmersed in the fluid) or liquid-liquid (drops of other liquid, bubles of gas) or simply a capacitor with two layers of different dielectric liquids, the ions may be stopped at the interface. These charges gives rise to what is called the interfacial polarization or the Maxwell-Wagner effect.

To illustrate this effect we consider a plane parallel capacitor, subject to an alternating electric potential difference $\phi(t) = Re(\phi e^{i\omega t})$, in which there are two layers of different conductivities σ_l and permittivities ε_l with $l = 1, 2$. According to the fourth Maxwell equation, we have in each fluid

$$\nabla \times \mathbf{H} = \mathbf{J}_l + i\omega\varepsilon_l\mathbf{E} = i\omega\varepsilon_l^*\mathbf{E}, \tag{2.85}$$

with i the imaginary number, $i = \sqrt{-1}$, ω the frequency of the applied potential, and

$$\varepsilon_l^* = \varepsilon_l - i\frac{\sigma_l}{\omega}. \tag{2.86}$$

Alternatively, we may write

$$\nabla \times \mathbf{H} = \mathbf{J}_l + i\omega\varepsilon_l \mathbf{E} = \sigma_l^* \mathbf{E}, \tag{2.87}$$

with $\sigma_l^* = \sigma_l + i\omega\varepsilon_l$ a complex conductivity.

Suppose that d_l is the thickness of dielectric l with $d_1 + d_2 = d$. We may view the system as two capacitors in serie with capacitances per unit area $C_l = \varepsilon_l^*/d_l$. The capacitance of the system is given by $1/C = 1/C_1 + 1/C_2$. Therefore

$$\frac{d_1 + d_2}{\varepsilon^*} = \frac{d_1}{\varepsilon_1^*} + \frac{d_2}{\varepsilon_2^*}, \tag{2.88}$$

and the permittivity may be expressed in the form

$$\varepsilon^* = \varepsilon_\infty + \frac{\varepsilon_s - \varepsilon_\infty}{1 + i\omega\tau} - i\frac{\varepsilon_0}{\omega\tau_o}, \tag{2.89}$$

with

$$\varepsilon_\infty = d\frac{\varepsilon_1\varepsilon_2}{(\varepsilon_1 d_2 + \varepsilon_2 d_1)}, \qquad \varepsilon_s = d\frac{\sigma_1\sigma_2(\varepsilon_1 d_2 + \varepsilon_2 d_1)}{(\sigma_1 d_2 + \sigma_2 d_1)^2}, \tag{2.90}$$

and

$$\tau = \frac{\varepsilon_1 d_2 + \varepsilon_2 d_1}{\sigma_1 d_2 + \sigma_2 d_1}, \qquad \tau_o = \frac{\varepsilon_0}{d}\frac{\sigma_1 d_2 + \sigma_2 d_1}{\sigma_1\sigma_2}. \tag{2.91}$$

The time τ is a combination of the individual relaxation time of each of the liquids $\tau_l = \varepsilon_l/\sigma_l$.

The factor $1 + i\omega\tau$ comes from the exponential relaxation of the interfacial charge density in the time domain, and it is similar to the relaxation processes of the polarization for the other mechanisms as postulated by Debye [19]. The last term is the contribution of the finite conductivity. This effect is of great practical importance whenever there are interfaces, and as mentioned above is the main mechanism responsable for dielectric losses in EHD of highly insulating liquids.

For liquids with a sufficiently high concentration of ions, this simple mechanism is greatly complicated due to the presence of electrical double layers, which contribute to a surface conductance along the interfaces present in our system. The polarization of these double layers may introduce a gigantic enhancement of the effective dielectric permittivity. The interested reader may find a description of these polarization mechanisms and related

REFERENCES

1. Raizer Y. P.: Gas Discharge Physics. Springer-Verlag, Berlin, 1991.
2. Bartnikas R. (editor).: Engineering Dielectrics. Vol III: Electrical Insulating Liquids, ASTM, Philadelphia, 1994.
3. Debye P.: Reaction rates in ionic solutions, Trans. Electrochem. Soc., 82(1942), 262–272.
4. Langevin P.: Recombinaison et mobilités des ions dans les gaz, Annales de Chimie et de Physique, 28(7) (1903), 433.
5. Fuoss R. M.: Ionic association. III. The equilibrium between ion pairs and free ions, J. Am. Chem. Soc., 80(5) (1958), 5059–5061.
6. Onsager L. Deviations from Ohm's law in weak electrolytes, Journal of Chemical Physics, 2 (1934), 599–615.
7. Thomson J. J. and Thomson G. P.: Conduction of Electricity through Gases, Cambridge University Press, third edition, 1928.
8. Gosse R. H. and Muller W. W.: Field emission, in: Encyclopedia of Physics XXI Springer-Verlag 1956, 232–303..
9. Halpern B. and Gomer R.: Field emission in liquids, J. Chem. Phys., 51(3) (1969), 1031–1047.
10. Halpern B. and Gomer R.: Field ionization in liquids, J. Chem. Phys., 51(3) (1969), 1048–1056.
11. Felici N. J.: A tentative explanation of the voltage-current characteristic of dielectric liquids. Journal of Electrostatics, 12 (1982), 165–172.
12. Alj A., Denat A., Gosse J. P. and Gosse B.: Creation of charge carriers in nonpolar liquids, IEEE Transactions on Electrical Insulation, EI-20(2) (1985), 221–231.
13. Gosse J. P., Gosse B. and Denat A.: La conduction électrique des liquides diélectriques, Revue General de l'Électricité, 10 (1985), 733–744.

14. Denat A., Gosse B. and Gosse J. P.: High field dc and ac conductivity of electrolyte solutions in hydrocarbons, Journal of Electrostatics, 11 (1982),179–194.

15. Nemancha M., Gosse J. P., Denat A. and Gosse B.: Temperature dependence of ion injection by metallic electrodes into non-polar liquids, IEEE Transactions on Electrical Insulation, EI-22(4) (1987), 459–465.

16. Scaife B. K. P.: Principles of Dielectrics. Clarendon Press, Oxford, 1989.

17. Onsager L. Electric moments of molecules in liquids, J. of the American Chemical Society, 58, (1936), 1486-1493.

18. Frölich H.: Theory of Dielectrics, Oxford Univ. Press, Oxford 1968.

19. Debye P.: Polar molecules, Dover, 1947

20. Russel W.B., Saville D.A. and Schowalter W.R.: Colloidal Dispersions. Cambridge University Press, Cambridge, UK, 1989.

21. Lyklema J.: Fundamentals of interface and colloid science. Vol. 2. Academic Press, 1995.

Chapter 3

MECHANICAL EQUATIONS

Ordinary fluids may be treated as a continuous media over a wide range of pressures and temperatures. In this chapter we present the fundamentals such as the deformation and the stress tensor required to understand any deformable fluid medium. This is followed by the introduction of the conservation laws for mass and momentum in a general setting. Two good texts in fluid mechanics, where the concepts given here are discussed in great detail, are [1], [2]. A separate section, following the argument of Landau & Lifshitz [3], considers the important problem of determining the electrical forces in dielectric fluids, with either a linear or a nonlinear polarization, in the context of thermodynamics. The boundary and jump conditions are stated for electrodes and interfaces, paying special attention to the capillary phenomena. Finally, some important nondimensional numbers found in EHD are discussed in the last section.

3.1 Mechanical equations

A fluid, liquid or gas, is composed of an enormous number of molecules. Under standard conditions (1 atmosphere, 273 K) the typical separation between molecules is of the order of $d_o \simeq 3.5 \times 10^{-10}$m in liquids and $10d_o$ in gases. Measuring instruments probe local volumes with linear dimensions several orders of magnitude larger than d_o, but still quite small by any macroscopic standard. Denoting this volume by δV, the mass density ρ, and the velocity $\mathbf{u}$, are given by:

$$\rho = \frac{\sum_{k=1}^{N} m_k}{\delta V}, \qquad \mathbf{u} = \frac{< \sum_{k=1}^{N} m_k \mathbf{u}_k >}{\sum_{k=1}^{N} m_k}, \tag{3.1}$$

where N is the number of molecules in volume δV, and m_k, $\mathbf{u}_k$, are the mass and velocity of each molecule respectively.

3.1.1 Continuum hypothesis

The fluid may be modeled as a continuum, where density and velocity are smooth functions of position, if the following inequalities hold:

$$(\delta V/N)^{1/3} \ll (\delta V)^{1/3} \ll l, \tag{3.2}$$

with l a typical scale of variation of any macroscopic quantity. In particular, if l is defined either as $\rho/\nabla_i\rho$, or as $u_j/\nabla_i u_j$ where $i, j = 1, 2, 3$, for any arbitrary chosen point in the fluid at a given time, the above inequality must be true for all points and over all time.

3.1.2 Kinematics of fluids

There are two ways to specify the position of a given element of fluid. In the first, called the Lagrangian description, the position, $\mathbf{x}$, of a particle at time t is specified as a function of its initial position, $\mathbf{x}(0) = \mathbf{x}_0$, at $t = 0$, i.e., $\mathbf{x} = \mathbf{x}(\mathbf{x}_0, t)$. In the second, called Eulerian and used in this text, the position of the particle is described as a function of the independent variables $\mathbf{x}$, and t. Any physical magnitude is also a function of these variables.

Consider how a given magnitude, $f(\mathbf{x}, t)$, varies in time for a fixed particle of fluid. This is defined as the material derivative of f, given by

$$\frac{df}{dt} = \frac{f(\mathbf{x} + d\mathbf{x}, t + dt) - f(\mathbf{x}, t)}{dt} = \frac{\partial f}{\partial t} + \mathbf{u} \cdot \nabla f, \tag{3.3}$$

where we have used the fact that $d\mathbf{x}(\mathbf{x}, t) = \mathbf{u}(\mathbf{x}, t)dt$ is the displacement of the particle during the time interval dt. In particular, for the acceleration of a particle of fluid, we have

$$\mathbf{a} = \frac{d\mathbf{u}}{dt} = \frac{\partial \mathbf{u}}{\partial t} + \mathbf{u} \cdot \nabla \mathbf{u}. \tag{3.4}$$

Local variations of the mass density and velocity fields (on the macroscopic scale) are given by $\nabla\rho$ and $\nabla\mathbf{u}$. The latter is usually decomposed in a symmetric and an antisymmetric part

$$\nabla\mathbf{u} = \frac{1}{2}\left(\nabla\mathbf{u} + \nabla\mathbf{u}^T\right) + \frac{1}{2}\left(\nabla\mathbf{u} - \nabla\mathbf{u}^T\right) = \mathbf{e}_i e_{ij} \mathbf{e}_i + \mathbf{e}_i \omega_{ij} \mathbf{e}_i, \tag{3.5}$$

where $\nabla\mathbf{u}^T$ is the transpose tensor, and $\mathbf{e}_i$, $i = 1, 2, 3$, denote the unit vectors of a cartesian system of coordinates. It follows that:

$$e_{ij} = \frac{1}{2}\left(\frac{\partial u_i}{\partial x_j} + \frac{\partial u_j}{\partial x_i}\right), \qquad \omega_{ij} = \frac{1}{2}\left(\frac{\partial u_i}{\partial x_j} - \frac{\partial u_j}{\partial x_i}\right). \tag{3.6}$$

Let us take a particle of fluid as the origin of a comoving frame of reference, and consider a neighbouring particle at distance $\Delta\mathbf{l}(0)$ at time $t = 0$. At time Δt it will be at position $\Delta\mathbf{l}(\Delta t) \simeq \Delta l(0) + (\nabla\mathbf{u} \cdot \Delta\mathbf{l})\Delta t$. In the limit $\Delta t \to 0$, replacing all increments by infinitesimals, and defining $\mathbf{n} = d\mathbf{l}/dl$, we have for the rate of dilatation of $d\mathbf{l}$ (see fig. 3.1.a)

$$\frac{\mathbf{n} \cdot \nabla\mathbf{u} \cdot \mathbf{n} dl dt}{dt} = (\mathbf{n} \cdot \mathbf{e} \cdot \mathbf{n} + \mathbf{n} \cdot \boldsymbol{\omega} \cdot \mathbf{n})dl = \mathbf{n} \cdot \mathbf{e} \cdot \mathbf{n} dl, \tag{3.7}$$

since $\mathbf{n}\cdot\boldsymbol{\omega}\cdot\mathbf{n} = \mathbf{e}_i n_i n_j \omega_{ij} \mathbf{e}_j = 0$ as $n_i n_j$ is a symmetric tensor and ω_{ij} is antisymmetric. Taking $dl = dx_i \mathbf{e}_i$ the rate of dilatation of volume dV is

$$\frac{dV(dt) - dV(0)}{dt} = (e_{11} + e_{22} + e_{33})dV(0) = e_{ii}dV(0) = (\mathrm{Tr}\ \mathbf{e})dV = (\nabla\cdot\mathbf{u})dV. \quad (3.8)$$

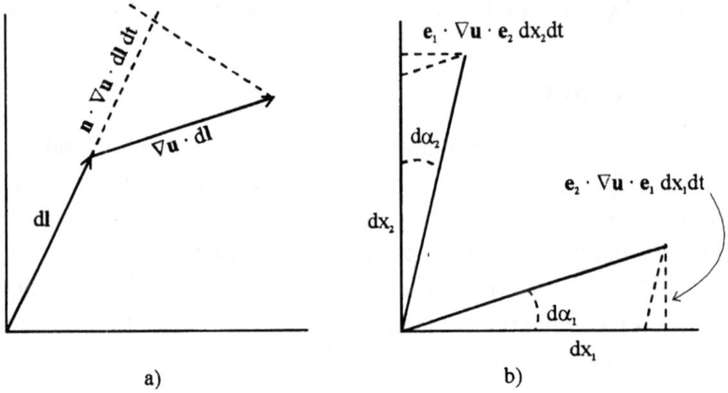

Fig. 3.1. Rate of deformation of an elementary area

From fig. 3.1.b, the rate of deformation of the elementary area $dx_1 \times dx_2$ is given by

$$\frac{d\alpha}{dt} = \frac{d\alpha_1}{dt} + \frac{d\alpha_2}{dt} = \frac{\mathbf{e}_2\cdot\nabla\mathbf{u}\cdot\mathbf{e}_1 dx_1 dt}{dx_1 dt} + \frac{\mathbf{e}_1\cdot\nabla\mathbf{u}\cdot\mathbf{e}_2 dx_2 dt}{dx_2 dt} = 2e_{12}. \quad (3.9)$$

If the tensor $\mathbf{e}$ is identically zero, the elements of fluid can not deform nor dilate, but they can still rotate as a rigid body. This is evident from the identity

$$\boldsymbol{\omega}\cdot d\mathbf{x} = \mathbf{e}_i w_{ij} dx_j = \frac{1}{2}(\nabla\times\mathbf{u})\times d\mathbf{x}, \quad (3.10)$$

which defines the angular velocity of rotation as $(\nabla\times\mathbf{u})/2$ with $\nabla\times\mathbf{u}$, known as the vorticity, defined in the usual way

$$\nabla\times\mathbf{u} = \mathbf{e}_i\, \varepsilon_{ijk}\, \omega_{jk}, \quad (3.11)$$

The tensor ε_{ijk} is the completely antisymmetric Levi-Civita tensor. In components, $\varepsilon_{ijk} = 1$ if ijk is an even permutation of 123, $\varepsilon_{ijk} = -1$ if ijk is an odd permutation of 123, and zero otherwise.

In conclusion, the tensor $\mathbf{e} = \mathbf{e}_i e_{ij} \mathbf{e}_j$ gives the rate of deformation and/or dilatation of the element of fluid, and $\boldsymbol{\omega} = \mathbf{e}_i \omega_{ij} \mathbf{e}_j$ gives the rotation of that element as a rigid body.

The mechanical equations that govern the dynamics of liquids are the equations for the conservation of mass and momentum.

3.1.3 Conservation of mass

The equation of mass is derived by considering an arbitrary fixed closed surface through which the fluid is flowing. Then, the decrease with time of the mass in the volume enclosed by the surface must be equal to the flux of mass through the surface:

$$\int_V \frac{\partial \rho}{\partial t} dV + \int_S \rho \mathbf{u} \cdot \mathbf{n} dS = 0. \tag{3.12}$$

Applying Gauss's theorem it follows that,

$$\int_V \left[\frac{\partial \rho}{\partial t} + \nabla \cdot (\rho \mathbf{u}) \right] dV = 0. \tag{3.13}$$

Since this is true for every control volume

$$\frac{\partial \rho}{\partial t} + \nabla \cdot (\rho \mathbf{u}) = 0. \tag{3.14}$$

Expanding the divergence term gives

$$\frac{\partial \rho}{\partial t} + \rho \nabla \cdot \mathbf{u} + (\mathbf{u} \cdot \nabla)\rho = 0, \tag{3.15}$$

or

$$\frac{d\rho}{dt} + \rho \nabla \cdot \mathbf{u} = 0. \tag{3.16}$$

For isentropic flows, pressure and density variations are related by

$$\frac{d\rho}{dt} = \left(\frac{\partial \rho}{\partial p} \right)_S \frac{dp}{dt} = \frac{1}{{c_s}^2} \frac{dp}{dt}, \tag{3.17}$$

where p is the pressure, S is the entropy and c_s is the velocity of sound in the fluid. Taking $\rho(l/\tau)^2$ as the scale for the pressure, we have $\Delta\rho/\rho \sim (v/c_s)^2$. Consequently, for fluid velocities much smaller than the velocity of sound, ρ may be taken as constant, and the mass conservation equation reduces to

$$\nabla \cdot \mathbf{u} = 0. \tag{3.18}$$

When this condition holds we say that the fluid is incompressible.

3.1.4 Conservation of momentum

Take an arbitrary fixed closed surface through which the fluid is flowing. The decrease with time of the linear momentum, in the volume enclosed by the surface, must be equal to the flux of momentum convected by the fluid through the surface, plus the rate of change of momentum due to volume and surface forces acting upon the fluid (Newton's second law),

$$\int_V \frac{\partial \rho \mathbf{u}}{\partial t} dV + \int_S \rho \mathbf{u}\mathbf{u} \cdot \mathbf{n} dS = \int_V \mathbf{f} dV + \int_S \mathbf{f}_S dS, \tag{3.19}$$

where $\mathbf{f}$ is the force per unit volume and $\mathbf{f}_S$ is the force per unit area acting on S.

Now, let us consider the forces exerting on a volume of fluid by the fluid surrounding it. Because the molecular forces are of short range, only the molecules in a layer the thickness of the order of ten times the molecular separation will be responsible for these forces. Macroscopically this distance is zero, and we may assume that these forces act on the surface.

The infinitesimal force $\delta\mathbf{f}$ acting on the element of surface, $\delta\mathbf{S} = \mathbf{n}\delta S$ with $\mathbf{n}$ the normal to the surface, is usually decomposed in the form

$$\delta\mathbf{f} = -p\mathbf{n}\delta S + \boldsymbol{\sigma}_{\mathbf{n}}\delta S. \tag{3.20}$$

By definition p is the thermodynamic pressure and it is taken to be positive when it is compressive. For fluids the value of p, at any point inside the fluid, is independent of the orientation of $\delta\mathbf{S}$. On the contrary, $\boldsymbol{\sigma}_{\mathbf{n}}$ depends on $\mathbf{n}$. However, knowledge of $\boldsymbol{\sigma}_{\mathbf{n}}$ on three mutually perpendicular planes is enough to determine $\boldsymbol{\sigma}_{\mathbf{n}}$ on any other plane. The classical demonstration is based on the equilibrium of an arbitrary tetrahedral volume, letting the volume of the tetrahedron approach zero [1]. In this way the volume forces are negligible compared to the surface forces, and denoting the unit vectors of a cartesian system of coordinates by $\mathbf{e}_i$, we have in this limit

$$\boldsymbol{\sigma}_{\mathbf{n}} = \boldsymbol{\sigma}_{\mathbf{e}_i}\mathbf{n}\cdot\mathbf{e}_i = \mathbf{n}\cdot\mathbf{e}_i\sigma_{ij}\mathbf{e}_j. \tag{3.21}$$

Conservation of the angular momentum for the tetrahedron leads in the same limit to the requirement that

$$\sigma_{ij} = \sigma_{ji}, \tag{3.22}$$

and therefore the state of the stresses in the fluid is characterised at each point by the mechanical stress tensor $\mathbf{T}^m$ with components

$$T_{ij}^{\mathrm{m}} = -p\delta_{ij} + \sigma_{ij}, \tag{3.23}$$

or

$$\mathbf{T}^{\mathrm{m}} = -p\boldsymbol{\delta} + \boldsymbol{\sigma}, \tag{3.24}$$

with $\boldsymbol{\delta}$ the unit tensor, and $\boldsymbol{\sigma}$ the viscous stress tensor. Equation (3.20) may be written as $\delta\mathbf{f} = \mathbf{T}^{\mathrm{m}}\cdot\mathbf{n}\delta S$.

The majority of fluids are characterised by the constitutive relation [2]

$$\boldsymbol{\sigma} = 2\eta\left[\mathbf{e} - \frac{1}{3}e_{ii}\boldsymbol{\delta}\right] + \lambda e_{ii}\boldsymbol{\delta}, \tag{3.25}$$

with $\mathrm{Tr}\,\mathbf{e} = e_{ii} = \nabla\cdot\mathbf{u}$ the trace of the rate of deformation tensor, and η and λ the first and second coefficients of viscosity. These fluids are referred to as Newtonian.

The surface force density is $\mathbf{f}_S = \mathbf{T}^{\mathrm{m}}\cdot\mathbf{n}$, and we have

$$\int_V \frac{\partial\rho\mathbf{u}}{\partial t}dV + \int_S \rho\mathbf{u}\mathbf{u}\cdot\mathbf{n}dS = \int_V \mathbf{f}dV + \int_S \mathbf{T}^{\mathrm{m}}\cdot\mathbf{n}dS. \tag{3.26}$$

Applying Gauss's theorem to the surface integrals we obtain

$$\frac{\partial \rho \mathbf{u}}{\partial t} + \nabla \cdot (\rho \mathbf{u}\mathbf{u}) = \mathbf{f} + \nabla \cdot \mathbf{T}^{\mathrm{m}}. \tag{3.27}$$

Using the mass conservation equation and the identities

$$\frac{\partial \rho \mathbf{u}}{\partial t} = \rho \frac{\partial \mathbf{u}}{\partial t} + \mathbf{u} \frac{\partial \rho}{\partial t}, \qquad \nabla \cdot (\rho \mathbf{u}\mathbf{u}) = \mathbf{u} \nabla \cdot (\rho \mathbf{u}) + \rho (\mathbf{u} \cdot \nabla) \mathbf{u}, \tag{3.28}$$

we can modify the left hand side of (3.27) as follows:

$$\frac{\partial \rho \mathbf{u}}{\partial t} + \nabla \cdot (\rho \mathbf{u}\mathbf{u}) = \rho \left[\frac{\partial \mathbf{u}}{\partial t} + \rho (\mathbf{u} \cdot \nabla) \mathbf{u} \right] = \rho \frac{d\mathbf{u}}{dt}. \tag{3.29}$$

Finally, equation (3.27) can be written as,

$$\rho \frac{d\mathbf{u}}{dt} = \nabla \cdot \mathbf{T}^{\mathrm{m}} + \mathbf{f}. \tag{3.30}$$

Substitution of the constitutive relation given above (equation 3.25) gives the Navier-Stokes equation for Newtonian fluids:

$$\rho \frac{d\mathbf{u}}{dt} = -\nabla p + \eta \nabla^2 \mathbf{u} + (\lambda + \frac{1}{3}\eta) \nabla (\nabla \cdot \mathbf{u}) + \mathbf{f}. \tag{3.31}$$

In this derivation it has been assumed that the viscosity coefficients η and λ are constants. If, in addition the fluid is incompressible, the Navier-Stokes equation reduces to

$$\rho \frac{d\mathbf{u}}{dt} = -\nabla p + \eta \nabla^2 \mathbf{u} + \mathbf{f}_e + \rho \mathbf{g}, \tag{3.32}$$

where $\mathbf{f}_e$ is the electrical force density and the last term is the gravitational force per unit volume.

3.2 Electrical forces

The problem of calculating the electrical forces in a fluid is a difficult problem. Any attempt to deduce this force from microscopic models and the subsequent average is of dubious use. Only for very dilute media, in which the molecules can be considered subjected only to the imposed field, can the force be calculated from first principles. But for dense gases and liquids this is not the case, and recourse to thermodynamics is necessary. This has the advantage that the solution is dependent on the detailed molecular structure only through the permittivity, which is assumed to be a known function of mass density and temperature.

Following the argument of Landau and Lifshitz [3], consider a layer of fluid confined between two parallel plates, a distance h apart, carrying the necessary surface charge density distribution for the generation of a uniform electric field in the layer. Assume also that the plates are perfect conductors, and are connected to a battery so that the potential between them is kept constant. Imagine now that the upper conductor is displaced quasistatically outwards a small distance $\boldsymbol{\xi}$, not necessarily in the direction of the normal $\mathbf{n} \equiv \mathbf{e}_z$ to the plate. Each point $\mathbf{x}$ is displaced to the position $\mathbf{x} + (z/h)\boldsymbol{\xi}$, and therefore,

$$\delta\phi = \phi(\mathbf{x} - (z/h)\boldsymbol{\xi}) - \phi(\mathbf{x}) = \frac{z}{h}\boldsymbol{\xi} \cdot \mathbf{E}, \tag{3.33}$$

and consequently

$$\delta\mathbf{E} = -\nabla\delta\phi = -\frac{1}{h}(\boldsymbol{\xi} \cdot \mathbf{E})\mathbf{n}. \tag{3.34}$$

Since we know the change of the electric field in the process, the appropriate thermodynamic potential per unit volume is $\tilde{F}$, defined by

$$\tilde{F} = F - \mathbf{E} \cdot \mathbf{D} = F_o - \mathbf{E} \cdot \mathbf{D} + \int_0^{\mathbf{D}} \mathbf{E} \cdot d\mathbf{D}, \tag{3.35}$$

with a differential given by:

$$\delta\tilde{F} = \delta F_o - \mathbf{D} \cdot \delta\mathbf{E}. \tag{3.36}$$

The work done by the fluid on the plate in this isothermal deformation at constant potential, per unit area of the plates, is equal to the decrease of $h\tilde{F}$. Or equivalently, the work done on the fluid by the plate is equal to the increase of $h\tilde{F}$. As a result, we have per unit area of the plates

$$\boldsymbol{\xi} \cdot \mathbf{T} \cdot \mathbf{n} = \delta(h\tilde{F}) = \tilde{F}\delta h + h\delta\tilde{F}, \tag{3.37}$$

with $\mathbf{T}$ the stress tensor acting upon the fluid, h the thickness of the layer, and $\tilde{F}$ the free energy per unit of volume, which is a function of temperature, mass density and the electric field. In our case the process is isothermal and

$$\begin{aligned}
\delta\tilde{F} &= \left(\frac{\partial\tilde{F}}{\partial\rho}\right)_{T,\mathbf{E}} \delta\rho + \left(\frac{\partial\tilde{F}}{\partial\mathbf{E}}\right)_{T,\rho} \cdot \delta\mathbf{E} \\
&= \left(\frac{\partial F_o}{\partial\rho}\right)_T \delta\rho + \int_0^{\mathbf{D}} \left(\frac{\partial\mathbf{E}}{\partial\rho}\right)_{T,\mathbf{D}} \cdot d\mathbf{D}\delta\rho - \mathbf{D} \cdot \delta\mathbf{E}.
\end{aligned} \tag{3.38}$$

Mass conservation implies $\delta(\rho h) = 0$, from which the variation of density with the thickness of the layer can be written as:

$$\delta\rho = -\frac{\rho}{h}\delta h = -\frac{\rho}{h}\xi_z = -\frac{\rho}{h}\boldsymbol{\xi} \cdot \mathbf{n}. \tag{3.39}$$

Substituting equations (3.34), (3.35), (3.38), and (3.39) in equation (3.37) we have

$$\begin{aligned}
\boldsymbol{\xi} \cdot \mathbf{T} \cdot \mathbf{n} &= \boldsymbol{\xi} \cdot \mathbf{n} \left\{ \left[F_o - \rho \left(\frac{\partial F_o}{\partial\rho}\right)_T \right] - \mathbf{E} \cdot \mathbf{D} + \int_0^{\mathbf{D}} \left[\mathbf{E} - \rho \left(\frac{\partial\mathbf{E}}{\partial\rho}\right)_{T,\mathbf{D}} \right] \cdot d\mathbf{D} \right\} \\
&+ (\boldsymbol{\xi} \cdot \mathbf{E})(\mathbf{n} \cdot \mathbf{D}).
\end{aligned} \tag{3.40}$$

Remembering that F_o/ρ is the free energy per unit of mass in the absence of the electric field, we have

$$\delta(F_o/\rho) = -(S_o/\rho)\delta T - p_o\delta(1/\rho), \tag{3.41}$$

where p_o is the thermodynamic pressure in the absence of the electric field, given by

$$p_o(\rho,T) = -\left[\frac{\partial(F_o/\rho)}{\partial(1/\rho)}\right]_T = -F_o - \rho\left(\frac{\partial F_o}{\partial\rho}\right)_T, \tag{3.42}$$

and we obtain

$$\mathbf{T} = \left\{-p_o - \mathbf{E}\cdot\mathbf{D} + \int_0^{\mathbf{D}}\left[\mathbf{E} - \rho\left(\frac{\partial\mathbf{E}}{\partial\rho}\right)_{T,\mathbf{D}}\right]\cdot d\mathbf{D}\right\} + \mathbf{DE}. \tag{3.43}$$

For a non linear medium, $\mathbf{E} = (\mathbf{D}-\mathbf{P})/\varepsilon_0$, and consequently,

$$\left(\frac{\partial\mathbf{E}}{\partial\rho}\right)_{T,\mathbf{D}} = -\frac{1}{\varepsilon_0}\left(\frac{\partial\mathbf{P}}{\partial\rho}\right)_{T,\mathbf{D}}. \tag{3.44}$$

For a linear medium

$$\left(\frac{\partial\mathbf{E}}{\partial\rho}\right)_{T,\mathbf{D}} = \mathbf{D}\left[\frac{\partial}{\partial\rho}\left(\frac{1}{\varepsilon}\right)\right]_T = -\frac{1}{\varepsilon^2}\left(\frac{\partial\varepsilon}{\partial\rho}\right)_T\mathbf{D}, \tag{3.45}$$

and we obtain the well known result

$$\mathbf{T} = \left[-p_o - \frac{1}{2}\mathbf{E}\cdot\mathbf{D} + \frac{1}{2}\mathbf{E}^2\rho\left(\frac{\partial\varepsilon}{\partial\rho}\right)_T\right]\boldsymbol{\delta} + \mathbf{DE}. \tag{3.46}$$

Therefore, for a linear medium the electrical part of the stress tensor is

$$\mathbf{T}^e = \varepsilon\mathbf{EE} - \frac{1}{2}\varepsilon(1-a)E^2\boldsymbol{\delta}, \tag{3.47}$$

with $a = (\rho/\varepsilon)(\partial\varepsilon/\partial\rho)_T$ the electrostriction parameter. In component form,

$$\mathbf{T}^e{}_{ij} = \varepsilon E_i E_j - \frac{1}{2}(1-a)E_k E_k\delta_{ij}. \tag{3.48}$$

In vacuum the electric tensor coincides with the electrostatic Maxwell tensor as should be the case.

Transforming the surface forces densities to volume forces densities by applying Gauss's theorem to the surface enclosing the volume of interest,

$$\mathbf{f}_E = \nabla\cdot\mathbf{T}^e = q\mathbf{E} - \frac{1}{2}E^2\nabla\varepsilon + \nabla p_{st}, \tag{3.49}$$

where we have introduced $p_{st} = (1/2)\varepsilon a E^2$ the electrostrictive pressure.

The term $q\mathbf{E}$, called the Coulomb force, is the force per unit volume on a medium containing free electric charge. This is the strongest EHD force term and usually dominates when d.c. electric fields are present. The next term, called the dielectric force, is due to the force exerted on a non-homogeneous dielectric liquid by an electric field. This is usually weaker than the Coulomb force and only dominates when an a.c. electric field with a period much shorter that the charge relaxation time and/or the ionic transit time, is applied. The electrostrictive force, being the gradient of a scalar is treated as a modification to the fluid pressure.

3.3 Boundary and jump conditions

Consider the physical system depicted in fig. 3.2.a

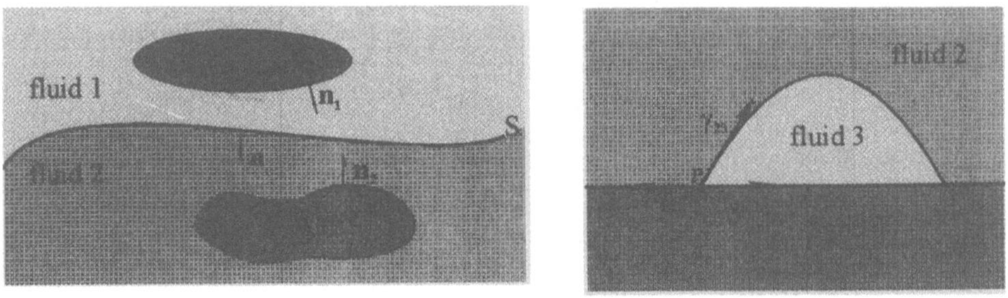

a) b)

Fig. 3.2. Mechanical system

In each solid surface, the impenetrability condition implies that

$$\mathbf{n}_i \cdot (\mathbf{u} - \mathbf{u}_{si}) = 0, \qquad i = 1, 2, \tag{3.50}$$

where $\mathbf{u}$, $\mathbf{u}_s$, are the fluid and solid velocities at the interface. If in addition the fluids are viscous, the tangential velocities have to be continuous also, a restriction known as the non-slip boundary condition,

$$\mathbf{n}_i \times (\mathbf{u} - \mathbf{u}_{si}) = 0, \qquad i = 1, 2. \tag{3.51}$$

We choose to define the geometry of the interface between the two fluids by the implicit function $F(\mathbf{r}, t) = 0$, parametrized in such a way that its position at time $t + dt$ is defined unambiguosly by $F(\mathbf{r} + v_{\mathrm{n}}\mathbf{n}\, dt, t + dt) = 0$, where $\mathbf{n} = \nabla F / |\nabla F|$ is the normal to the surface and v_{n} the normal velocity. With this parametrization, cavitation is excluded at the interface. The actual fluid particles of the surface may have tangential velocities that are not included in this geometrical definition. From this definition, and by Taylor-series expansion to first order, the kinematic condition inmediately follows:

$$v_n \mathbf{n} \cdot \nabla F + \frac{\partial F}{\partial t} = 0. \tag{3.52}$$

Before considering the equilibria of forces at the interface between the two fluids, and at the contact line defined by the set of points at which the three phases meet, we need to discuss capillary phenomena.

3.3.1 Surface tension

Consider an interface between a gas and a liquid. Molecules tend to attract each other through forces induced by the fluctuating induced dipole-dipole interactions (Van der Waals forces). A molecule sitting on the surface of the liquid will suffer an inward force toward the liquid, because the attraction due to the large number of molecules in the liquid phase is much stronger than the attraction due to the small number of gas molecules. As a consequence, the surface tends to be as small as possible. The energy, W, needed to create a unit area of new surface is denoted as γ, and is called the coefficient of surface tension

$$\delta W = \gamma \delta S, \tag{3.53}$$

and is taken to be positive. Now consider the surface between two liquids A and B. A molecule A in the close vicinity of the interface suffers a inward attractive force, f_A, due to molecules of liquid A, and an outward force, f_B, due to molecules of type B. If f_A is greater than f_B both liquids will be immiscible, and the interface will tend to a minimum. Again the coefficient of interfacial tension is the energy needed to create a unit of interface area. It is clear than this coefficient has to be smaller than in the case of a surface between a gas and a liquid, since the number of molecules at each side of the interface are comparable. In the case that f_A is less than f_B the liquids will be miscible, and we say that the coefficient of interfacial surface tension is negative.

The surface tension may also be interpreted as a force per unit length, as is evident from fig. 3.3

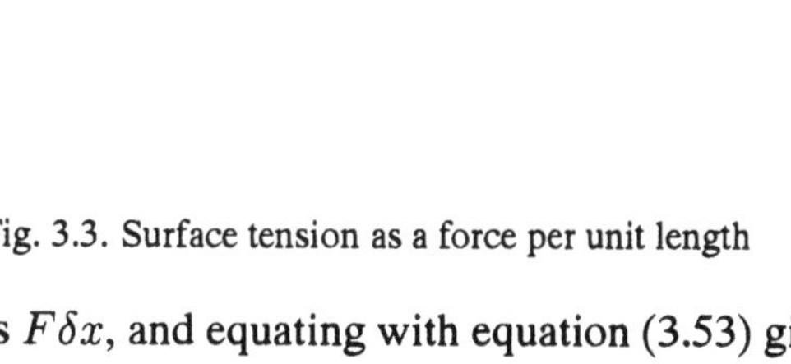

Fig. 3.3. Surface tension as a force per unit length

The work done by F is $F\delta x$, and equating with equation (3.53) gives:

$$\delta W = \gamma \delta S = \gamma(\delta \cdot 1) = F \delta x, \tag{3.54}$$

from which we deduce that the force per unit length is $F = \gamma$.

3.3.2 Contact line

At point P, equilibrium of forces (see fig. 3.2.b) requires that:

$$\gamma_{12} = \gamma_{13} + \gamma_{32}\cos\theta. \tag{3.55}$$

If $|\gamma_{12}| = |\gamma_{13}| + |\gamma_{32}|$ equilibrium is not possible. If fluid 2 is a liquid and fluid 3 a gas, the drop on a solid spreads on the solid surface as a monolayer.

The boundary of the drop on the solid defines the contact line, and two boundary conditions may be realized. Either the angle θ is fixed, and the contact line is free to move, or else, the contact line is fixed, and the angle is free. The latter condition is usually met when the liquid encounters an end wedge of the solid, or it is pinned by chemical action on a given line.

3.3.3 Equilibria of forces at the interface

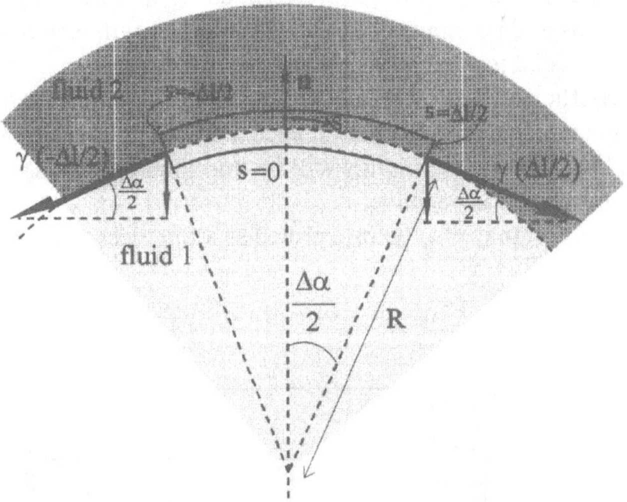

Fig. 3.4. Force equilibria at the interface

Consider a small volume of height Δ, width Δl and unit length perpendicular to the paper, that encloses the surface at time t (as shown in fig. 3.4). Taking the limit $\Delta \to 0$, the equilibria of normal surface forces, per unit length perpendicular to the paper, is given by,

$$\mathbf{n}\cdot[\mathbf{T}^m + \mathbf{T}^e]\cdot\mathbf{n}\Delta l \simeq \gamma(\Delta l)\times\frac{\Delta\alpha}{2} + \gamma(-\Delta l)\times\frac{\Delta\alpha}{2} \simeq \gamma(0)\Delta\alpha \simeq \gamma\frac{1}{R}\Delta l, \tag{3.56}$$

and in the limit $\Delta l \to 0$,

$$\mathbf{n}\cdot[\mathbf{T}^m + \mathbf{T}^e]\cdot\mathbf{n} = \gamma\frac{1}{R}. \tag{3.57}$$

For an arbitrary surface with radii of curvature, R_1, R_2, it results

$$\mathbf{n}\cdot[\mathbf{T}^m+\mathbf{T}^e]\cdot\mathbf{n}=\gamma\left(\frac{1}{R_1}+\frac{1}{R_2}\right)=\gamma\nabla\cdot\mathbf{n}. \tag{3.58}$$

The equilibria of tangential forces gives

$$\mathbf{s}\cdot[\mathbf{T}^m+\mathbf{T}^e]\cdot\mathbf{n}\Delta l=\gamma(\Delta l/2)\times 1+\gamma(-\Delta l/2)\times 1\simeq\frac{\partial\gamma}{\partial s}\Delta l, \tag{3.59}$$

and, again in the limit $\Delta l\to 0$, we have

$$[\mathbf{T}^m+\mathbf{T}^e]\cdot\mathbf{n}-(\mathbf{n}\cdot[\mathbf{T}^m+\mathbf{T}^e]\cdot\mathbf{n})\,\mathbf{n}=\nabla_s\gamma, \tag{3.60}$$

with $\nabla_s=(\boldsymbol{\delta}-\mathbf{nn})\cdot\nabla$, where $\boldsymbol{\delta}=\mathbf{e}_i\mathbf{e}_i$ is the unit dyad. To conclude, the equilibria of forces may be stated in the form

$$[\mathbf{T}^m+\mathbf{T}^e]\cdot\mathbf{n}=\gamma\nabla\cdot\mathbf{n}+\nabla_s\gamma. \tag{3.61}$$

Is the surface tension dependent on the electric field?

The answer is no if we refer to the coefficient γ defined by equation (3.61). The origin of the confusion is due to the fact that there are electrical forces acting at interfaces that behave exactly in the same way as the capillary forces. The classical example is the case of a electrical double layer with dimensions on the order of nanometers, that from the macroscopic point of view can be considered as surface dipole distribution on a given surface, with dipoles parallel to $\mathbf{n}$, the normal to the interface.

Consider such a charged electrical double layer on a plane surface with unit length perpendicular to the paper (see fig. 3.5.a). The electrical forces acting on the volume enclosed by the two surfaces plus the two closing lateral surfaces are calculated using the Maxwell stress tensor. The important point is that there is a net force acting at any of the lateral surfaces due to the electric field. As it is evident from fig. 3.5.a we have at each lateral surface S_l that the force is

$$\mathbf{n}_l\cdot(\varepsilon\mathbf{E}_s\mathbf{E}_s-\frac{1}{2}\varepsilon E_s^2\boldsymbol{\delta})\times\Delta=-\frac{1}{2}E_s^2\Delta\mathbf{n}_l=-\gamma_e\mathbf{n}_l, \tag{3.62}$$

tangential to the surface and in the opposite direction to the surface tension force. The coefficient, γ_e, is the electrical energy stored in the surface per unit area. For a curved surface and a nonuniform dipole distribution, we will have from fig. 3.5.b, that the equilibria of forces for the interface is given by

$$[\mathbf{T}^m+\mathbf{T}^e]\cdot\mathbf{n}=(\gamma-\gamma_e)\nabla\cdot\mathbf{n}+\nabla_s(\gamma-\gamma_e). \tag{3.63}$$

Thus γ_e plays a role similar to the surface tension, but acts in the opposite direction. As it is independent of the direction of the electric field, the "effective surface tension" given by $\gamma-\gamma_e$ will be always smaller than the actual surface tension γ. A classical example is the double electrical layer between mercury and water that is responsible of the decreasing of "effective surface tension" observed in mercury drops immersed in water, and subjected to a potential drop of a few milivolts, an effect known as electrocapillarity. For very insulating liquids this double layer may extend well beyond the interface and electrocapillarity is absent.

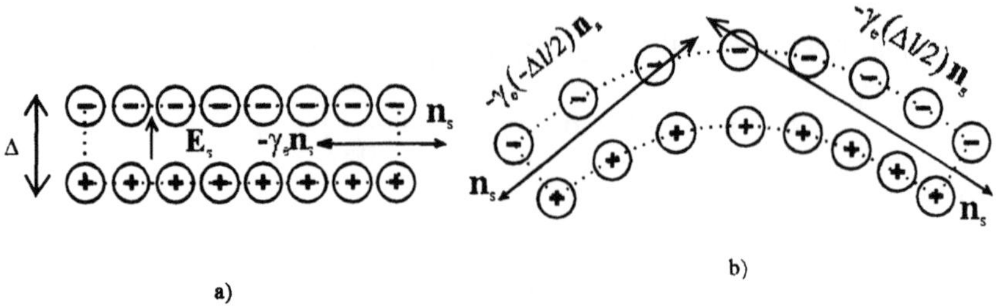

Fig. 3.5. Electrocapillarity: electrical double layer

3.4 Nondimensional numbers in EHD

The fluid equilibria and motions are influenced by electrical forces. The influence in the bulk is governed by the Navier-Stokes equation, and at the interfaces by the jump conditions. In turn, the motion modifies the charge density distribution, both in the bulk and at the interfaces, through the convective term of the current density in the corresponding charge conservation equation. The fields are determined by the space charge and surface charge distributions, and the motion may substantially change the value of the electric field. In general, the importance of the principal electrical and mechanical effects depends on the ratio of charge decay times and electrical and mechanical time constants, or equivalently of the pressure ratios or forces ratios of mechanical and electrical origin. An interesting discussion of these effects has been done by Crowley in [4]. Also Melcher [5] describes the time constants more relevant to EHD. Here, we follow an alternative approach, based on the nondimensionalization of the EHD equations.

Before entering into the discussion of the nondimensional numbers, let us summarize the EHD laws for the simple case of homogenous isothermal incompressible fluids subjected to gravity and electric forces. In the bulk of the fluid we have

$$\nabla \cdot (\varepsilon \mathbf{E}) = q, \qquad \nabla \times \mathbf{E} = 0, \qquad \frac{\partial q}{\partial t} + \nabla \cdot \mathbf{J} = 0, \tag{3.64}$$

$$\nabla \cdot \mathbf{u} = 0, \qquad \rho \frac{d\mathbf{u}}{dt} = -\nabla p + \eta \nabla^2 \mathbf{u} + q\mathbf{E} - \frac{1}{2} E^2 \nabla \varepsilon + \nabla p_{st} + \rho \mathbf{g}, \tag{3.65}$$

together with the jump conditions at the interface

$$\mathbf{n} \cdot [\varepsilon \mathbf{E}] = q_s, \qquad \mathbf{n} \times [\mathbf{E}] = 0, \qquad \frac{\partial q_s}{\partial t'} + v_n (\nabla \cdot \mathbf{n}) q_s + \nabla_s \cdot \mathbf{K}_s + \mathbf{n} \cdot [\mathbf{J}'] = 0, \tag{3.66}$$

$$v_n \mathbf{n} \cdot \nabla F + \frac{\partial F}{\partial t} = 0, \qquad \mathbf{n} \cdot [\mathbf{u}] = 0, \tag{3.67}$$

$$\mathbf{n} \cdot [2\eta \mathbf{e}] \cdot \mathbf{n} - [p] - \frac{1}{2} \left[\varepsilon (1 - a) E^2 \right] + \left[\varepsilon E_n^2 \right] = \gamma \left(\frac{1}{R_1} + \frac{1}{R_2} \right), \tag{3.68}$$

$$\mathbf{s}_i \cdot [2\eta \mathbf{e}] \cdot \mathbf{n} + q_s \mathbf{s}_i \cdot \mathbf{E} = 0, \qquad i = 1, 2, \tag{3.69}$$

plus the electrical and mechanical boundary conditions at two metallic fixed electrodes explained in section 1.4, and 3.3 above. To these equations we must add the constitutive equation for the density current **J**.

3.4.1 Charge decay

We identify first some of the typical times and dimensionless numbers related to the evolution of charge density in EHD.

Ohmic regime

For the ohmic regime we may put $\mathbf{J} = \sigma\mathbf{E} + q\mathbf{u}$ with $\mathbf{u}$ the liquid velocity, in the charge conservation equation and using Gauss's law, this equation, in nondimensional form, becomes

$$\frac{dq}{dt} + \frac{\tau}{\tau_e}q = 0 \tag{3.70}$$

where dq/dt stands for $\partial q/\partial t + \mathbf{u}\cdot\nabla q$, the material derivative, and σ has been assumed to be constant. The scale τ depends on the processes taking place in our physical system. For a medium at rest, $\tau = \tau_e$ and charge relaxes exponentially, $q(t) = q(0)e^{-t}$, with t measured in units of $\tau_e = \varepsilon/\sigma$, the charge relaxation time. On the other hand, τ may be related to the period, $T = 2\pi/\omega$, of an imposed a.c. field. Then, for sufficiently high values of the frequency, $dq/dt \simeq 1/\omega\tau_e \simeq 0$, and the presence of the charge density may be of no relevance to the dynamics of the system.

If τ is related to a mechanical time the ratio, τ_e/τ, is usually called the electric Reynolds number R_E. For large values of R_E the charge appears to be frozen and, if it is initially zero, will be always zero. In more general situations τ will be a combination of electrical and mechanical times.

Unipolar injection regime

Consider now the case of unipolar injection and suppose, for simplicity, that only one ionic species, taken to be positive, is injected in a perfectly insulating liquid, $\sigma = 0$. The constitutive law for the current density is

$$\mathbf{J} = q\mu\mathbf{E} - D\nabla q + q\mathbf{u}, \tag{3.71}$$

If λ and l denote the spatial scales of variation for charge density and electrical potential, V, the applied electrical potential difference, q_0 the charge density at the injector, the dimensionless charge conservation equation can be written as,

$$\frac{dq}{dt} + \left(\frac{l}{\lambda}\right)\left(\frac{\tau}{\tau_\imath}\right)\mathbf{E}\cdot\nabla q - \left(\frac{\tau}{\tau_d}\right)\nabla^2 q + \left(\frac{\tau}{\tau_q}\right)q^2 = 0 \tag{3.72}$$

where $\tau_\imath = l^2/\mu V$ is the ionic transit time, $\tau_d = \lambda^2/D$ the molecular ionic diffusion time and $\tau_C = \varepsilon/\mu q_0$ is the charge relaxation time associated to Coulombian repulsion.

The ratio between diffusion and drift is $(D/\mu V)(l/\lambda) \sim (0.025/V)(l/\lambda)$, where Einstein's relation $D/\mu = k_B T/e$ has been taken into account for an ambient temperature of $T \simeq 300$K. As can be seen, diffusion will be negligible compared to drift unless the potential difference is of the order of 0.025 volts for $\lambda \sim l$, or else very high gradients are present ($\lambda \ll l$). Usually this will only happen in thin layers, either boundary layers or internal layers, where diffusion is required to match the outer or mainstream solutions of the diffusion-free problem. In most practical situations $\lambda \sim l$ and diffusion may be safely neglected.

Notice that, if diffusion is neglected, the dimensional charge conservation equation has an exact solution $q = q_0/(1+t/\tau_C)$ with τ_C the algebraic bulk relaxation time. This solution is valid on $d\mathbf{r}/dt = \mu\mathbf{E} + \mathbf{u}$, so that unless a given element of liquid can be traced via a particle line to a source of charge, it will support no bulk charge density. The algebraic charge relaxation time τ_C depends on the initial level of charge, and after several decay times, the charge density reaches a limiting value $\varepsilon/\mu t$ independent of the initial conditions.

The ratio between drift and Coulombian repulsion is given by the nondimensional number $C = \tau_C/\tau_i = q_0 l^2/\varepsilon V$. For low (large) values of C the characteristic time that enters into the problem is the transit time τ_i (Coulombian repulsion time τ_C). Small values of C indicate that the electric field is determined by the external power supply connected to the electrodes in the fluid. On the contrary, for high values of C the electric field is strongly dependent on the charges which are either inside or on the surface of the fluids.

Again, if τ is associated with the motion of the fluid, we can construct an appropriate electric Reynolds number R_E as the ratio of the mechanical time to the algebraic relaxation time, and for high values of this ratio the charge may be considered to be frozen, i.e. the liquid behaves as a good insulator. On the contrary, for small values of R_E, the liquid may be considered to be conductor.

3.4.2 Electrohydrostatics

There may exist solutions to the EHD equations that are time independent. The simplest one is the hydrostatic solution, $\mathbf{u} = 0$. For this solution to be possible, the following conditions must be satisfied

$$q_s \mathbf{s}_i \cdot \mathbf{E} = 0 \quad i = 1,2 \quad \text{(at the interface)} \tag{3.73}$$

$$\nabla \times \left(q\mathbf{E} - \frac{1}{2}E^2 \nabla\varepsilon + \rho\mathbf{g} \right) = 0 \quad \text{(in the bulk)}, \tag{3.74}$$

or, equivalently,

$$q\mathbf{E} - \frac{1}{2}E^2 \nabla\varepsilon - \rho\mathbf{g} = -\nabla \left(p_e + \rho\mathbf{g} \cdot \mathbf{r} \right) \tag{3.75}$$

that defines p_e. Notice that the electrostriction pressure is not included in p_e since the curl of a gradient is identically zero.

The conditions (3.73) are automatically satisfied for perfectly insulating liquids for which is $q_s = 0$, or perfect liquid conductors for which $\mathbf{s}_i \cdot \mathbf{E} = 0 \quad (i = 1,2)$. In general, these conditions, plus condition (3.74) in the bulk, are met only in highly symmetrical geometries. In these cases we have for each phase

$$\mathbf{T}^m = -\nabla p, \tag{3.76}$$

$$p_i + p_{ei} + \rho_i \mathbf{g} \cdot \mathbf{r} - \frac{1}{2}\varepsilon a E^2 = \Pi_i = f(t) \quad i = 1, 2. \tag{3.77}$$

The equilibrium equation at the interface is

$$[\Pi] - [\rho]\mathbf{g} \cdot \mathbf{r} - [p_e] - \frac{1}{2}[\varepsilon(E_n^2 - E_s^2)] = \gamma\left(\frac{1}{R_1} + \frac{1}{R_2}\right), \tag{3.78}$$

where $\mathbf{r}$ is evaluated at the interface through $F(\mathbf{r}) = 0$. This equation together with the Poisson equation for the potential, the surface charge conservation equation, and the appropriate boundary conditions define the equilibrium hydrostatic state $\mathbf{E}$, p, $F(\mathbf{r}) = 0$, $\mathbf{u} = 0$. The electrostriction term has disappeared in equation (3.78) since the bulk and surface terms cancel each other.

In the absence of interfaces, we have $p \sim \rho g l$ with l the typical length of the system in the vertical direction. Since $p_e \simeq \varepsilon E_0^2$, with E_0 the typical magnitude of the electric field, the relevance dimensionless ratio is $\varepsilon E^2/\rho g l$. Due to the fact that breakdown occurs for values of the electric fields ranging from 3 (air) to 30 MV/m (insulating oils), l considered to be the height that a liquid can climb on the Earth, varies from centimetres to meters.

Consider those situations for which the relevant forces are the interfacial ones. The order of magnitude of these forces acting at the interface, is from (3.78) γ/R for the capillary ones, with R the minimum radius of curvature; $\varepsilon E_0^2/2$ for the electrical terms, with E_0 the typical field at the interface; $[\rho]gL$ for the gravitational force, where L is the length scale for the interface along the gravity acceleration. Therefore, the system behavior is determined by two of the following nondimensional numbers: the Bond number $\mathrm{Bo} \equiv [\rho]gLR/\gamma$, defined by the ratio between the gravitational force and the capillary one; the Electric Bond Number $\mathrm{B_e} \equiv [\varepsilon E^2]R/\gamma$, which is the analog using the electrical force instead of the gravitational one; and the number $\mathrm{W_e} \equiv [\varepsilon E^2]/[\rho]gL$, defined by the ratio of the electrical force to the gravitational one. This parameter governs the equilibria when the capillary effects are small compared to the electrical and gravitational forces.

Examples of these equilibria are plane interfaces perpendicular to $\mathbf{g}$, and drops either free [6],[7], or captive (conical points [7],[8], sessile or pendants drops [9], and liquid bridges[10],[11]).

3.4.3 Dynamics

For simplicity, let us consider two limiting situations. First, that the dynamics are governed by electrical interfacial forces. Secondly, that there are no interfaces and the dynamics are governed by forces acting in the bulk.

Interfacial dynamics

Here, the dynamics are governed by the normal and tangential stresses given by equations (3.68), and (3.69). In the bulk, the velocity field is governed by the Navier-Stokes equation, which in nondimensional form is

$$\frac{\partial \mathbf{u}}{\partial t} + \frac{\tau}{\tau_m}(\mathbf{u} \cdot \nabla)\mathbf{u} = -\frac{p_o \tau}{\rho u_o l}\nabla p + \left(\frac{\tau}{\tau_\nu}\right)\nabla^2 \mathbf{u} + \frac{g\tau}{u_o}\mathbf{e}_z, \tag{3.79}$$

where, u_o and p_o are the characteristic values of velocity and pressure, $\tau_m = l/u_o$ is the mechanical transport time, $\tau_\nu = l^2/\nu$ the viscous relaxation time, and τ a typical time depending on the physical processes that govern the velocity. For example, for a pulsating flow created by a oscillating plate, τ will be the imposed period of the plate. For a steady flow $\tau = \infty$, and the first term vanishes. Usually, $\tau = \tau_m$, and the ratio of the inertia terms to the viscous term is given by $Re = lu_o/\nu$, the Reynolds number. We also have $p_o = \rho u_o^2$, and $g\tau/u_o$ is the inverse of the Froude number, $Fr = u_o/(gl)^{(1/2)}$, raised to the power two.

When only the normal stresses are different from zero, and $Re \gg 1$, the fluid may be considered as ideal, and the electric Bond number, and W_e or a combination of both, will govern the dynamics of the system. Well known examples are linear waves and instabilities in horizontal interfaces and jets [12]. Nonlinear waves on charged free surfaces are analysed by Zhakin in chapter 8. If the effects due to viscosity need to be considered, a new nondimensional number appears, the Ohnesorge Number $\mathrm{Oh} \equiv [\rho\nu]u_0R/\gamma h_0$, which compares the viscous terms with the capillary ones.

When the tangential electrical stresses are different from zero and dominates the dynamics, it is readily seen from equation (3.69) that the velocity u_0 is determined through stresses $[\rho\nu]u_0/h \sim q_s E_s$, and the magnitude and changes of q_s are important (this requires a solution of eq. (3.66). Viscosity must always be considered when the liquids are neither perfectly conducting nor perfect insulators, since it is the only mechanism that balances the electrical stresses at the interface. See [13] and references therein for a detailed analysis of viscous drops subjected to electric fields.

An example of a steady-state solution is *Quincke's rotor* [5]. This is a solid, highly insulating, cylinder immersed in a viscous, conducting, liquid. An electric field, uniform and of value E_o at large distances of the cylinder, is imposed perpendicular to its axis. Due to the symmetry of the system, there is an equilibrium state, but this state is unstable. A slight rotation produces an asymmetrical distribution of charge. In this case, there is a net torque on the cylinder and it begins to rotate. Assuming that the system is viscous dominated, the angular velocity may be estimated as

$$\rho\nu\frac{u_o}{R} \sim \rho\nu\Omega \sim q_s E_s. \tag{3.80}$$

The tangential electric field is, in this system, of the same order as the external field, whereas the surface charge density can be obtained from the surface charge conservation equation. For steady conditions the latter can be written as $\Omega(\partial q_s/\partial\theta) + [\sigma E_r] = 0$, and from here, the order of magnitude for q_s is $q_s \sim (\sigma E_o)/\Omega$. Finally, we obtain

$$\Omega^2 \sim \frac{\sigma E_o^2}{\rho\nu} = \frac{1}{t_{ev}\tau_e} \tag{3.81}$$

with $t_{ev} = \rho\nu/\varepsilon E_o^2$ the electroviscous relaxation time (see below), and $\tau_e = \sigma/\varepsilon$ the charge decay time. For corn oil ($\varepsilon \simeq 3.1 \times 10^{-11}$, $\sigma \simeq 5 \times 10^{-11}$, $\rho\nu \simeq 5 \times 10^{-2}$ in SI units) and an applied electric field around 1 kV/cm, the resulting rotation period is of order of one second. This phenomenon has been observed in practice, and is responsible for the rotation of small particles, immersed in insulating liquids subjected to dc fields.

Bulk dynamics

Here, the dynamics are governed by Navier-Stokes equation, which is, written in dimensionless form, given by:

$$\frac{\partial \mathbf{u}}{\partial t} + \frac{\tau}{\tau_m}(\mathbf{u}\cdot\nabla)\mathbf{u} = -\frac{p_o\tau}{\rho u_o l}\nabla p + \frac{\tau}{\tau_\nu}\nabla^2\mathbf{u} + \left(\frac{q_0 E_0 \tau}{\rho u_o l}\right) q\mathbf{E} - \left(\frac{\varepsilon_o {E_0}^2 \tau}{\rho u_o l}\right)\frac{1}{2}E^2\nabla\varepsilon$$
$$-\left(\frac{\varepsilon_o {E_0}^2 \tau}{\rho u_o l}\right)\nabla\left(\frac{1}{2}a\varepsilon E^2\right) + \frac{g\tau}{u_o}\mathbf{e}_z, \qquad (3.82)$$

with ε_o a characteristic reference value of ε, and a the electrostriction parameter. In addition to the classical nondimensional numbers in fluid mechanics, we have two new electromechanical parameters that may be written in a variety of ways depending on the physical mechanisms that determine the velocity field, the charge density and the gradient of permittivity. Since $q_0 \sim \varepsilon_o E/l$, both electromechanical parameters are of the same order of magnitude. However, the Coulomb force is typically much larger than the dielectric force; for example, when the variation of the permittivity is induced by a temperature gradient is $\nabla\varepsilon = 10^{-3}\Delta T\varepsilon$, which is much smaller than ε, for the imposed temperature differences, which are typically of a few degrees centigrades. In general, only for those situations for which q_0 is negligible, and $\nabla\varepsilon$ different from zero, the dielectric force is the dominant electric force.

Viscous flows: consider a viscous dominated stationary flow driven by an externally imposed pressure, p_o. The relevant time scale is $\tau = \tau_\nu$, and the velocity scale u_o is determined by balancing the first two terms of the right hand side of equation (3.82), $p_o\tau_\nu/\rho u_o l = 1$, which gives $u_o = \eta u_o/l$. On the other hand, in case where the flow is driven by the electrical forces, the typical scale for the velocity is obtained by equating the electromechanical parameter to 1, which gives $u_o = \varepsilon_o E_o^2 l/\eta$. The ratio of these two velocities is equal to the ratio of pressures, and this dimensionless ratio $p_o/\varepsilon_o E_o^2$ determines whether the electrical forces are dominant or not. For small values of this parameter the velocity field is not affected by the electric forces, while for high values the electric field determines the dynamic evolution of the velocity.

Consider the transient motion of a liquid driven solely by the electrical forces. In the viscous regime, the second term is of the same order as the third. Equating both terms, we find a natural time scale $\tau_{ev} = \eta/\varepsilon E^2$, called the electroviscous scale, which plays a significant role in these transient viscous regimes [5],[14].

Inertial flows: in the inertial regime, the most frequently encountered in nature or in industrial processes, the acceleration term is comparable to the electric force. In other terms, this is tantamount to saying that the electrical energy is converted to kinetic energy of the liquid. This implies that $(1/2)\varepsilon E^2 \simeq (1/2)\rho u_o^2$, and consequently that the typical liquid velocity is given by $u_o \simeq (\varepsilon_o/\rho)^{1/2}E_o = \mu_h E_o$. The liquid velocity is proportional to the electric field, and the constant, $\mu_h = (\varepsilon_o/\rho)^{1/2}$ is called the hydrodynamic mobility. It is normal to define a dimensionless number $M = (\varepsilon/\rho)^{1/2}/\mu$ as the ratio of two mobilities, the hydrodynamic mobility, and μ, the true ionic mobility. Taking into account *Walden's rule,* i.e. $\mu\eta \sim$ constant, we see that M is proportional to viscosity. Since $v \simeq M\mu E$ and for liquids M may vary from 4 to 400, ions are entrained by the motion of the liquid, contrary to the case of gases where M is less than one. For charged liquids, the time of flight of ions may be much larger than the algebraic charge relaxation time or the ionic transit time, and the final charge distribution may be strongly affected by the fluid motion. As a consequence, the electric field is modified, which in turn changes the fluid velocity, thus making these type of problems particularly difficult. See chapter 11 by Watson and chapters 15 to 18 by Atten where the dynamics of charged liquids are analysed in detail.

REFERENCES

1. Batchelor G. K.: An Introduction to Fluid Dynamics, Cambridge University Press, Cambridge, UK, 1967.

2. Landau L. M. and Lifshitz E. M.: Mecanique des Fluides , MIR, Moscow 1989.

3. Landau L. M. and Lifshitz E. M.: Electrodynamique des Milieux Continus, MIR, Moscow 1990.

4. Crowley J. M.: Dimensionless Ratios in Electrohydrodynamics, in: Handbook of Electrostatics, Marcel-Dekker, 1994, 99-119.

5. Melcher J. R.: Continuum Electromechanics, The MIT Press, Cambridge, Massachusetts, 1981.

6. Rayleigh, Lord: On the equilibrium of liquid conducting masses charged with electricity, Phil. Mag. 14 (1882) 184.

7. Taylor G.I.: Disintegration of water drops in an electric field, Proc. R. Soc. London A280 (1964), 347.

8. Ramos A. and Castellanos A.: Conical points in liquid-liquid interfaces subjected to electric fields, Physics Letters A184 (1994), 268-272.

9. Wohlhuter F.K. and Basaran O.A.: Shapes and stability of pendant and sessile drops in an electric field, J. Fluied Mechanics 235 (1992), 481-510.

10. González H., McCluskey F.M.J., Castellanos A. and Barrero A.: Stabilization of dielectric liquid bridges by electric fields in the absence of gravity, Journal of Fluid Mechanics 206 (1989), 545-561.

11. Ramos A., González H. and Castellanos A.: Experiments on dielectric liquid bridges subjected to axial electric fields, Physics of Fluids 6 (1994), 3206-08.

12. Melcher J. R.: Field-coupled surface waves, The MIT Press, Cambridge, Massachusetts, 1964.

13. Saville D.A.: Electrohydrodynamics: the Taylor-Melcher leaky dielectric model, Annual Review Fluid Mechanics 29 (1997), 27-64.

14. Tobazéon R.: Electrohydrodynamic instabilities and electroconvection in the transient and a.c. regime of unipolar injection in insulating liquids: a review, Journal of Electrostatics 15 (1984), 359-384.

Chapter 4

THE TEMPERATURE EQUATION

In this chapter we carry out the classical procedure of thermodynamics of irreversible process [1],[2], to obtain the equation of the temperature in electrohydrodynamics. First, we review the thermodynamics of dielectric fluids at thermodynamic equilibrium [3],[4]. The internal electric energy of polar fluids depends on temperature through the permittivity. Given that temperature and internal energy are intimately connected, we split the internal electrical energy for polar fluids into two terms. The term that depends on temperature is added to the term for the internal energy of the fluid in the absence of the electric field to form the total internal energy. The remainder term, which is of the form $(1/2)\varepsilon_{np}E^2$, where ε_{np} is the permittivity at frequencies above the microwave region but below the short infra red region, is ascribed to the macroscopic electrical energy. The permittivity ε_{np} only depends on the ionic and electronic molecular polarizabilities, and therefore it is independent of temperature.

The assumption of local equilibrium allow us to extend the Gibbs relation to non-equilibrium processes, and the rate of variation of entropy can be calculated once the internal energy has been identified, since the production of the latter is equal to minus the production of kinetic, gravitational potential, and electrical energies (first principle of thermodynamics). The linearity between fluxes and forces, together with the Onsager reciprocity relations are used to determine the constitutive equations for the heat flux and electrical current density.

The equation for the temperature is then derived from the entropy equation. The physics behind the terms of electrical origin is discussed, and their orders of magnitude estimated. Finally, an important simplification of the EHD equations, referred to as the Boussinesq approximation in fluid dynamics, is extended to the case of EHD.

4.1 Internal energy

For a fluid consisting of structureless point particles of mass m_k, with interaction energy negligible compared to their kinetic energy, the total energy, U per unit volume averaging over a small volume δV is given by

$$U = \frac{< \sum_k \frac{1}{2} m_k \mathbf{u}_k^2 >}{\delta V} = \frac{1}{2}\rho \mathbf{u}^2 + \frac{< \sum_k \frac{1}{2} m_k {\mathbf{u}'_k}^2 >}{\delta V} = U_m + \rho U_o, \tag{4.1}$$

with U_m the kinetic energy per unit volume and U_o the internal energy per unit mass. For this simple fluid (ideal monatomic gas) the concept of temperature is introduced through the relation $\rho U_o = (3/2)Nk_BT$ with N the number of particles per unit volume, $k_B = 1.38 \times 10^{-23}\mathrm{JK}^{-1}$ Boltzmann's constant, and T the absolute temperature.

For ideal gases we have for the energy per unit mass

$$U_o = c_V T + U_o^o, \qquad \frac{p}{\rho} = \frac{k_B T}{m}, \tag{4.2}$$

with U_o^o constant for any given gas, m the molecular mass, p the pressure, and c_V the specific heat at constant volume. This specific heat is the energy added in the form of heat (keeping the volume constant) to increase the absolute temperature of the unit of mass by one degree.

For liquids, at equilibrium, the interaction energy between molecules is not negligible compared to their kinetic energy, and the exact expression of the internal energy as a function of temperature is not known. The specific heat at constant volume still gives the increase of internal energy as no external work is done in the process. However, c_V is not amenable to direct measurement, as enormous pressures develop in a liquid which is heated at constant density. What is easily measurable is the specific heat at constant pressure, c_p, which is related to c_V through the thermodynamic relation [5]

$$c_p - c_V = \frac{T\alpha^2}{\rho \kappa_T}, \tag{4.3}$$

with $\alpha = -(1/\rho)(\partial\rho/\partial T)_p$ the coefficient of thermal expansion, and $\kappa_T = (1/\rho)(\partial\rho/\partial p)_T$ the coefficient of isothermal compressibility. Since liquids expand very little, most of the heat transferred to the body at constant pressure goes to the internal energy, and therefore $c_p \simeq c_V$. It is found that c_p is less dependent on temperature than c_V, and in addition most of the processes for liquids are at constant pressure. It is then more convenient use for liquids the pressure instead of mass density as the independent variable.

In the range of temperatures and pressures of interest in most problems of EHD we may assume as the equation of state for the density of liquids [5] to be,

$$\rho(T,p) = \rho(T_o, P_o)[1 - \alpha(T - T_o) + \kappa_T(p - p_o)] \simeq \rho(T_o, P_o)[1 - \alpha(T - T_o)], \tag{4.4}$$

since $\kappa_T(p-p_o)$ represents only a one per cent even for $p-p_o = 100$ atmospheres. Therefore we will assume that density and permittivity are functions only of density and temperature, and independent of pressure.

4.2 Thermodynamics of dielectric fluids

In this section we introduce the basic concepts of thermodynamics for dielectric fluids at equilibrium, that is, for fluids at rest in the absence of gravity. In addition, all electric process are performed quasistatically with no flow of electric current, and polarization is in equilibrium with the electric field.

In classical thermodynamics it is normal to refer all magnitudes to the unit of mass. Since $dV = (1/\rho)dm$, the thermodynamic potentials per unit mass are obtained from those defined per unit volume by dividing by ρ. Thus, we have for the free energy per unit mass

$$F = F_o + \frac{\mathbf{D}^2}{2\rho\varepsilon}, \tag{4.5}$$

which has the differential:

$$dF = dF_o - \frac{1}{2\rho}\left(\frac{\partial\varepsilon}{\partial T}\right)_\rho E^2 dT + \frac{1}{\rho^2}\left[-\frac{1}{2}\mathbf{E}\cdot\mathbf{D} - \frac{1}{2}\rho\left(\frac{\partial\varepsilon}{\partial\rho}\right)_T E^2\right] d\rho + \frac{1}{\rho}\mathbf{E}\cdot d\mathbf{D}. \tag{4.6}$$

From the relation

$$dF = -SdT + \frac{p}{\rho^2}d\rho + \frac{1}{\rho}\mathbf{E}\cdot d\mathbf{D}, \tag{4.7}$$

we immediately obtain for the entropy per unit mass,

$$S = S_o + \frac{1}{2\rho}\left(\frac{\partial\varepsilon}{\partial T}\right)_\rho E^2 = S_o - \frac{D^2}{2\rho}\left[\frac{\partial}{\partial T}\left(\frac{1}{\varepsilon}\right)\right]_\rho, \tag{4.8}$$

and for the pressure

$$p = p_o - \frac{1}{2}\mathbf{E}\cdot\mathbf{D} - \frac{1}{2}\rho\left(\frac{\partial\varepsilon}{\partial\rho}\right)_T E^2. \tag{4.9}$$

This pressure is not the same as the pressure in equation (3.46) of chapter 3, since the latter was deduced at constant potential. Using $\tilde{F} = F_o - (1/2\rho)\varepsilon E^2$ instead of F, in the above thermodynamic relations, we obtain the pressure given in equation (3.46).

From the relation $U = F + TS$, the internal energy per unit mass is

$$U = U_o + \frac{D^2}{2\varepsilon\rho} + \frac{1}{2\rho}T\left(\frac{\partial\varepsilon}{\partial T}\right)_\rho E^2, \tag{4.10}$$

with U_o the internal energy in the absence of electric field.

An important relation is Gibbs relation

$$TdS = dU - \delta W = dU - \frac{p}{\rho^2}d\rho - \frac{1}{\rho}\mathbf{E}\cdot d\mathbf{D}. \tag{4.11}$$

We will now apply this relation to nonpolar and polar fluids.

Nonpolar fluids

For nonpolar fluids $U = U_o + (D^2/2\varepsilon\rho) = U_o + (1/2\rho)\varepsilon E^2$, since $(\partial\varepsilon/\partial T)_\rho = 0$, and we have for the differential of the internal energy

$$dU = dU_o + \frac{1}{\rho}\mathbf{E}\cdot d\mathbf{D} - \frac{1}{2\rho^2}\mathbf{E}\cdot\mathbf{D}d\rho - \frac{1}{2\rho}E^2\left(\frac{\partial\varepsilon}{\partial\rho}\right)_T d\rho. \tag{4.12}$$

Substituting this expression in the Gibbs relation, taking into account the expression for the pressure given by equation (4.9) we obtain

$$TdS = dU_o - \frac{p_o}{\rho^2}d\rho, \tag{4.13}$$

which is also valid in the absence of electric field. This is obvious since $S = S_o$.

Polar fluids

We have seen in section one that the temperature and the internal energy are closely associated magnitudes, and that the electrical energy also depends on temperature through the dielectric constant. It is then convenient to separate the electrical internal energy for polar fluids in two terms; one containing all temperature dependence and the other depending only on mass density, in the way

$$U_e = \frac{1}{2\rho}\left[(\varepsilon - \varepsilon_{np}) + T\left(\frac{\partial(\varepsilon-\varepsilon_{np})}{\partial T}\right)_\rho\right]E^2 + \frac{1}{2\rho}\varepsilon_{np}E^2, \tag{4.14}$$

where ε_{np} is the dielectric constant measured for frequencies well above the rotational frequencies of the molecules, so that the permanent dipoles are frozen in the electric field, but still below the vibrational ionic and atomic frequencies. The dielectric constant ε_{np} is only a function of the density. Formally, ε_{np} can be considered as the limit of p_o going to zero with p_o the permanent dipole of the molecules. Therefore ε_{np} is given by the corresponding formulae for gases and liquids, but with the polarizability α equal to the sum of the electronic and ionic polarizabilities. Thus, for polar gases

$$\varepsilon_{np} - \varepsilon_0 = \left[\frac{N_A(\alpha_e + \alpha_i)}{M}\right]\rho, \tag{4.15}$$

while for polar liquids, we have the equivalent of a Clausius-Mossotti relation

$$\frac{\varepsilon_{np} - \varepsilon_0}{\varepsilon_{np} + 2\varepsilon_0} = \left[\frac{N_A(\alpha_e + \alpha_i)}{3M\varepsilon_0}\right]\rho. \tag{4.16}$$

From

$$\mathbf{D} = \varepsilon\mathbf{E} = \varepsilon_{np}\mathbf{E} + (\varepsilon - \varepsilon_{np})\mathbf{E} = \mathbf{D}_{np} + \mathbf{P}_p = (\varepsilon_0\mathbf{E} + \mathbf{P}_{np}) + \mathbf{P}_p, \tag{4.17}$$

it is useful to define the following quantities: $\mathbf{D}_{np} = \varepsilon_{np}\mathbf{E}$ and $\mathbf{P}_p = (\varepsilon - \varepsilon_{np})\mathbf{E}$.

The sum of the internal energy in the absence of the electric field and the internal electrical energy may be written as

$$U = U_o + \frac{1}{2\rho}E^2\left[\frac{\partial}{\partial T}(\varepsilon - \varepsilon_{np})T\right]_\rho + \frac{D_{np}^2}{2\rho\varepsilon_{np}} = \tilde{U}_o + \tilde{U}_e, \tag{4.18}$$

with $\tilde{U}_o$ defined as the new internal energy

$$\tilde{U}_o = U_o + \frac{1}{2\rho}E^2\left[\frac{\partial}{\partial T}(\varepsilon - \varepsilon_{np})T\right]_\rho, \tag{4.19}$$

which includes a contribution from the electric field.

On the other hand, we identify $\tilde{U}_e = D_{np}^2/2\rho\varepsilon_{np}$ as the macroscopic electrical energy. The differential of the total energy is:

$$dU = d\tilde{U}_o + d\tilde{U}_e = d\tilde{U}_o + \frac{1}{\rho}\mathbf{E}\cdot d\mathbf{D}_{np} - \frac{1}{2\rho^2}\mathbf{E}\cdot\mathbf{D}_{np}d\rho - \frac{1}{2\rho}E^2\left(\frac{\partial\varepsilon_{np}}{\partial\rho}\right)_T d\rho. \tag{4.20}$$

The Gibbs relation for polar fluids is written as

$$TdS = dU - \frac{p}{\rho^2}d\rho - \frac{1}{\rho}\mathbf{E}\cdot d\mathbf{D} = d\tilde{U}_o + d\tilde{U}_e - \frac{p}{\rho^2}d\rho - \frac{1}{\rho}\mathbf{E}\cdot d\mathbf{D}_{np} - \frac{1}{\rho}\mathbf{E}\cdot d\mathbf{P}_p \tag{4.21}$$

and after substitution of equation (4.9) we obtain

$$TdS = d\tilde{U}_o - \frac{1}{\rho^2}\left\{p_o - \frac{1}{2}(\varepsilon - \varepsilon_{np})E^2 - \frac{1}{2}E^2\rho\left[\frac{\partial(\varepsilon - \varepsilon_{np})}{\partial\rho}\right]_T\right\}d\rho - \frac{1}{\rho}\mathbf{E}\cdot d\mathbf{P}_p, \tag{4.22}$$

as the Gibbs relation for polar fluids at thermodynamic equilibrium.

4.3 Energy conservation equation

For dielectric fluids the total energy is not simply the sum of the kinetic and potential energy due to intermolecular interactions, although it may be still possible to identify several types of well defined macroscopic energy densities. An example of these are homogeneous fluids in gravitational and electric fields where, in addition to kinetic energy, we have gravitational potential energy, and electric energy. The remaining energy is called the internal energy, and it is associated with the hidden atomic degrees of freedom, and molecular interactions, However, when any one of these macroscopic energies depend on temperature, as happens in our case with the internal electrical energy of polar liquids, the identification of the internal energy is not simple, and a choice must be made.

In a process the internal energy may change either because the macroscopic forces have done work on the constraints of the system and/or because energy at a microscopic scale has flowed into or out of the boundary of the system. The latter form of energy flow is called heat flow and its detailed expression depends on the fluid properties and the process involved.

Since the total energy must be conserved (first principle of thermodynamics) the total energy content within an arbitrary fixed volume V can only change if energy flows into (or out of) the volume through its boundary. In EHD this conservation law is expressed for nonpolar fluids as

$$\int_V \frac{\partial(U_m + U_g + U_e + \rho U_o)}{\partial t} dV + \int_S (\mathbf{J}_m + \mathbf{J}_g + \mathbf{J}_e + \mathbf{J}_o) \cdot \mathbf{n} dS = 0. \tag{4.23}$$

with $U_m = (1/2)\rho v^2$, $U_g = \rho g z$, $U_e = (1/2)\mathbf{E} \cdot \mathbf{D}$, and ρU_o are the kinetic, potential, electrical and internal energy per unit volume, and $\mathbf{J}_m$, $\mathbf{J}_g$, $\mathbf{J}_e$, $\mathbf{J}_o$ the corresponding flux densities. For polar fluids the same relation holds putting $\tilde{U}_e$ and $\tilde{U}_o$ in the place of U_e and U_o respectively.

Using the conservation laws for each of the macroscopic energy densities, to obtain the relations

$$\int_V \frac{\partial U_k}{\partial t} dV + \int_S \mathbf{J}_k \cdot \mathbf{n} dS = \int_V P_k dV, \qquad (k = m, g, e), \tag{4.24}$$

where P_k represent the sources or sinks per unit volume of the energy U_k, we have for the internal energy

$$\int_V \frac{\partial \rho U_o}{\partial t} dV + \int_S (\rho U_o \mathbf{u} + \mathbf{J}'_o) dS = -\int_V (P_m + P_g + P_e) dV, \tag{4.25}$$

where we have conveniently distinguished the internal energy convected by the fluid, $\rho U_o \mathbf{u}$, from the flow of energy measured by an observer comoving with the element of fluid, $\mathbf{J}'_o$. As this relation holds for any volume we have, using the divergence theorem,

$$\frac{\partial \rho U_o}{\partial t} + \nabla \cdot (\rho U_o \mathbf{u}) + \nabla \cdot \mathbf{J}'_o = -P_m - P_g - P_e \tag{4.26}$$

Using the mass conservation equation and the identities

$$\frac{\partial \rho U_o}{\partial t} = \rho \frac{\partial U_o}{\partial t} + U_o \frac{\partial \rho}{\partial t}, \tag{4.27}$$

$$\nabla \cdot (\rho U_o \mathbf{u}) = U_o \nabla \cdot (\rho \mathbf{v}) + \rho(\mathbf{u} \cdot \nabla) U_o, \tag{4.28}$$

we have

$$\rho \frac{\partial U_o}{\partial t} + \rho(\mathbf{u} \cdot \nabla) U_o = -\nabla \cdot \mathbf{J}'_o - P_m - P_g - P_e, \tag{4.29}$$

or equivalently

$$\rho \frac{dU_o}{dt} = -\nabla \cdot \mathbf{J}'_o - P_m - P_g - P_e. \tag{4.30}$$

We now calculate the source terms for the kinetic energy, and the potential energy.

4.3.1 Kinetic energy conservation equation

Multiplying the Navier-Stokes equation (see section 3.2.4) by $\mathbf{u}$ we obtain

$$\rho\mathbf{u}\cdot\frac{d\mathbf{u}}{dt}=\mathbf{u}\cdot(\nabla\cdot\mathbf{T}^m)+q\mathbf{u}\cdot\mathbf{E}-\frac{1}{2}E^2\mathbf{u}\cdot\nabla\varepsilon+\mathbf{u}\cdot\nabla p_{st}+\rho\mathbf{g}\cdot\mathbf{u}, \tag{4.31}$$

with p_{st} the electrostriction pressure, given by

$$p_{st}=\frac{1}{2}\mathbf{E}^2\rho\left(\frac{\partial\varepsilon}{\partial\rho}\right)_T=\frac{1}{2}a\varepsilon\mathbf{E}^2, \tag{4.32}$$

where $a=(\rho/\varepsilon)(\partial\varepsilon/\partial\rho)_T$ is the electrostriction coefficient. Equation (4.31) can be written as,

$$\rho\mathbf{u}\cdot\frac{d\mathbf{u}}{dt}=\mathbf{u}\cdot(\nabla\cdot\mathbf{T}^m)+\mathbf{u}\cdot\mathbf{f}_e+\rho\mathbf{g}\cdot\mathbf{u}, \tag{4.33}$$

with $\mathbf{f}_e$ the electrical force density.

The first term of the right hand side can be re-written:

$$\mathbf{u}\cdot(\nabla\cdot\mathbf{T}^m)=\nabla\cdot(\mathbf{T}^m\cdot\mathbf{u})-\mathbf{T}^m:\nabla\mathbf{u}, \tag{4.34}$$

with $\mathbf{T}^m:\nabla\mathbf{u}=T^m_{ij}(\partial u_i/\partial x_j)$. We have that:

$$\mathbf{T}^m:\nabla\mathbf{u}=(-p_o\boldsymbol{\delta}+\boldsymbol{\sigma}):\nabla\mathbf{u}=\left\{-p_o\boldsymbol{\delta}+2\eta\left[\mathbf{e}-\frac{1}{3}(\mathrm{Tr}\ \mathbf{e})\boldsymbol{\delta}\right]+\lambda(\mathrm{Tr}\ \mathbf{e})\boldsymbol{\delta}\right\}:\nabla\mathbf{u}, \tag{4.35}$$

with p_o the pressure in the absence of the electric field. Putting $\nabla\mathbf{u}=\mathbf{e}+\boldsymbol{\omega}$ and taking into account that $\boldsymbol{\omega}:\boldsymbol{\delta}=0$ and $\mathbf{e}:(\mathrm{Tr}\ \mathbf{e})\boldsymbol{\delta}=(\mathrm{Tr}\ \mathbf{e})^2=(\nabla\cdot\mathbf{u})^2$ it is easy to show that

$$\mathbf{T}^m_{ij}e_{ij}=-p_o\nabla\cdot\mathbf{u}+2\eta\left[e_{ij}-\frac{1}{3}(\nabla\cdot\mathbf{u})\delta_{ij}\right]^2+\lambda(\nabla\cdot\mathbf{u})^2=\frac{p_o}{\rho}\frac{d\rho}{dt}+\Phi. \tag{4.36}$$

The term $(p_o/\rho)d\rho/dt$ represents the increase in the internal energy due to compression of the fluid, and is zero for an incompressible fluid. The second term, Φ, is the rate at which energy is dissipated, irreversibly, by viscosity. For an incompressible fluid Φ is given by

$$\Phi=2\eta e_{ij}^2. \tag{4.37}$$

We have finally for the kinetic energy density

$$\frac{\partial U_m}{\partial t}+\nabla\cdot(U_m\mathbf{u}-\mathbf{T}^m\cdot\mathbf{u})=\mathbf{u}\cdot\mathbf{f}_e+\rho\mathbf{g}\cdot\mathbf{u}-\frac{p_o}{\rho}\frac{d\rho}{dt}-\Phi, \tag{4.38}$$

and therefore $\mathbf{J}_m=U_m\mathbf{u}-\mathbf{T}^m\cdot\mathbf{u}$ is the kinetic energy density flux, and

$$P_m=\mathbf{u}\cdot\mathbf{f}_e+\rho\mathbf{g}\cdot\mathbf{u}-\frac{p_o}{\rho}\frac{d\rho}{dt}-\Phi, \tag{4.39}$$

is the source term.

4.3.2 Gravitational potential energy conservation equation

The potential energy of an element of fluid at height z is $U_g = \rho g z$. We have

$$\frac{\partial \rho g z}{\partial t} = gz\frac{\partial \rho}{\partial t} = -gz\nabla \cdot (\rho \mathbf{u}) = -\nabla \cdot (\rho g z \mathbf{u}) + \rho \mathbf{u} \cdot \nabla(gz). \tag{4.40}$$

Since $\nabla(gz) = -\mathbf{g}$, the final expression for the conservation of the potential energy is given by

$$\frac{\partial U_g}{\partial t} + \nabla \cdot (U_g \mathbf{u}) = P_g, \tag{4.41}$$

with

$$P_g = -\rho \mathbf{g} \cdot \mathbf{u}. \tag{4.42}$$

4.3.3 Internal energy conservation equation for nonpolar fluids

Let us consider the conservation law for the electrical internal energy U_e of nonpolar fluids. As discussed in chapter 2 for these fluids the permittivity is only a function of the mass density, i.e. $\varepsilon = \varepsilon(\rho)$.

The internal electrical energy per unit volume is

$$U_e = \frac{1}{2}\mathbf{E} \cdot \mathbf{D} = \frac{\mathbf{D}^2}{2\varepsilon}, \tag{4.43}$$

and therefore

$$\frac{\partial}{\partial t}\left(\frac{1}{2}\mathbf{E} \cdot \mathbf{D}\right) = \mathbf{E} \cdot \frac{\partial \mathbf{D}}{\partial t} - \frac{1}{2}\mathbf{E}^2\frac{\partial \varepsilon}{\partial t}. \tag{4.44}$$

From the fourth Maxwell equation

$$\mathbf{E} \cdot \frac{\partial \mathbf{D}}{\partial t} = \mathbf{E} \cdot (\nabla \times \mathbf{H} - \mathbf{J}) = -\mathbf{E} \cdot \mathbf{J}' - q\mathbf{E} \cdot \mathbf{u} + \mathbf{E} \cdot \nabla \times \mathbf{H}. \tag{4.45}$$

The following equation holds

$$\nabla \cdot (\mathbf{E} \times \mathbf{H}) = \mathbf{H} \cdot \nabla \times \mathbf{E} - \mathbf{E} \cdot \nabla \times \mathbf{H} = -\mathbf{E} \cdot \nabla \times \mathbf{H}, \tag{4.46}$$

since in EHD $\nabla \times \mathbf{E} = 0$.

For a given fluid element we also have that

$$\frac{d\varepsilon}{dt} = \left(\frac{\partial \varepsilon}{\partial \rho}\right)_T \frac{d\rho}{dt}, \tag{4.47}$$

and consequently,

$$\begin{aligned}\frac{1}{2}\mathbf{E}^2\frac{\partial \varepsilon}{\partial t} &= \frac{1}{2}\mathbf{E}^2\left(\frac{d\varepsilon}{dt} - \mathbf{u} \cdot \nabla\varepsilon\right) = \frac{1}{2}\mathbf{E}^2\left[\left(\frac{\partial \varepsilon}{\partial \rho}\right)_T \frac{d\rho}{dt} - \mathbf{u} \cdot \nabla\varepsilon\right] \\ &= \frac{1}{2}\mathbf{E}^2\rho\left(\frac{\partial \varepsilon}{\partial \rho}\right)_T \left(\frac{1}{\rho}\frac{d\rho}{dt}\right) - \frac{1}{2}\mathbf{E}^2\mathbf{u} \cdot \nabla\varepsilon. \end{aligned} \tag{4.48}$$

Using the mass conservation equation in the form $(1/\rho)(d\rho/dt) = -\nabla \cdot \mathbf{u}$, and the identity $\nabla \cdot (p_{st}\mathbf{u}) = p_{st}\nabla \cdot \mathbf{u} + \mathbf{u} \cdot \nabla p_{st}$ where p_{st} is the electrostriction pressure given by

$$p_{st} = \frac{1}{2}\mathbf{E}^2\rho\left(\frac{\partial \varepsilon}{\partial \rho}\right)_T, \tag{4.49}$$

we arrive at the conservation law for the internal electric energy

$$\frac{\partial U_e}{\partial t} + \nabla \cdot (\mathbf{E} \times \mathbf{H} - p_{st}\mathbf{u}) = -\mathbf{E} \cdot \mathbf{J}' - \mathbf{f}_e \cdot \mathbf{u}. \tag{4.50}$$

Here $\mathbf{E} \times \mathbf{H}$ is the Poynting vector, which, together with $-p_{st}\mathbf{u}$, gives us the flux of internal electrical energy. $\mathbf{J}' = \mathbf{J} - q \cdot \mathbf{u}$ is the current density measured in the comoving frame of reference of the fluid element, and $\mathbf{f}_e$ is the electric force density given by

$$\mathbf{f}_e = q\mathbf{E} - \frac{1}{2}E^2\nabla\varepsilon + \nabla p_{st}. \tag{4.51}$$

The source term for the electrical internal energy of nonpolar fluids is then

$$P_e = -\mathbf{E} \cdot \mathbf{J}' - \mathbf{f}_e \cdot \mathbf{u}. \tag{4.52}$$

Putting the source terms for kinetic, potential and electrical energy densities into the right hand side of equation (4.30), we obtain

$$\rho\frac{dU_o}{dt} = -\nabla \cdot \mathbf{J}'_o + \Phi + \mathbf{E} \cdot \mathbf{J}' + \frac{p_o}{\rho}\frac{d\rho}{dt}. \tag{4.53}$$

This is the energy conservation equation for nonpolar fluids.

4.3.4 Internal energy conservation equation for polar liquids

As discussed in section 4.2, the inseparable nature of the internal degrees of freedom of thermal and electrical origin, forced us to define a new internal energy per unit mass $\tilde{U}_o$, and a new electrical energy, independent of temperature, defined as

$$\tilde{U}_e = \frac{D_{np}^2}{2\varepsilon_{np}}. \tag{4.54}$$

Now, we determine a conservation law for the newly defined electrical energy, along the same lines as we did for nonpolar liquids. We have that

$$\frac{\partial}{\partial t}\left(\frac{D_{np}^2}{2\varepsilon_{np}}\right) = \mathbf{E} \cdot \frac{\partial \mathbf{D}_{np}}{\partial t} - \frac{1}{2}E^2\frac{\partial \varepsilon_{np}}{\partial t}. \tag{4.55}$$

Putting the fourth Maxwell equation in the form,

$$\nabla \times \mathbf{H} = \mathbf{J} + \frac{\partial \mathbf{D}_{np}}{\partial t} + \frac{\partial \mathbf{P}_p}{\partial t}, \tag{4.56}$$

multiplying this equation by $\mathbf{E}$, and using the well known relations

$$\mathbf{J} = \mathbf{J}' + q\mathbf{u}, \qquad \frac{\partial}{\partial t} = \frac{d}{dt} - \mathbf{u}\cdot\nabla, \qquad \mathbf{P}_p = \mathbf{P}'_p, \tag{4.57}$$

we obtain

$$\mathbf{E}\cdot\frac{\partial \mathbf{D}_{np}}{\partial t} = -\nabla\cdot(\mathbf{E}\times\mathbf{H}) - \mathbf{E}\cdot\mathbf{J}' - q\mathbf{E}\cdot\mathbf{u} - \mathbf{E}\cdot\frac{d\mathbf{P}'_p}{dt} + \mathbf{E}(\mathbf{u}\cdot\nabla)\mathbf{P}_p. \tag{4.58}$$

On the other hand, substituting ε by ε_{np} in (4.48), we arrive at the equation

$$\frac{1}{2}E^2\frac{\partial \varepsilon_{np}}{\partial t} = \frac{p_{st}^{np}}{\rho}\frac{d\rho}{dt} - \frac{1}{2}E^2\nabla\varepsilon_{np}\cdot\mathbf{u}, \tag{4.59}$$

with the partial electrostriction pressure, p_{st}^{np} defined as

$$p_{st}^{np} = \frac{1}{2}\mathbf{E}^2\rho\left(\frac{\partial\varepsilon_{np}}{\partial\rho}\right)_T. \tag{4.60}$$

Substituting (4.58), (4.59) in equation (4.55) gives

$$\frac{\partial\tilde{U}_e}{\partial t}+\nabla\cdot(\mathbf{E}\times\mathbf{H}) = -\mathbf{E}\cdot\mathbf{J}'-\mathbf{E}\cdot\frac{d\mathbf{P}'_p}{dt}+\mathbf{E}(\mathbf{u}\cdot\nabla)\mathbf{P}_p-q\mathbf{E}\cdot\mathbf{u}-\frac{p_{st}^{np}}{\rho}\frac{d\rho}{dt}+\frac{1}{2}E^2\nabla\varepsilon_{np}\cdot\mathbf{u}. \tag{4.61}$$

The production term for the new electrical energy is

$$P_{\tilde{e}} = -\mathbf{E}\cdot\mathbf{J}' - q\mathbf{E}\cdot\mathbf{u} - \mathbf{E}\cdot\frac{d\mathbf{P}'_p}{dt} + \mathbf{E}(\mathbf{u}\cdot\nabla)\mathbf{P} - \frac{p_{st}^{np}}{\rho}\frac{d\rho}{dt} + \frac{1}{2}E^2\nabla\varepsilon_{np}\cdot\mathbf{u}. \tag{4.62}$$

Using the identity

$$\nabla p_{st}\cdot\mathbf{u} = \nabla\cdot(p_{st}\mathbf{u}) - p_{st}\nabla\cdot\mathbf{u} = \nabla\cdot(p_{st}\mathbf{u}) + \frac{p_{st}}{\rho}\frac{d\rho}{dt}, \tag{4.63}$$

together with

$$\begin{aligned}\frac{1}{2}E^2\nabla(\varepsilon-\varepsilon_{np})\cdot\mathbf{u} - \mathbf{E}(\mathbf{u}\cdot\nabla)\mathbf{P}_p &= \frac{1}{2}E^2u_i\frac{\partial}{\partial x_i}(\varepsilon-\varepsilon_{np})\\ -E_k\left(u_i\frac{\partial}{\partial x_i}\right)(\varepsilon-\varepsilon_{np})E_k &= -\frac{1}{2}E^2u_i\frac{\partial}{\partial x_i}(\varepsilon-\varepsilon_{np})\\ -(\varepsilon-\varepsilon_{np})u_i\frac{\partial}{\partial x_i}\left(\frac{1}{2}E^2\right) &= -\nabla\cdot\left[\frac{1}{2}(\varepsilon-\varepsilon_{np})E^2\mathbf{u}\right] - \frac{1}{2\rho}(\varepsilon-\varepsilon_{np})E^2\frac{d\rho}{dt},\end{aligned} \tag{4.64}$$

we get

$$\begin{aligned}P_m + P_g + P_{\tilde{e}} &= \nabla\cdot\left[\frac{1}{2}(\varepsilon-\varepsilon_{np})E^2\mathbf{u} + p_{st}\mathbf{u}\right]\\ &- \frac{1}{\rho}\left\{p_o - \frac{1}{2}(\varepsilon-\varepsilon_{np})E^2 - \frac{1}{2}E^2\rho\left[\frac{\partial(\varepsilon-\varepsilon_{np})}{\partial\rho}\right]_T\right\}\frac{d\rho}{dt} - \Phi\\ &- \mathbf{E}\cdot\mathbf{J}' - \mathbf{E}\cdot\frac{d\mathbf{P}'_p}{dt}.\end{aligned} \tag{4.65}$$

The first term of the right hand side is in the form of a divergence, and it may therefore be included in the flux of kinetic, potential and electrical energy. Only the remaining four terms represent a true production of the internal energy.

According to the conservation of total energy, the corresponding equation for the new internal energy, $\tilde{U}_o$, is given by

$$\begin{aligned}\rho\frac{d\tilde{U}_o}{dt} &= -\nabla\cdot\mathbf{J}'_{\tilde{o}} + \frac{1}{\rho}\left\{p_o - \frac{1}{2}(\varepsilon-\varepsilon_{np})E^2 - \frac{1}{2}E^2\rho\left[\frac{\partial(\varepsilon-\varepsilon_{np})}{\partial\rho}\right]_T\right\}\frac{d\rho}{dt}\\ &+ \Phi + \mathbf{E}\cdot\mathbf{J}' + \mathbf{E}\cdot\frac{d\mathbf{P}'_p}{dt},\end{aligned} \tag{4.66}$$

where $\mathbf{J}'_{\tilde{o}}$ is the flux of $\tilde{U}_o$. This is the appropiate equation for the conservation of electrical internal energy of polar fluids.

4.4 Entropy production in dielectric fluids

To be able to determine the entropy production in fluids we must assume that the continuum hypothesis holds and that simultaneously each element of volume δV is in thermodynamic equilibrium, even though globally, the system may be far from thermodynamic equilibrium. This will be true if there are enough collisions over the time δt among the molecules of δV to attain thermodynamic equilibrium. This time, δt, must be infinitesimal from the macroscopic point of view, i.e. quite small compared to any typical scale of variation of the macroscopic quantities, but very large compared to the time between molecular collisions. If τ_c is the typical collision time between molecules, it must be $\tau_c \ll \delta t \ll \tau$ with τ any of the characteristic time of variation of the different subprocesses ocurring in the system. When these conditions hold we say that there is local equilibrium in the system, and we may define the usual thermodynamic quantities for each of the fluid elements. Obviously, these thermodynamic quantities vary now from point to point of our system.

To calculate the entropy production we postulate that the Gibbs relation holds locally. We obtain the law of conservation for the entropy, substituting in Gibbs relation the rate of change of the internal energy, that we have obtained from the phenomenological mechanical and electrical equations that describe dynamic processes out of thermodynamic equilibrium.

4.4.1 Entropy production for nonpolar fluids

In equilibrium thermodynamics the total differential of the entropy per unit mass is given by the Gibbs relation

$$TdS = dU_o - \frac{p_o}{\rho^2}d\rho. \tag{4.67}$$

The hypothesis of local equilibrium implies that

$$T\frac{dS}{dt} = \frac{dU_o}{dt} - \frac{p_o}{\rho^2}\frac{d\rho}{dt}, \tag{4.68}$$

is valid for a given mass element.

Substituting equation (4.53) in this equation, after multiplying by ρ, dividing by T, and reordering the terms, we obtain the equation for the entropy per unit volume

$$\rho\frac{dS}{dt} = -\frac{1}{T}\nabla\cdot\mathbf{J'}_o + \frac{1}{T}\Phi + \frac{1}{T}\mathbf{E}\cdot\mathbf{J'}. \tag{4.69}$$

It is convenient to re-write this equation in the form

$$\rho\frac{dS}{dt} = -\nabla\cdot\left(\frac{\mathbf{J'}_o}{T}\right) - \frac{1}{T^2}\mathbf{J'}_o\cdot\nabla T + \frac{1}{T}\Phi + \frac{1}{T}\mathbf{E}\cdot\mathbf{J'}. \tag{4.70}$$

From this equation it follows that the expression for the entropy flux is

$$\mathbf{J'}_S = \frac{\mathbf{J'}_o}{T}, \tag{4.71}$$

and the entropy production is

$$P_S = \mathbf{J'}_o\cdot\nabla\left(\frac{1}{T}\right) + \frac{\Phi}{T} + \mathbf{J'}\cdot\left(\frac{\mathbf{E}}{T}\right). \tag{4.72}$$

The entropy production is zero if the system is in thermodynamic equilibrium. For irreversible processes it can be expressed in the form

$$P_S = \sum_k J_k X_k, \tag{4.73}$$

with J_k the thermodynamic fluxes, and X_k the thermodynamic forces. In our case

$$P_S = \mathbf{J'}_o\cdot\nabla\left(\frac{1}{T}\right) + \frac{1}{T}2\eta\left[e_{ij} - \frac{1}{3}(\nabla\cdot\mathbf{u})\delta_{ij}\right]^2 + \frac{1}{T}\lambda(\nabla\cdot\mathbf{u})^2 + \mathbf{J'}\cdot\left(\frac{\mathbf{E}}{T}\right). \tag{4.74}$$

The viscous dissipation term is split into two terms of different tensorial order. The first one is driven by $(e_{ij} - \frac{1}{3}\nabla\cdot\mathbf{u}\delta_{ij})$ and is a tensor of second order, and the second by $\nabla\cdot\mathbf{u}$ and is a scalar (tensor of order zero). The first term depends on the dynamic viscosity and the second one on the bulk viscosity. We know that for isotropic systems, fluxes and thermodynamic forces of different tensorial character do not couple (Curie symmetry principle). Therefore both the dynamic viscosity and the bulk viscosity must be positive, since each term must be independently greater than zero.

It is also known empirically that, in many experimental conditions, the irreversible fluxes are linear functions of the thermodynamic forces. Moreover, for fluxes and forces of the same tensorial nature, in general

$$J_k = \sum_i L_{ik}X_k. \tag{4.75}$$

In our case we must have

$$\mathbf{J'}_o = L_{11}\nabla\left(\frac{1}{T}\right) + L_{12}\left(\frac{\mathbf{E}}{T}\right), \tag{4.76}$$

and

$$\mathbf{J}' = L_{21}\nabla\left(\frac{1}{T}\right) + L_{22}\left(\frac{\mathbf{E}}{T}\right). \tag{4.77}$$

Since entropy production must be positive

$$\mathbf{J}'_o \cdot \nabla\left(\frac{1}{T}\right) + \mathbf{J}' \cdot \left(\frac{\mathbf{E}}{T}\right) > 0, \tag{4.78}$$

and consequently the following inequalities must be satisfied

$$L_{11} > 0, \qquad L_{22} > 0, \qquad (L_{12} + L_{21})^2 < 4L_{11}L_{22}. \tag{4.79}$$

In addition, according to the reciprocity theorem of Onsager

$$L_{12} = L_{21}, \tag{4.80}$$

and from the last relation of (4.79) we have

$$L_{12} = L_{21} < \sqrt{L_{11}L_{22}}. \tag{4.81}$$

Therefore,

$$\mathbf{J}'_o = L_{11}\nabla\left(\frac{1}{T}\right) + L_{12}\left(\frac{\mathbf{E}}{T}\right), \tag{4.82}$$

$$\mathbf{J}' = L_{12}\nabla\left(\frac{1}{T}\right) + L_{22}\left(\frac{\mathbf{E}}{T}\right). \tag{4.83}$$

The off-diagonal terms are responsible for thermoelectric effects. Since these terms are documented only for ionic melts and molten semiconductors of low conductivity such as selenium [6], and these liquids behave as liquids in the ohmic regime, we will restrict our considerations to ohmic liquids. However, we must keep in mind that, in EHD, fluids are poor conductors, and molten semiconductors of high conductivity such as molten silicon, are outside the realm of EHD.

In many practical situations, for ordinary fluids in the absence of electric fields, $\mathbf{J}'_o$, the heat flow is given by Fourier law $\mathbf{J}'_q = -k\nabla T$ with $k = L_{11}/T^2$ the thermal conductivity of the fluid. It should be noted that this law is a phenomenological one, and that, for example, in mixtures we have a contribution to the flow of internal energy due to the gradient of the concentration in addition to the gradient of T. This is not difficult to understand since a flux of different molecules can transport internal energy in various ways depending on the processes that are taking into account.

On the other hand, for isothermal systems and low values of the electric field $\mathbf{E}$, the current density, $\mathbf{J}'$, is given by Ohm's law $\mathbf{J}' = \sigma\mathbf{E} = \sigma T(\mathbf{E}/T)$ with σ the electrical conductivity of the fluid. When both a thermal gradient and an electric field are present we have

$$\mathbf{J}'_o = kT^2\nabla\left(\frac{1}{T}\right) + L_{12}\left(\frac{\mathbf{E}}{T}\right), \tag{4.84}$$

$$\mathbf{J}' = L_{12}\nabla\left(\frac{1}{T}\right) + \sigma T\left(\frac{\mathbf{E}}{T}\right), \tag{4.85}$$

where now both k and σ are functions of the temperature and the electric field simultaneously. Following Landau and Lifshitz [3], we put $L_{12} = \sigma\alpha T^2$, where α is referred to as the thermoelectric coefficient.

The final expressions for the flux densities are

$$\mathbf{J}'_o = kT^2\nabla\left(\frac{1}{T}\right) + \alpha\sigma T^2\left(\frac{\mathbf{E}}{T}\right) = -k\nabla T - \alpha\sigma T\mathbf{E}, \tag{4.86}$$

$$\mathbf{J}' = \sigma\alpha T^2\nabla\left(\frac{1}{T}\right) + \sigma T\left(\frac{\mathbf{E}}{T}\right) = -\sigma\alpha\nabla T + \sigma\mathbf{E}. \tag{4.87}$$

If the system is isolated electrically, so that no electrical current can flow in the stationary state and the fluid is not isothermal, then $\mathbf{J}' = 0$, and an electric field appears in the liquid, given by

$$\mathbf{E} = \alpha\nabla T. \tag{4.88}$$

Some measurements exist for molten semiconductors of low conductivity such as selenium [6], and for molten salts. The values reported for the thermoelectric coefficient are $\alpha \le 10^{-4}\mathrm{VK}^{-1}$ The electrical potential difference generated is at most 0.3 V, so usually the contribution of this term to the electrical current is quite small, although it may be the cause of certain instabilities in heated semiconductor melts [7].

For molten semiconductors at ambient temperature the ratio of both terms in equation (4.86) is

$$\frac{k\nabla T}{\alpha\sigma\mathbf{E}} \sim \frac{kT}{\phi_o\alpha\sigma} \sim 10^3 \gg 1, \tag{4.89}$$

where we have taken $k \sim 100\ \mathrm{W\,m^{-1}\,K^{-1}}$, $\phi_o \sim 1$ V, and $\sigma \simeq 3\times10^{-4}\ \mathrm{S\,m^{-1}}$.

Experimental values of coefficient α are not available for ordinary dielectric liquids. Due to the small values for the thermoelectric coefficient and the electrical conductivity for ordinary insulating liquids, these cross effects seems to be negligible. So far no physical phenomena related to these terms have been reported in the literature.

4.4.2 Entropy production for polar fluids

The local equilibrium hypothesis allow us to extend Gibbs relation at thermodynamic equilibrium

$$TdS = d\tilde{U}_o - \frac{1}{\rho^2}\left\{p_o - \frac{1}{2}(\varepsilon-\varepsilon_{np})E^2 - \frac{1}{2}E^2\rho\left[\frac{\partial(\varepsilon-\varepsilon_{np})}{\partial\rho}\right]_T\right\}\frac{d\rho}{dt} - \frac{1}{\rho}\mathbf{E_{eq}}\cdot d\mathbf{P}'_p, \tag{4.90}$$

to irreversible processes. We have denoted the electric field in the last term with the subscript *eq* to indicate that its value is in equilibrium with the value of $\mathbf{P}'_p$ (remember that the process is done quasistatically in equilibrium thermodynamics, which implies that the polarization vector follows the electric field with no time lag).

Dividing by dt, substituting for the internal energy from equation (4.66), multiplying the equation by ρ/T, and reordering, we have the entropy conservation equation for polar fluids

$$\rho\frac{dS}{dt} = -\frac{1}{T}\nabla\cdot\mathbf{J}'_{\bar{o}} + \frac{1}{T}\mathbf{E}\cdot\mathbf{J}' + \frac{1}{T}\Phi + \frac{1}{T}(\mathbf{E}-\mathbf{E}_{eq})\cdot\frac{d\mathbf{P}'_p}{dt}. \tag{4.91}$$

Putting the first term of the right hand side in the form of a divergence plus the term $\mathbf{J}'_{\bar{o}}\cdot\nabla(1/T)$ we obtain for the entropy per unit volume.

$$\rho\frac{dS}{dt} + \nabla\cdot\mathbf{J}'_S = P_S, \tag{4.92}$$

with the entropy flux given by

$$\mathbf{J}'_S = \frac{\mathbf{J}'_{\bar{o}}}{T}, \tag{4.93}$$

and the entropy production by

$$P_S = \mathbf{J}'_{\bar{o}}\cdot\nabla\left(\frac{1}{T}\right) + \mathbf{J}'\cdot\left(\frac{\mathbf{E}}{T}\right) + \frac{\Phi}{T} + \frac{d\mathbf{P}'_p}{dt}\cdot\left[\frac{(\mathbf{E}-\mathbf{E}_{eq})}{T}\right], \tag{4.94}$$

that must be positive according to the Second Principle of Thermodynamics. This expression also satisfies the requirement that it is zero if the system is in thermodynamic equilibrium. As in the case of nonpolar fluids, the entropy is a bilinear form of the flux densities and thermodynamic forces.

Effects related to the off-diagonal elements of the matrix of coefficients have not been described for polar fluids in the literature to the knowledge of the author. Therefore we will suppose that to a first approximation the matrix of coefficients is diagonal.

For simplicity we assume that Fourier's law, and Ohm's law are valid for the heat flux and the electrical current density respectively. Along the same lines as de Groot and Mazur [1], we assume that

$$\frac{d\mathbf{P}'_p}{dt} = \frac{L_{33}}{T}(\mathbf{E}-\mathbf{E}_{eq}). \tag{4.95}$$

This relation is strictly true for a perfectly insulating isothermal fluid in hydrostatic equilibrium.

Since $\mathbf{P}'_p = \mathbf{P}_p = (\varepsilon - \varepsilon_{np})\mathbf{E}_{eq} = \chi\mathbf{E}_{eq}$, where χ is the electrical susceptibility associated to $\mathbf{P}'_p$, the differential equation satisfied by the polarization vector is

$$\frac{d\mathbf{P}'_p}{dt} + \frac{1}{\tau_P}\mathbf{P}'_p = \frac{\chi}{\tau_P}\mathbf{E}, \tag{4.96}$$

with

$$\tau_P = \frac{\chi T}{L_{33}}. \tag{4.97}$$

For an applied electric field $\mathbf{E}(t) = \mathbf{E}_0 h(t)$ with $\mathbf{E}_0$ constant and $h(t)$ the Heaviside function, $h(t) = 1$ if $t > 0$ and $h(t) = 0$ if $t < 0$, the solution to this equation is

$$\mathbf{P'}_p(t) = \chi \mathbf{E}_0 \left(1 - e^{-\frac{t}{\tau_P}}\right), \tag{4.98}$$

for $t > 0$ and zero for $t < 0$. The time constant τ_P indicates the characteristic time needed by the polarization vector to attain its equilibrium value with the applied electric field, and is called the polarization relaxation time.

In the general case of a time-dependent electric field $\mathbf{E}(t)$, equation (4.96) is solved in the frequency domain expanding $\mathbf{P'}_p$ and $\mathbf{E}$ into Fourier integrals. We obtain

$$-i\omega \mathbf{P'}_p(\omega) = -\frac{1}{\tau_P}\mathbf{P'}_p(\omega) + \frac{\chi}{\tau_P}\mathbf{E}(\omega). \tag{4.99}$$

Defining the complex electrical susceptibility in the frequency domain

$$\chi(\omega) = \frac{\chi}{1 - i\omega\tau_P}, \tag{4.100}$$

is

$$\mathbf{P'}_p(\omega) = \chi(\omega)\mathbf{E}(\omega), \tag{4.101}$$

and from the convolution theorem, we have in the time domain

$$\mathbf{P'}_p(t) = \int_{-\infty}^{\infty} \chi(\tau)\mathbf{E}(t-\tau)d\tau, \tag{4.102}$$

with

$$\chi(\tau) = \int_{-\infty}^{+\infty} \chi(\omega)e^{-i\omega\tau}\, d\omega, \tag{4.103}$$

In this way we have recovered the Debye theory of dielectric relaxation for the case of one relaxation time. To obtain the general theory of dielectric relaxation that includes a set of relaxation times we need to consider the internal degrees of freedom related to the polarization and the corresponding production of entropy must now take into account the internal variables, with the associated fluxes and thermodynamic forces.

4.5 Equation for the temperature in EHD

The equation for the temperature is obtained from the entropy equation. The entropy defined by $S = -(\partial F/\partial T)_{\rho,\mathbf{D}}$ is function of temperature, mass density and displacement vector. We have

$$dS = \left(\frac{\partial S}{\partial T}\right)_{\rho,\mathbf{D}} dT + \left(\frac{\partial S}{\partial \rho}\right)_{T,\mathbf{D}} d\rho + \left(\frac{\partial S}{\partial \mathbf{D}}\right)_{T,\rho} \cdot d\mathbf{D}, \tag{4.104}$$

with

$$\left(\frac{\partial S}{\partial T}\right)_{\rho,\mathbf{D}} = \left(\frac{\partial S_o}{\partial T}\right)_{\rho} - \frac{D^2}{2\rho}\left[\frac{\partial^2}{\partial T^2}\left(\frac{1}{\varepsilon}\right)\right]_{\rho} = \frac{c_V}{T} - \frac{D^2}{2\rho}\left[\frac{\partial^2}{\partial T^2}\left(\frac{1}{\varepsilon}\right)\right]_{\rho}. \tag{4.105}$$

and

$$\left(\frac{\partial S}{\partial \mathbf{D}}\right)_{T,\rho} = -\frac{\mathbf{D}}{\rho}\left[\frac{\partial}{\partial T}\left(\frac{1}{\varepsilon}\right)\right]_{\rho}. \tag{4.106}$$

Also, from $\partial^2 F/\partial T\partial\rho = \partial^2 F/\partial\rho\partial T$ we get

$$\left(\frac{\partial S}{\partial \rho}\right)_{T,\mathbf{D}} = -\frac{1}{\rho^2}\left(\frac{\partial p}{\partial T}\right)_{\rho,\mathbf{D}}. \tag{4.107}$$

Collecting terms we obtain

$$dS = \left\{\frac{c_V}{T} - \frac{D^2}{2\rho}\left[\frac{\partial^2}{\partial T^2}\left(\frac{1}{\varepsilon}\right)\right]_{\rho}\right\} dT - \frac{1}{\rho^2}\left(\frac{\partial p}{\partial T}\right)_{\rho,\mathbf{D}} d\rho - \frac{1}{\rho}\left[\frac{\partial}{\partial T}\left(\frac{1}{\varepsilon}\right)\right]_{\rho} \mathbf{D}\cdot d\mathbf{D}, \tag{4.108}$$

where $c_V = T(\partial S_o/\partial T)_\rho$ is the specific heat at constant volume.

We may want to choose the pressure p instead of the mass density ρ as independent variable for the entropy, since most processes in liquids take place at constant pressure. To this end, we introduce the Gibbs free energy, $G = F + (p/\rho)$, with differential

$$dG = -SdT + \frac{1}{\rho}dp + \frac{1}{\rho}\mathbf{E}\cdot d\mathbf{D}. \tag{4.109}$$

The entropy, given by $S = -(\partial G/\partial T)_{p,\mathbf{D}}$, is a function of temperature, pressure, and displacement vector. We have

$$dS = \left(\frac{\partial S}{\partial T}\right)_{p,\mathbf{D}} dT + \left(\frac{\partial S}{\partial p}\right)_{T,\mathbf{D}} dp + \left(\frac{\partial S}{\partial \mathbf{D}}\right)_{T,p}\cdot d\mathbf{D}. \tag{4.110}$$

From $\partial^2 G/\partial T\partial p = \partial^2 G/\partial p\partial T$ we obtain that $(\partial S/\partial p)_{T,\mathbf{D}} = -\alpha/\rho$, with α the coefficient of thermal expansion. Therefore

$$dS = \left[\frac{c_p}{T} - \frac{D^2}{2}\left(\frac{\partial}{\partial T}\left\{\frac{1}{\rho}\left[\frac{\partial}{\partial T}\left(\frac{1}{\varepsilon}\right)\right]_{\rho}\right\}\right)_{p}\right] dT - \frac{\alpha}{\rho}dp - \frac{1}{\rho}\left[\frac{\partial}{\partial T}\left(\frac{1}{\varepsilon}\right)\right]_{\rho}\mathbf{D}\cdot d\mathbf{D}, \tag{4.111}$$

where $c_p = T(\partial S_o/\partial T)_p$ is the specific heat at constant pressure, and $\rho = \rho(p, T)$.

We are now ready to write the temperature equation.

4.5.1 Nonpolar fluids

For nonpolar fluids $S = S_o$, and equation (4.111) reduces to

$$dS = \frac{c_p}{T}dT - \frac{\alpha}{\rho}dp. \tag{4.112}$$

Dividing this equation by dt, and multiplying by ρT, after substitution in equation (4.69) we obtain the equation for the temperature

$$\rho c_p \left[\frac{\partial T}{\partial t} + \mathbf{u}\cdot\nabla T\right] = \alpha T\frac{dp}{dt} - \nabla\cdot\mathbf{J'}_o + \mathbf{E}\cdot\mathbf{J'} + \Phi, \tag{4.113}$$

with the viscous dissipation Φ given by

$$\Phi = 2\eta\left[e_{ij} - \frac{1}{3}(\nabla\cdot\mathbf{u})\delta_{ij}\right]^2 + \lambda(\nabla\cdot\mathbf{u})^2. \tag{4.114}$$

In the case of a fluid in the ohmic regime with the thermoelectric phenomena included, the heat due to the second and third terms of (4.113) is

$$-\nabla\cdot\mathbf{J'}_o + \mathbf{E}\cdot\mathbf{J'} = \nabla\cdot(\kappa\nabla T) + \frac{J'^2}{\sigma} - T(\mathbf{J'}\cdot\nabla\alpha), \tag{4.115}$$

with $\kappa = k - \sigma\alpha^2 T$. It has been assumed, for simplicity, that $\nabla\cdot\mathbf{J'} = 0$, recovering in this way the well known result for solid metallic conductors (see Landau and Lifshitz [3]).

In most situations the thermoelectric effects are negligible. If the velocity of the fluid is much less than the sound velocity, the fluid may be treated as incompressible, and we have

$$\rho c_p \left[\frac{\partial T}{\partial t} + \mathbf{u}\cdot\nabla T\right] = k\nabla^2 T + \Phi + \mathbf{E}\cdot\mathbf{J'}, \tag{4.116}$$

where Φ reduces now to

$$\Phi = 2\eta e_{ij}^2 \tag{4.117}$$

4.5.2 Polar fluids

It is convenient to distinguish between polar gases and polar liquids. For polar gases we may use for the entropy the variables T, ρ and $\mathbf{D}$. Multiplying equation (4.108) by ρT, dividing by dt and substituting in equation (4.91) we obtain the temperature equation in the form:

$$\begin{aligned}\rho\left\{c_V - T\frac{D^2}{2\rho}\left[\frac{\partial^2}{\partial T^2}\left(\frac{1}{\varepsilon}\right)\right]_\rho\right\}\frac{dT}{dt} &= \frac{T}{\rho}\left(\frac{\partial p}{\partial T}\right)_{\rho,\mathbf{D}}\frac{d\rho}{dt} - \nabla\cdot\mathbf{J'}_o + \Phi \\ &+ \mathbf{E}\cdot\mathbf{J'} + T\left[\frac{\partial}{\partial T}\left(\frac{1}{\varepsilon}\right)\right]_\rho \mathbf{D}\cdot\frac{d\mathbf{D}}{dt}.\end{aligned} \tag{4.118}$$

For $\varepsilon = \varepsilon(\rho)$ it is easy to show that this equation reduces to the equation of nonpolar fluids (4.113).

For polar liquids it is more convenient to use the entropy as a function of pressure. Since ρ and ε are both independent of pressure , the entropy is a function of temperature and $\mathbf{D}$ only. Consequently, for processes at constant pressure the temperature equation is

$$\rho\left[c_p - T\frac{D^2}{2}\left(\frac{\partial}{\partial T}\left\{\frac{1}{\rho}\left[\frac{\partial}{\partial T}\left(\frac{1}{\varepsilon}\right)\right]_\rho\right\}\right)_p\right]\frac{dT}{dt} = -\nabla\cdot\mathbf{J}'_{\tilde{o}} + \mathbf{E}\cdot\mathbf{J}' + \Phi + (\mathbf{E}-\mathbf{E}_{eq})\cdot\frac{d\mathbf{P}'_p}{dt} + T\left[\frac{\partial}{\partial T}\left(\frac{1}{\varepsilon}\right)\right]_\rho \mathbf{D}\cdot\frac{d\mathbf{D}}{dt}. \quad (4.119)$$

To complete the temperature equation for polar fluids we assume that $\mathbf{J}'_{\tilde{o}} = -\tilde{k}\nabla T$, the flux of internal energy $\tilde{U}_o$, with $\tilde{k} = \tilde{k}(T,\varepsilon,E)$, that will depend on the temperature, the electrical permittivity and the electric field. Here, we have also assumed, for simplicity, that thermoelectric effects are negligible. To the knowledge of the author, any difference between the conduction of heat by molecular diffusion in the absence and in the presence of an electric field has not been reported in the literature. It is probable that the difference is quite small and we may safely put k instead of $\tilde{k}$ in the equation for the temperature for polar fluids.

An estimation of the order of magnitude of second term of the left hand side of equation (4.119), along the lines discussed in section 1.5, shows that it is negligible compared to the first, since $c_p \sim 10^3\mathrm{JK}^{-1}$. As a result, we have

$$\rho c_p\frac{dT}{dt} = -\nabla\cdot\mathbf{J}'_{\tilde{o}} + \mathbf{E}\cdot\mathbf{J}' + \Phi + (\mathbf{E}-\mathbf{E}_{eq})\cdot\frac{d\mathbf{P}'_p}{dt} + T\left[\frac{\partial}{\partial T}\left(\frac{1}{\varepsilon}\right)\right]_\rho \mathbf{D}\cdot\frac{d\mathbf{D}}{dt}. \quad (4.120)$$

The last term of the right hand side may be linked[1] to the reversible variations of temperature with the electric field, and it is similar to the first term of the right hand side of equation (4.118). In a quasistatic adiabatic process with electrically isolating walls (no flow of dissipative electrical current) and uniform electric and temperature fields, it follows from equation (4.111) or equation (4.120) that,

$$\rho c_p dT - T\left[\frac{\partial}{\partial T}\left(\frac{1}{\varepsilon}\right)\right]_\rho \mathbf{D}\cdot d\mathbf{D} = 0, \quad (4.121)$$

which is the equation obtained in chapter 1 for these processes. Even for intense fields this term, at ambient temperature, is only responsible for a variation of order 10^{-4} Kelvins. If the system is in contact with a thermal reservoir the heat flux through the surface (the surface integral of $\mathbf{J}'_0$), is given by the volume integral of the last term of (4.118) or (4.120).

[1]Personal communication of A. Ramos

4.6 The Boussinesq approximation in Electrohydrodynamics

In many practical situations thermal gradients are produced by imposing a temperature difference across the boundaries or, more rarely, due to Joule heating and/or dielectric heating. For moderate heating the equation of state for the mass density may be approximated by $\rho = \rho_0[1 - \alpha(T - T_0)]$, with ρ_0 the reference density at temperature T_o and α the volume expansion coefficient evaluated at the reference temperature. As a result density differences are much smaller than the mean density and the effects of adiabatic temperature gradient and compressibility can be ignored. The density is taken as constant at its reference value ρ_o, except in the buoyancy term of Navier-Stokes equation.

The permittivity ε is also assumed to vary linearly with temperature, $\varepsilon = \varepsilon_0[1 - e_1(T - T_0)]$, with e_1 the relative derivative of the permittivity with respect to temperature. At ambient temperature $e_1 \sim 10^{-3}$. For frequencies below the microwave region, which are typical in EHD, the molecular polarization is in equilibrium with the field, and dielectric heating is not present. The last term of equations (4.118) or (4.120) is also negligible when the imposed temperature difference is of a few Kelvin. Since under the Boussinesq approximation the fluid properties η, k, c_V and α are assumed to be constants, and the contribution from viscous dissipation is neglected, the electrohydrodynamic equations reduce to

$$\nabla \cdot \mathbf{D} = q, \tag{4.122}$$

$$\nabla \times \mathbf{E} = 0, \tag{4.123}$$

$$\nabla \cdot \mathbf{J}' + \frac{\partial q}{\partial t} = 0, \tag{4.124}$$

$$\nabla \cdot \mathbf{u} = 0, \tag{4.125}$$

$$\rho_0 \frac{d\mathbf{u}}{dt} = -\nabla p^* + \eta_0 \nabla^2 \mathbf{u} + q\mathbf{E} - \frac{1}{2} E^2 \nabla \varepsilon + \rho \mathbf{g}, \tag{4.126}$$

$$\rho_0 C_p \frac{dT}{dt} = k\nabla^2 T + \mathbf{E} \cdot \mathbf{J}', \tag{4.127}$$

with $p^* = p - p_{st}$.

A further simplification is achieved for very highly insulating liquids, since the currents are now of the order of μA or less, and the contribution of Joule heating to the energy equation may be neglected. The energy equation is decoupled from the other equations and therefore the solution depends only upon the boundary and initial conditions imposed on the temperature T. We have

$$\rho c_V \left[\frac{\partial T}{\partial t} + \mathbf{u} \cdot \nabla T \right] = k\nabla^2 T. \tag{4.128}$$

However, the temperature can not be ignored as its distribution affects the dynamic behavior of the system, through the buoyancy and electric forces. In addition the temperature also affects the charge density distribution through the variations induced in the permittivity, the conductivity and the mobility. Reference [8] has a discussion of these effects in some detail.

REFERENCES

1. De Groot S.R. and Mazur P.: Non-Equilibrium Thermodynamics, North Holland, Amsterdan, 1969.

2.- Kreuzer H.J.: Non-Equlibrium Thermodynamics and its Statistical Foundations, Clarendon Press, Oxford, 1981.

3. Landau L. M. and Lifshitz E. M.: Electrodynamique des Milieux Continus, MIR, Moscow 1989.

4. Panofsky W. K. H. and Phillips M.: Classical Electricity and Magnetism, Addison-Wesley, Reading, USA, 1962.

5. Callen H. B.: Thermodynamics, John Wiley, NY, 1960.

6. Cutler M.: Liquid Semiconductors. Academic Press, N.Y., 1977.

7. Eidel'man E.D.: Convection under thermoelectric field in liquid semiconductors. Zh. Eksp. Theor. Fiz., 103 (1993), 1633-1643 (Sov. Fis. JETP, 76, (1993), 802).

8. Pontiga F. and Castellanos A.: Physical mechanisms of instability in a liquid layer subjected to an electric field and a thermal gradient. Physics of Fluids A, 6 (1994), 1684-1701.

PART II

CONDUCTION PHENOMENA IN DIELECTRIC LIQUIDS, EHD MODELS, LINEAR AND NONLINEAR EFFECTS ON THE CHARGED INTERFACE, EHD PUMPING OF DIELECTRIC LIQUIDS

A.I. Zhakin
Kursk State Technical University, Kursk, Russia

Abstract

The aim of this part is to describe the basic conduction processes in dielectric liquids (DL), to consider the linear and non linear deformation of charged interface in a long wave approach and discuss the main problems on EHD pumping. This part consists of five chapters. In chapter [5] we consider the general theories on the dissociation of ion pairs, pair-ion recombination and discuss the ion-complex formation in high electric field. In chapter [6] we give a short review on conduction phenomena in DL and associated conduction models in EHD. Chapter [7] is devoted to the general investigation of conductivity processes in DL on the basis of the multi-ion dissociation-injection EHD model. The non-linear analysis of free surface shapes and non-linear EHD waves in initially plane charged interfaces in the long wave approach is given in chapter [8]. Here we study EHD solitons (dimples, bumps, etc.) on the charged free surface. In chapter [9] two problems are considered: electrokinetic pumping by a high voltage travelling wave and ion-drag pumping.

Chapter 5

CLASSIC THEORIES OF ION RECOMBINATION AND DISSOCIATION IN LIQUIDS

5.1 Introduction

Liquid dielectrics, as weak electrolytes, have a very low concentration of ionic components. That is why considering the recombination process we usually take into account only pair interactions. The recombination rate is determined by the slowest process of ionic approach which is controlled by ion diffusion and migration in the electric field. Neutral components recombine only due to diffusion. For ionic components it is necessary to take into account electrostatic interaction and their migration in an external field.

The first calculation of the recombination rate between neutral components had been performed by Smoluchovski [1, 2] and for ionic ones by Debye [3]. The same result was obtained by Langevin, considering the ionic recombination in high density gases.

After the recombination of two ions, an ion pair is generally formed, a structure suggested by Bjerrum [2, 4].

For low dielectric permittivity of the parent liquids and high concentration of ionic components there is the possibility to form ionic triplets due to interaction between ion pairs and monoions. This process accounts for the so-called anomalous conductivity, which was considered by Fuoss and Krauss [5].

In this section we will present the general concepts on ion recombination following the Debye - Smoluchovski method and then obtain the expression for the recombination rate and the conditions under which the recombination depends, or not, on the external electric intensity. Then the Bjerrum concept of ion pairs will be discussed. The electrochemical reactions describing the process of an ionic triplet formation are considered. Finally, the Onsager theory on the dependence of the dissociation constant on the electric field is discussed in detail.

5.2 General theory of pair recombination in solutions

Let us consider a finite volume V where there are two species of particles. Let us introduce for the i-th particle the mobility μ_i, the diffusion coefficient D_i, the total number of particles N_i, the force $\mathbf{F}_i$ and the two - particle distribution function $f_{ij} = f_{ij}(\mathbf{r}_1,\mathbf{r}_2)$ where $\mathbf{r}_1(\mathbf{r}_2)$ is the radius-vector of particle 1(2). The function f_{12} determines the probability dW_{12} of finding the particles 1, 2 in the elementary volumes dV_1, dV_2:

$$dW_{12} = f_{12}N_1N_2dV_1dV_2/V^2 = f_{12}n_1n_2dV_1dV_2,\ \ n_i = N_i/V. \tag{5.1}$$

Since at $r = |\mathbf{r}_1 - \mathbf{r}_2| \rightarrow \infty$ it takes place $dW_{12} \rightarrow n_1n_2dV_1dV_2$ then

$$f_{12} \rightarrow 1 \text{ at } r = |\mathbf{r}_1 - \mathbf{r}_2| \rightarrow \infty. \tag{5.2}$$

The velocity of i-th particle is defined as

$$\mathbf{u}_i = -D_i\nabla_i f_{ij} + f_{ij}(\mu_i/e)\mathbf{F}_i. \tag{5.3}$$

Here (and later) the lower number indexes denote the differential operations on the variables $\mathbf{r}_1$, $\mathbf{r}_2$.

The current of particles 2 to an arbitrary particle 1 is expressed as

$$\mathbf{i}_{21} = n_2\mathbf{u},\ \mathbf{u} = \mathbf{u}_2 - \mathbf{u}_1 = D_i\nabla_1 f_{12} - D_2\nabla_2 f_{21} + (\mu_2/e_2)\mathbf{F}_2 - (\mu_1/e_1)\mathbf{F}_1,$$

where $\mathbf{u}$ is the relative velocity between the two particles.

The recombination rate is defined as the number of particles' collisions per unit time and volume:

$$\dot{\xi}_r = (dN/dt)/V,\ \ dN = N_1\int_S \mathbf{i}_{21}\cdot\mathbf{n}\, dS\, dt \tag{5.4}$$

where S is the spherical surface of arbitrary radius $R \gg R_1$ surrounding the central ion (its centre is determined by the vector $\mathbf{r}_1$, R_i is the characteristic size of i-th particle) and $\mathbf{n}$ is the internal normal to S. From (5.4) it follows:

$$\dot{\xi}_r = \alpha\, n_1 n_2,\ \ \ \alpha = \int_S \mathbf{u}\cdot\mathbf{n}\, dS\, dt \tag{5.5}$$

The functions f_{12}, f_{21} are determined by the stationary diffusion equation

$$\nabla\cdot\mathbf{i}_{21} = 0 \tag{5.6}$$

In the symmetrical case when the forces $\mathbf{F}_i$ depend on the difference $\mathbf{r} = \mathbf{r}_2 - \mathbf{r}_1$ it is $f_{12} = f_{21} = f(\mathbf{r})$ and the function $f(\mathbf{r})$ is determined by the equation

$$\nabla \cdot \{ -(D_1 + D_2)\nabla f + f[(\mu_2/e_2)\mathbf{F}_2 - (\mu_1/e_1)\mathbf{F}_1] \} = 0 \tag{5.7}$$

with the boundary condition (5.2). The recombination constant is expressed as

$$\alpha = \int_S \{ -(D_1 + D_2)\nabla f + f[(\mu_2/e_2)\mathbf{F}_2 - (\mu_1/e_1)\mathbf{F}_1] \} \cdot \mathbf{n}dS \tag{5.8}$$

A simple method to calculate the recombination constant at strong interaction between particles was proposed by Zhakin [6]. This method is based on the equation of relative motion of two particles

$$\dot{\mathbf{r}} = (\mu_2/e_2)\mathbf{F}_2 - (\mu_1/e_1)\mathbf{F}_1, \qquad \mathbf{r} = \mathbf{r}_2 - \mathbf{r}_1. \tag{5.9}$$

For the recombination constant it follows

$$\alpha = \int_{S+} [(\mu_2/e_2)\mathbf{F}_2 - (\mu_1/e_1)\mathbf{F}_1)] \cdot \mathbf{n}dS, \tag{5.10}$$

where S_+ is the part of the sphere S starting from which all particles will get into its centre.

5.3 Debye – Smoluchovski – Langevin theories of pair recombination in the absence of an external electric field. The influence of an external field on recombination

In the absence of an external electric field the interaction between ions is caused by Coulomb electrostatic forces

$$\mathbf{F}_1 = e_1 e_2 \mathbf{r} / (4\pi\varepsilon r^3), \quad \mathbf{F}_2 = -e_1 e_2 \mathbf{r} / (4\pi\varepsilon r^3),$$

where e_1, e_2 are the absolute values of ion charges. Then equation (5.7) takes the form

$$\frac{d}{dr}\left(r^2 \frac{df}{dr}\right) + l_B \frac{df}{dr} = 0, \qquad l_B = e_1 e_2 / (4\pi\varepsilon k_B T). \tag{5.11}$$

Here the known relationship $D_i = \phi_i \mu_i$, $\phi_i = k_B T/e_i$, $i = 1,2$ has been used. Besides the boundary condition (5.2) the function f must satisfy the following condition

$$\text{at} \quad r \to R_{12}: \quad f \to 0, \tag{5.12}$$

where $R_{12} = R_1 + R_2$ and R_{12} is the effective radius of the i-th ion. The condition (5.12) implies that recombinating ions approach from infinite distance. From (5.11), (5.12) it follows

$$f = [\,\exp(l_B/R_{12}) - \exp(l_B/r)\,] \,/\, [\,\exp(l_B/R_{12}) - 1]. \tag{5.13}$$

Using (5.8) we get the Debye result

$$\alpha = 4\pi\,(D_1 + D_2)\, l_B \,/\, [\,1 - \exp(-l_B/R_{12})\,], \tag{5.14}$$

At $l_B/R_{12} << 1$ from (5.14) it follows the Smoluchovski formula

$$\alpha = 4\pi\,(D_1 + D_2)\, R_{12}\,. \tag{5.15}$$

At $l_B/R_{12} >> 1$ with closing relations $D_i = \phi_i\mu_i$ from (5.11) we obtain the Langevin correlation

$$\alpha = (e_2\mu_1 + e_1\mu_2) \,/\, \varepsilon. \tag{5.16}$$

Formula (5.15) responds to reactions of neutral components and (5.16) is valid at small ionic radii.

The distance l_B was introduced by Bjerrum; this is the distance at which the Coulombian electrostatic energy interaction between two ions of opposite polarity equals their thermal energies.

Langevin formula (5.16) can be obtained using the simple theory given by (5.9), (5.10). In that case the equation (5.9) has the form

$$\dot{\mathbf{r}} = -\,\mathbf{u} - A\mathbf{r}/r^3, \quad A = (\mu_1 e_2 + \mu_2 e_1) \,/\, 4\pi\varepsilon, \quad \mathbf{u} = (\mu_2 + \mu_1)\mathbf{E},$$

where $\mathbf{u}$ is the relative velocity between recombinating ions at $r \to \infty$. From (5.10) it follows

$$\alpha = \pi d_*^2 u, \qquad u = (\mu_1 + \mu_2)E,$$

where d_* is the cross-section of the recombination process which is expressed as

$$d_* = 2(A/u)^{1/2}.$$

From these formulae we obtain again the Langevin correlation (5.16). Therefore at small cross-section values, $d_* < R_{12}$, it is

$$\alpha = \pi R_{12}^2 (\mu_1 + \mu_2)E,$$

that is, the recombination constant is a linear function of electrical strength. Hence the Langevin correlation at high electric field is valid under the condition

$$d_* = (A/u)^{1/2} \sim (e/\varepsilon E)^{1/2} > R_{12}, \qquad (5.17)$$

where e is the proton charge.

5.4 Bjerrum conception of ion pairs

From (5.13) it is seen that at small distance between ions $r < l_B$ the distribution function has an asymptotic form

$$f = C \exp(l_B/r) \ , \qquad C = \text{const.}$$

The probability of localizing particle 1 in a spherical layer enclosing particle 2 is defined as $dP = 4\pi r^2 f\, dr$.
Hence, the function $W = 4\pi r^2 f$ is the radial distribution function. This function has the minimum at

$$r_B = l_B/2 = e_1 e_2/(8\pi \varepsilon k_B T). \qquad (5.18)$$

The value r_B is called the Bjerrum radius. Bjerrum suggested that a pair of ions at a distance $r < r_B$ should be considered as "bound" to each other.

5.5 Fuoss – Krauss theory of anomalous conductivity due to ionic triplets

In classical electrochemistry the electrolyte conductivity is defined by the coefficient $\Lambda = \sigma / N_0$, where

$$\sigma = (e_1\mu_1 + e_2\mu_2)\, n$$

is the ohmic conductivity coefficient and N_0 is the initial concentration of a dissociating salt (ion pairs). For the reaction

$$A^+B^- \underset{\alpha_r}{\overset{\alpha_d}{=}} A^+ + B^-, \qquad (5.19)$$

bearing in mind the equilibrium state, we have

$$\alpha_r n^2 = \alpha_d N, \quad n + N = N_0.$$

At small dissociation degrees $n/N_0 << 1$ it follows $n = (\alpha_d N_0/\alpha_r)^{1/2}$, and hence

$$\Lambda \sim N_0^{-1/2}$$

Experiments have shown that in non-aequous solvents such as acetone, nitrobenzene, etc., there is an anomalous conductivity with

$$\Lambda \sim C_1 N_0^{-1/2} + C_2 N_0^{1/2}, \tag{5.20}$$

where C_1, C_2 are constants.

To explain this effect, Fuoss and Krauss suggested to complement the reaction (5.19) with [5]

$$A^+B^- + A^+ \underset{k_3^+}{\overset{\alpha_{12}^+}{=}} A^+B^-A^+, \qquad A^+B^- + B^- \underset{k_3^-}{\overset{\alpha_{12}^-}{=}} B^-A^+B^-. \tag{5.21}$$

Note that the ion-dipole interaction is much smaller than the ion-ion one. Therefore, it must be taken into consideration the recombination of ion triplets, for example, accordingly to reactions

$$A^+B^-A^+ + B^- = A^+B^-A^+B^- = 2\,A^+B^-,$$

$$B^-A^+B^- + A^+ = A^+B^-A^+B^- = 2\,A^+B^-,$$

$$A^+B^-A^+ + B^-A^+B^- = A^+B^-A^+B^-A^+B^- = 3\,A^+B^-.$$

It seems that at small concentration of ion pairs A^+B^- the most important reactions can be

$$A^+B^-A^+ + B^- \xrightarrow{\alpha_{13}^-} 2\,A^+B^-, \quad B^-A^+B^- + A^+ \xrightarrow{\alpha_{13}^-} 2\,A^+B^-, \tag{5.22}$$

$$A^+B^-A^+ + B^-A^+B^- \xrightarrow{\alpha_{13}^-} 3\,A^+B^-.$$

To reveal the basic feature of the formation of ionic triplets we will consider a symmetrical case $\alpha_{12}^{+} = \alpha_{12}^{-} = \alpha_{12}$, $k_3^{+} = k_3^{-} = k_3$, and denote

$$n_1 = [A^+], \quad n_2 = [B^-], \quad n_3 = [A^+B^-A^+], \quad n_4 = [B^-A^+B^-], \quad N = [A^+B^-] \tag{5.24}$$

In this case $n_1 = n_2$, $n_3 = n_4$. At the equilibrium state we get

$$n_1 + n_3 + N = N_0, \qquad \alpha_d N - \alpha_r n_1^2 - \alpha_{13} n_1 n_2 = 0, \tag{5.25}$$

$$\alpha_{12} n_1 N - k_3 n_3 - \alpha_{13} n_1 n_3 - \alpha_{33} n_3^2 = 0.$$

Let us consider a typical case of small dissociation degrees

$$n_1 / N_0 \ll 1, \qquad n_3 / N_0 \ll 1. \tag{5.26}$$

Then Eq. (5.25) yields

$$n_3 = K_3 N_0 n_1, \quad n_1 = [\alpha_d N_0 / (\alpha_r + \alpha_{13} N_0 / K_3)]^{1/2}, \tag{5.27}$$

$$K_3 = k_3 / \alpha_{12}.$$

Hence the inequalities (5.26) are valid under conditions

$$N_0 / K_3 < 1, \quad [\alpha_d / (\alpha_r N_0)]^{1/2} \ll 1. \tag{5.28}$$

The conductivity coefficient in this case is expressed as

$$\sigma = e\,[\mu_1 + \mu_2 + (\mu_3 + \mu_4) N_0 / K_3]\, n_1, \tag{5.29}$$

where n_1 was defined in (5.27) and μ_j (j=1,..,4) are the mobility of ions A^+, B^-, $A^+B^-A^+$, $B^-A^+B^-$ respectively.

Using (5.29) it is easy to show the trend (5.20).
In conclusion we point out that the complex-ion formation is negligibly small under conditions

$$\alpha_{13} < \alpha_r, \qquad N_0 / K_3 \ll 1, \tag{5.30}$$

An important question in the determination of the constant of equilibrium K_3 is revealed.

5.6 On the determination of the equilibrium constants

Let us consider the determination of the equilibrium constant K of the following reactions

$$AX \underset{\alpha}{\overset{k}{=}} A + X \qquad (5.31)$$

We will assume that the effects of the reactions (5.31) which take place between ionic components and solvent (interaction between molecules A, X, AX and those of parent liquid) are negligibly small and to be shown up in liquid polarization. So in this case it is possible to use the general theory of chemical reactions in gaseous phase where the equilibrium constant of the reactions (5.31) is expressed as

$$K_2 = k/\alpha = Q_A Q_X / Q_{AX} \qquad (5.32)$$

where Q_A, Q_X, Q_{AX} are the statistical sums of the molecules A, X, AX respectively which can be written as

$$Q_j = g_j\, Q^j_{in} / v^j_q, \qquad j = A, X \qquad (5.33)$$

$$Q_{AX} = g_{AX} Q^A_{in}\, Q^B_{in} \exp(D_{AX}/k_B T)/v_q^{AX}$$

where g_j, g_{AX} are the number of the spinous states, Q^j_{in} is the statistical sums of the internal energy states,

$$v^j_q = (2\pi\, h^2 / m_j\, k_B T)^{3/2}, \qquad j = A, X, AX$$

are the quantum volumes (m_j denotes the molecular mass) and D_{AX} is the interaction energy between A and X in the molecule AX.

Using (5.33), from (5.32) it follows

$$K_2 = A \exp(-D_{AX} / k_B T) \qquad (5.34)$$

where $A = (g_A\, g_B / g_{AB})\, (v^{AB}_q / v^A_q\, v^B_q)$ depends only on temperature and for ionic components is expressed as

$$A = (m\, k_B T / 2\pi\, h^2)^{3/2},\ m = m_A m_B / (m_A + m_B) \qquad (5.35)$$

Let us consider the application of the formula (5.34) to various physical situations.

5.6.1 Equilibrium between ionic pairs and free ions in an external electric field

In this case the reaction (5.31) is that of (6.1). Interaction between ions A^+, B^- is determined by the electrostatic energy

$$U(r) = -(e^2/4\pi\varepsilon)/r - eEr\cos\theta, \qquad (5.36)$$

where $r = |\mathbf{r}|$, **r** is the vector connecting ions B^- and A^+, and θ is the angle between **r** and **E**.

Following Bjerrum we assume that ionic pair radius r_* is defined by the following condition

$$\frac{d}{dr}[r^2 \exp(-U(r)/k_BT)] = 0 \qquad (5.37)$$

leading up to equation

$$r_*^2 \cos\theta + (r_* - r_B)\, l_E = 0, \qquad (5.38)$$

$$r_B = e^2/(8\pi\varepsilon k_BT), \quad l_E = 2k_BT/(eE).$$

So we obtain for r_*

$$r_* = r_B f(s,x), \quad f(s,x) = (\sqrt{s+4sx} - s)/2s, \qquad (5.39)$$

$$x = \cos\theta, \qquad s = l_E/r_B$$

and for the dissociation energy

$$D_{AB} = U(r_*) - U(R_{12}) = D^0_{AB} - \Delta U_E,$$

$$D^0_{AB} = e^2/(4\pi\varepsilon R_{12}) - 2k_BT,$$

$$\Delta U_E = 4[e^2/(8\pi\varepsilon r_*) - k_BT] = 4(1/f(s,x) - 1)k_BT \qquad (5.40)$$

where D_{AB}^{o} is the dissociation energy in the absence of an external electric field. From (5.40) it is seen that the external electric field reduces the dissociation energy, so at low fields when $s \to \infty$ and $f \to 1$ it follows $\Delta U_E \to 0$, and at high fields when $s \to 0$, it is $U_E \sim s^{-1/2} \sim E^{1/2}$.

Using (5.40) from (5.34) we obtain

$$K_2 = K_{20} \exp(\Delta U_E / k_B T) \tag{5.41}$$

where K_{20} is the equilibrium constant in the absence of the external electric field.

The expression (5.41) determines the equilibrium constant when dissociation and recombination take place at a certain orientation of the ionic pair, that is at fixed angle θ which varies in the limits $0 < \theta < \pi/2$. Since the recombination rate does not depend on ionic pair orientation and electric intensity, then expression (5.41) determines the dependence of the dissociation rate α_d in the angle θ and electric strength

$$\alpha_d(\theta) = \alpha_r K_{20} \exp(\Delta U_E / k_B T)$$

The statistical average procedure gives the following result

$$\alpha_d = \langle \alpha_d(\theta) \rangle = \alpha_{d0} F_1(p),$$

$$F_1(p) = [\,(2\sqrt{1+4p^2} - 1) \exp(2(\sqrt{1+4p^2} - 1)) - 1] / 8p^2,$$

$$p = \sqrt{S} = (r_B/l_E)^{1/2} = e\sqrt{eE/4\pi\varepsilon}\;/\,(2\,k_B T)$$

At high field ($p >> 1$) we get the following asymptotic expression

$$F(p) = (2p)^{-1} \exp(4p) \tag{5.42}$$

and at low field ($p \to 0$), we have $F(p) = 1$.

5.6.2 Equilibrium between ionic triplets and ionic pairs and monoions

Calculation of the equilibrium constant K_3 of the reaction

$$A^+B^-A^+ \underset{\alpha_{12}}{\overset{k_3}{=}} A^+B^- + A^+$$

will be carried out under the following assumptions:

1. The ions A^+, B^- in the ionic pair A^+B^- are bounded,
2. Ionic pairs can be considered as "rigid" dipoles **p**, $p = |\mathbf{p}| = e\, l_{12}$, where l_{12} is the mean "length" of the ionic pair.

In this case the energy of ion A^+ in the electric field of the dipole **p** and the external field **E** can be expressed as

$$U(r, \theta, \theta_1) = e^2 / (4\pi\varepsilon)(1/r - 1/r_1) - eEr\cos\theta, \tag{5.43}$$

$$r_1^2 = l_{12}^2 + r^2 - 2rl_{12}\cos(\theta - \theta_1),$$

where the distances r, r_1 and angles θ, θ_1 are shown in Fig. 5.1. We do not take into account the energy of the dipole in the external field – **p·E** because it is much smaller than the other components.

Using (5.37) we obtain the following equation for the ionic triplet radius r_*:

$$r_*^3 \cos\theta + l_E\,[r_*^2 - r_{id}^2 \cos(\theta_1 - \theta)\,] = 0 \tag{5.44}$$

where l_E is defined in (5.38) and

$$r_{id} = [ep\,/\,(4\pi\,\varepsilon\,k_BT)\,]^{1/2} = (2\,r_B\,l_{12})^{1/2},\;\; p = e\,l_{12} \tag{5.45}$$

In the absence of the external electric field ($l_E = \infty$) we get

$$r_{A0} = r_{id}\,\sqrt{\cos(\eta)}\,, \qquad \eta = \theta_1 - \theta. \tag{5.46}$$

This formula shows that ionic triplets exist only at $0 < \eta < \pi/2$. From the physical point of view this condition means that the positive ion A^+ is attracted to the negative charge side of the ionic pair A^+B^-. Hence, we can conclude that both angles are confined in the interval $[0, \pi/2]$. The value r_{id} determines the typical length of the ionic triplets.

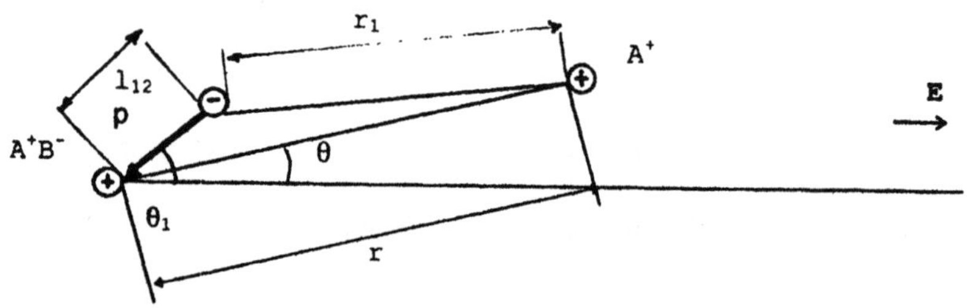

Fig. 5.1, Interaction between ionic pair A^+B^- and free ion A^+.

In the presence of an external electric field it will be $r_* < r_{id}$ and at high field when $r_* << r_{id}$ we can put

$$r_* = \xi_E\,[\cos(\theta_1 - \theta)\,/\,\cos\theta]^{1/3}, \qquad \xi_E = [l_E\,r_{id}^2]^{1/3} \tag{5.47}$$

It is seen that the ionic triplet (dimension) density decreases with an increase in the field strength as $r_* \sim E^{-1/3}$.

Introducing the characteristic length of the ionic triplets l_{13} we obtain analogously (5.40) the dissociation theory

$$D_{ABA} = U(r_*, \theta, \theta_1) - U(l_{12}, \theta, \theta_1) = D^0_{ABA} - \Delta U_E, \tag{5.48}$$

$$D^0_{ABA} = e^2/(4\pi\varepsilon)\,[1/r_{13} - 1/r_{1*0} + 1/l_{13} - 1/r_{*0}],$$

$$\Delta U_E = eEr_* \cos\theta + e^2/(4\pi\varepsilon)\,[1/r_{1*} - 1/r_{1*0} + 1/r_{*0} - 1/r_*],$$

$$r_{13} = [l_{12}^2 + l_{13}^2 - 2\, l_{12} l_{13} \cos(\theta_1 - \theta)\,]^{1/2},$$

$$r_{1*0} = [\, l_{12}^2 + r_{*0}^2 - 2\, r_{*0} l_{12} \cos(\theta_1 - \theta)\,],$$

$$r_{1*} = [l_{12}^2 + r_*^2 - 2\, r_* l_{12} \cos(\theta_1 - \theta)\,]^{1/2}.$$

Polarization data show that the mean ionic pair length is about 100 nm (see subsection 1.2). If we assume this value for l_{12} then for non-polar liquids at room temperature it follows $r_{*0} \sim r_{1*0} \sim r_{id} \sim 500$ nm while the magnitudes of l_E, r_* at low field E = 1 kV/cm have the order $l_E = 5 \cdot 10^4$ nm, $r_* \sim r_{*0}$. At high field when $r_* \ll r_{*0}$ we can put

$$\Delta U_E / k_B T = 2\,(r_*/l_E) \cos\theta + r_{*0}^2 / (2\, r_*^2) \cos(\theta_1 - \theta),$$

where r_* is determined by (5.47). This expression shows that the strong influence of the electric field on the ion-triplet decay is to be observed under condition $r_{id} \sim l_E$, that is, at E ~ 100 kV/cm the dissociation triplet rate is expressed as

$$k_3 = A\alpha_{12} \int_0^{\pi/2} \int_0^{\pi/2} W \exp[-(D^0_{ABA} - \Delta U_E)/k_B T] \sin\theta \sin\theta_1\, d\theta\, d\theta_1,$$

where A is determined by the formula (5.35) and $W = W(\theta, \theta_1)$ determines the probability dP of dipole and ion orientations in the spherical angles $d\Omega = \sin\theta\, d\theta d\varphi$, $d\Omega_1 = \sin\theta_1\, d\theta_1 d\varphi$: $dP = W d\Omega d\Omega_1$. The function $W(\theta, \theta_1)$ is determined by the Boltzmann factor, at low field W = 1 and at high field W is close to a δ-function. So at high field we have

$$k_3 = k_{30} \exp[\,2.5\,(r_{id}/l_E)^{2/3}],$$

$$k_{30} = A\alpha_{12} \exp(-D^0_{ABA} / k_B T),$$

$$D^0_{ABA} = e^2/(4\pi\varepsilon)\,[1/(l_{13} - l_{12}) - 1/l_{13} - 1/(r_{id} - l_{12}) + 1/r_{id}].$$

So at high field the dissociation constant increases as $k_3 \sim \exp(E^{2/3})$.

5.7 Recombination of ion pairs and monoions at high electric fields

Let us calculate the coefficient α_{12}^- in the first reaction of (5.22) using the equations (5.9), (5.10), which in this case can be represented as

$$\dot{\mathbf{r}} = -\mu_2 (\mathbf{E} + \mathbf{E}_d) - \mu_2 (\mathbf{p}\cdot\nabla)\, \mathbf{E}_i\, ;$$

$$\mathbf{E}_i = \mathbf{e r} / 4\pi\varepsilon r^3, \;\; \mathbf{E} = -\nabla\, (\mathbf{p}\cdot\mathbf{r}/4\pi\varepsilon r),$$

where **E**, $\mathbf{E}_d$, $\mathbf{E}_i$ are the intensities of the external, dipole and ion B^- electric fields, **p** is the dipole momentum of the ion pair A^+B^-. It can be shown that under conditions

$$E \gg e/4\pi\varepsilon r_*^2, \;\; r_* = (p/4\pi\varepsilon E)^{1/3}, \qquad \mathbf{p} \uparrow\uparrow \mathbf{E},$$

the recombination constant α_{12}^- is expressed as

$$\alpha_{12}^- = 3\pi\mu_2\, (p/4\pi\varepsilon)^{2/3} E. \tag{5.49}$$

Assuming $p = \beta_d E$, where β_d is the dipole polarization constant, we obtain

$$\alpha_{12}^- = 3\pi\mu_2\, (\beta_d/4\pi\varepsilon)^{2/3} E. \tag{5.50}$$

Hence at high electric field the process of the triplet-ion formation can be important.

5.8 Onsager theory: introduction

In this section we intend to discuss the Onsager theory on the increment of the dissociation rate of ion pairs by external electric field [7]. At first, we will consider the general theory of dissociation of ion pairs, then following to Onsager we obtain the dependence of the dissociation constant on the electrical intensity.

5.8.1 General theory of the dissociation of ion pairs into free ions at high external electric fields

Let us define the probability of finding a particle 2 at a distance r from a particle 1 in an elementary volume dV as

$$dW = F_{21} dV. \tag{5.51}$$

This definition allows to consider that F_{21} is a density of particles 2 in the vicinity of the particle 1. Hence a probable current of particles can be written in the form

$$\mathbf{i}_k = -D_k \nabla_k F_{kj} + F_{kj} \mu_k \mathbf{F}_k. \tag{5.52}$$

The number of the ion pair decays at the time range dt is expressed as

$$dN = N_{12} \int_S \mathbf{i} \cdot \mathbf{n}\, dS\, dt, \quad \mathbf{i} = \mathbf{i}_2 - \mathbf{i}_1, \tag{5.53}$$

where N_{12} is the total number of ion pairs in the volume V (see section 2 of this chapter). Using the general formula (5.4) we can find for the dissociation rate

$$\dot{\xi}_d = \alpha_d N, \qquad N = N_{12}/V, \qquad \alpha_d = \int_S \mathbf{i} \cdot \mathbf{n}\, dS, \tag{5.54}$$

$$\mathbf{i} = -D_2 \nabla_2 F_{21} + D_1 \nabla_1 F_{12} + F_{21} (\mu_2/e_2)\, \mathbf{F}_2 - F_{12} (\mu_1/e_1)\, \mathbf{F}_1.$$

The functions F_{21} are determined from the continuity equation

$$\nabla_2 \cdot \mathbf{i} = 0, \tag{5.55}$$

and the boundary and integral conditions follow from the relationship (5.51)

$$\int_{V_{AB}} F_{21}\, dV_2 = 1, \qquad F_{21} \to 0 \qquad \text{at} \qquad r \to \infty, \tag{5.56}$$

where V_{AB} is the "volume" of the ionic pair.

If the forces $\mathbf{F}_k$ depend on the difference $\mathbf{r} = \mathbf{r}_2 - \mathbf{r}_1$ then the Eq. (5.55) has the form

$$\nabla \cdot \{ -(D_1 + D_2) \nabla f + f [(\mu_2/e_2)\, \mathbf{F}_2 - (\mu_1/e_1)\, \mathbf{F}_1] \} = 0, \tag{5.57}$$

where $f = F_{21}$ and

$$\alpha_d = \int_S [-(D_1 + D_2) \nabla f + f ((\mu_2/e_2)\, \mathbf{F}_2 - (\mu_1/e_1)\, \mathbf{F}_1)] \cdot \mathbf{n} dS. \tag{5.58}$$

Let us consider as an example the calculation of the dissociation constant α_{d0} in the absence of an external electric field. In this case the equation (5.57) has the form (5.11) and, with the closing relations (5.56) we find

$$f = [\exp(l_B/r) - 1]/A, \quad A = 4\pi \int_{R_{12}}^{l_B} [\exp(l_B/r) - 1]\, r^2 dr. \qquad (5.59)$$

Using (5.58) we obtain

$$\alpha_{d0} = (D_1 + D_2)\, l_B / \left[\int_{R_{12}}^{l_B} \exp(l_B/r)\, r^2 dr - (l_B^3 - R_{12}^{\ 3})/3 \right]$$

and at small ion radii $R_{12} << l_B$ this expression reduces to

$$\alpha_{d0} = 3\, [\, (D_1 + D_2)\, l_B/R_{12}^3\,]\, \exp(-\, l_B/R_{12}). \qquad (5.60)$$

Under condition $R_{12} << l_B$ from (5.14), (5.60) it follows the known result for the ionic dissociation constant K_2 obtained by Fuoss [8]

$$K_2^{-1} = K_{ass} = \alpha_r/\alpha_{d0} = (4\pi/3)\, R_{12}^3 \exp(l_B/R_{12})$$

5.8.2 Onsager calculations on the dependence of the dissociation constant on the electric field

In the presence of an electric field the equation (5.57) takes the form

$$\nabla^2 f - \nabla f \cdot \nabla \Pi = 0, \qquad (5.61)$$

$$\Pi = 2\, (r_B/r + \beta r \cos\theta), \qquad \beta = r_B/r_E^2,$$

$$r_E^2 = [\, (e_1\mu_2 + e_2\mu_1)/4\pi\varepsilon\, (\mu_1 + \mu_2)\, E],$$

where r_B is the Bjerrum distance (5.18), r_E is the Onsager distance; at the distance r_E the force due to external electric field E equals to the attractive force between two ions of opposite polarity, and θ is the angle between **r** and **E**.

To solve this equation it is convenient to take the transformation

$$f = r^{-1} \exp(-\beta\, r + \beta\, r\, x + 2r_B/r)\, u(r, x), \qquad x = \cos\theta, \qquad (5.62)$$

where u (r, x) is the unknown function being determined by the equation

$$r^2 \partial^2 u/\partial r^2 - 2(\beta r^2 + r_B)\, \partial u/\partial r + (1 - x^2)\, \partial^2 u/\partial x^2 - 2x\, \partial u/\partial x + 2\beta r_B\, (1 + x)\, u = 0.$$

Let us change to the new variable s = 1/r. Then the last equation can be put in the form

$$s^2\partial^2 u/\partial s^2 + 2(\beta + r_B s^2 + s)\,\partial u/\partial s + 2r_B su + (1 - x^2)\,\partial^2 u/\partial x^2 - 2x\,\partial u/\partial x + 2r_B\beta(1 - x)u = 0. \tag{5.63}$$

We will seek the solution of this equation in the form

$$u = \int_{M_1}^{M_2} K(t,x)\exp(-st)\,dt, \tag{5.64}$$

where M_1, M_2, K (t, x) are unknown numbers and function. The integral (5.64) will satisfy Eq. (5.63) under conditions

$$t(t - 2r_B)\,\partial K^2/\partial t^2 + 2(t - r_B)\,\partial K/\partial t - 2\beta tK - \xi(\xi - 2)\,\partial^2 K/\partial \xi^2 - 2(\xi - 1)\,\partial K/\partial \xi + 2\beta r_B \xi K = 0, \tag{5.65}$$

$$\exp(-st)(t - 2r_B)\,t\,[K - \partial K/\partial t]\Big|_{t=M_1}^{t=M_2} = 0 \tag{5.66}$$

with $\xi = 1 + x$. Finally, after changing the variables $\eta = t/r_B$, Eq. (5.65) is reduced to

$$[L(\eta) - L(\xi)]\,K(\eta, \xi) = 0, \tag{5.67}$$

$$L(\eta) = \eta(\eta - 2)\,\partial^2/\partial\eta^2 + 2(\eta - 1)\,\partial/\partial\eta - 2\beta r_B\eta, \quad \eta = t/r,$$

$$L(\xi) = \xi(\xi - 2)\,\partial^2/\partial\xi^2 + 2(\xi - 1)\,\partial/\partial\xi - 2\beta r_B\xi, \quad \xi = 1 + x.$$

Let us seek a partial solution in the form

$$K = (\eta, \xi) = H_1(z), \quad z = \eta\xi. \tag{5.68}$$

Then for the function $K(\eta, \xi)$ from (5.67) it follows

$$z H_1'' + H_1' - \beta r_B H_1 = 0.$$

The general solution of this equation has the form

$$H_1(z) = C_1 I_0(2\sqrt{\beta r_B z}) + C_2 K_0(2\sqrt{\beta r_B z}), \tag{5.69}$$

where I_0, K_0, are the modified Bessel functions of order zero. The function I_0 is finite at $z \to 0$ while $K_0 \to 0$ at $z \to \infty$.

Another partial solution can be obtained in the form

$$K(\eta, \xi) = H_2(z), \qquad z = \eta + \xi - 2. \tag{5.70}$$

For the function $H_2(z)$ we have

$$z H_2'' + 2 H_2' - 2\beta r_B H_2 = 0$$

with general solution

$$H_2(z) = C_3 z^{-1/2} I_1(8\beta r_B z)^{1/2} + C_4 z^{-1/2} K_1(8\beta r_B z)^{1/2}. \tag{5.71}$$

There are two possible solutions which satisfy the condition (5.66), assuming $M_1 = 0$, $M_2 = 2r_B$ and $M_1 = 2r_B$, $M_2 = \infty$. The conditions of a divergency in the integral (5.64) and the smoothness on the variables θ, r lead to the following set of partial solutions of Eq. (5.63)

$$u = u_1 = C_3 \int_0^{l_B} I_0\left(\sqrt{8\beta r_B t}\cos\frac{\theta}{2}\right) \exp(-t/r)\, dt, \tag{5.72}$$

$$u = u_2 = C_2 \int_0^{l_B} \frac{I_1\left(\sqrt{8\beta r_B (t/r - 2\sin^2\theta/2)}\right)\exp(-t/r)}{\sqrt{8\beta r_B (t/r - 2\sin^2\theta/2)}}\, dt\,, \tag{5.73}$$

where C_1, C_2 are arbitrary constants.

Substituting (5.72) into (5.62) we get the Onsager solution [7] which can be represented as

$$f = g\, s^{-1} \exp\left[-2p^2 s (1 - \cos\theta) + 1/s\right] \int_0^1 I_0(\mu\sqrt{t}) \exp(-t/s)\, dt\,, \tag{5.74}$$

$$\mu = 4p\cos(\theta/2), \qquad s = r/l_B, \qquad p = r_B/r_E.$$

The constant g is determined by the integral condition in (5.56). Using (5.56), (5.77), (5.78) we can determine the constant g from the following relationship

$$2\pi \int_0^{\pi} \int_{R_{12}}^{r_*(\theta)} f r^2 \sin\theta\, d\theta dr = 1. \tag{5.75}$$

where $r_*(\theta)$ is the radius of the ion pair determined by (5.39).

Let us consider the calculation of the dissociation constant taking into account the Onsager solution (5.74). Using the result obtained by Mason and Mcilroy [9] we can represent (5.74) as

$$f = g\left[\exp \Pi - H\right], \tag{5.76}$$

$$\Pi = 1/s + 4p^2 s \cos\theta,\ \ H = \exp(-\mu^2 s/4)\sum_{n=0}^{\infty}(\mu s/2)^n I_n(\mu),$$

$$s = r/l_B, \qquad p = r_B/r_E, \qquad \mu = 4p\cos(\theta/2).$$

where I_n is the modified Bessel function of the order n. Substituting (5.76) into (5.58) yields

$$\alpha_d = 2\pi g(D_1 + D_2)\, l_B \int_0^{\pi} (\partial H/\partial s - H\,\partial\Pi/\partial s)\, s^2 \sin\theta\, d\theta. \qquad (5.77)$$

As it was shown by Mason and Mcilroy [9] this integral does not depend on the distance r. Hence we can put $s \to 0$ in (5.81), yielding the Onsager formula

$$\alpha_d = 4\pi g(D_1 + D_2)\, l_B F(p), \qquad (5.78)$$

$$F(p) = \frac{I_1(4p)}{2p}, \quad p = r_B/r_E = \left|\frac{(\mu_1+\mu_2)E}{4\pi\varepsilon(\mu_1 e_2 + \mu_2 e_1)}\right|^{1/2} \frac{e_1 e_2}{2k_B T}.$$

At first sight the result (5.78) coincides with Onsager's [7], but taking into account (5.75), (5.76) we can see that the parameter g depends on the electric intensity E. In this case the dependence α_d from E can be quite different from Onsager's result. Putting g = 1/A we obtain

$$A = 8\pi l_B^3 \int_0^1 \int_\delta^{\bar r} s\,u \exp(-4p^2(1-u^2) + 1/s) \int_0^1 I_0(4pu\sqrt{t}) \exp(-t/s)\, dt\, du\, ds,$$

$$\delta = R_{12}/l_B, \qquad p = r_B/r_E, \qquad \bar r = r_*(\theta)/l_B, \quad u = \cos(\theta/2). \quad (5.79)$$

where $r_*(\theta)$ is the function defined in (5.39). We recall that this function determines the distribution of the ion position in the ionic pair (see Fig. 5.2).

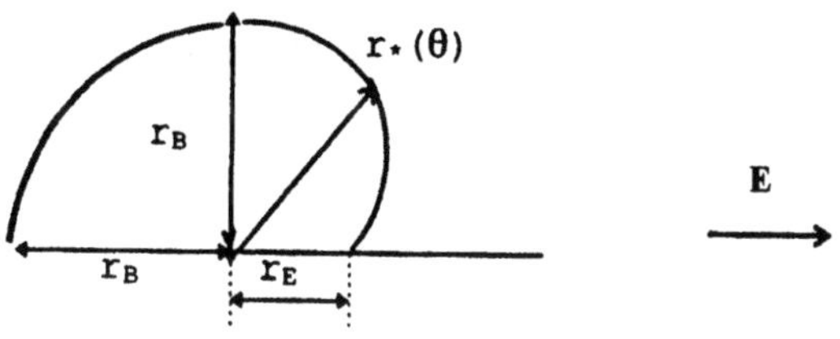

Fig. 5.2. The ion distribution in ionic pair at high field

At high field when p >> 1 the integral (5.79) has the following asymptotic solutions:
at $r_E >> R_{12}$ and $p >> 1$:

$$g^{-1} = A = (\pi/3)\, R_{12}^3 \exp(l_B/R_{12})\, p^{-2}, \tag{5.80}$$

at $r_E \geq R_{12}$ and $p >> 1$:

$$g^{-1} = A = (\pi/8)\, l_B^3\, (1 - R_{12}^2/r_E^2)\, F(p) \tag{5.81}$$

where F(p) is the Onsager factor defined in (5.78).

Finally at $p << 1$ the expression (5.78) coincides with the former result (5.60). We point out that in both limiting cases (5.80), (5.81), the dissociation constant increases with the increase of electric intensity but the asymptotic behaviour of α_d with E differs from Onsager's result and the one obtained by the statistical sum method.

References.

1. Smoluchovski, M.V.: Phys. Zeits. 15 (1916), 593.
2. Amdur, L., Hammes, G.G.: Chemical kinetics, McGraw-Hill Book Co., New York 1965.
3. Debye, P.: Reaction rates in solutions, Trans. Electrochem. Soc., 82 (1942), 262-272.
4. Bjerrum, N.: Kgl. Danske Vid. Selkab. Math—Fys. Medd., 7 (1926), 9.
5. Fuoss, A.M., Krauss. C.A.: J. Am. Chem. Soc: 55 (1933), 1019; 55 (1933), 2387.
6. Zhakin, A.I.: On calculation of pair-recombination rate in weak concentrated solutions, Magnitnaja Gydrodinamika, Latvija Acad. Nauka Press, Riga, 1 (1991), 63-70.
7. Onsager, L.: Deviation from Ohm's law in Weak Electrolytes, J. Chem. Phys. 2 (1934), 599-615.
8. Fuoss, R.M.: Ionic Association III, The equilibrium berween ion pairs and free ions, J. Am. Chem. Soc., 80 (1958), 5059-5061.
9. Mason, D.P., Mcilroy, D.K.: On the theory of Wien dissociation for weak electrolytes, Physics, 82A (1976), 463-476.

Chapter 6

CONDUCTION MODELS IN DIELECTRIC LIQUIDS

6.1 Introduction

Conduction in gases and dielectric liquids (DL) has been reviewed many times. Comprehensive treatments on conduction in gases can be found in the works of Thompson and Thompson [1], and as for DL in the review by Gallagher [2], Adamczevski [3] and Felici [4]. In this section we shall devote our discussion to some of the major controversial issues concerning the natural conductivity due to special ionizing admixtures. The latter has important significance to Electrohydrodynamics applications [5, 6, 7]. Then we shall consider conduction models in dielectric liquids.

6.2 General conduction characteristics of DL

The main conduction parameters of DL are the concentration of impurity ions n_0 and electronacceptive neutral admixture molecules c_O. Usually these concentration values are related by the relationship $n_0 = (K_2 \cdot c_0)^{1/2}$, where K_2 is the equilibrium constant of the dissociation - recombination reactions. The typical magnitudes of $n_0 = 10^{12}cm^{-3}$ are quite small compared with the molecular concentration of the parent liquids $c_0 = 10^{23}cm^{-3}$.

The concentration n_0 responds for a kind of charge carrier. At small values of n_0 (usually $n_0 < 10^8$ cm^{-3} or conductivity coefficient $\sigma < 10^{-18}$ S/m) charge carriers are free electrons which are called fast electrons. At high impurity concentrations ($n_0 > 10^{11}$ cm^{-3}) conductivity is caused by ions because electrons are trapped by the neutral electronacceptor molecules. As EHD effects are usually observed in DL with high level of saturation on impurities ($\sigma \geq 10^{-14}$ S/m) therefore only ionic carriers can be taken into account.

Another important question is the ionic structure. It is relevant to recall here the difference between aqueous and non-aqueous ionic solutions. First of all in aqueous solutions it is necessary to take into account the effect of an ionic atmosphere radius which is determined by the Debye length

$$\lambda_D = (\varepsilon k_B T / 2e^2 n_0)^{1/2}.$$

Let us introduce the mean distance between ions $n_0^{-1/3}$. The effect of ionic atmosphere appears under condition $\lambda_D \ll n_0^{-1/3}$, which is generally valid in aqueous electrolytes. For DL usually it is $\lambda_D \gg n_0^{-1/3}$ and the effect of ionic atmosphere can be neglected. Note that this conclusion is the consequence of the small ionic concentration n_0. Furthermore in this case it can be taken into account only pair interaction in ionic and neutral components.

Pair interaction between ions leads to ion pairs formation. We have considered in detail the theory of ion pairs in chapter 5. The ion pair A^-B^+ is two ions A^+, B^- bounded by the electrostatical ion-ion interaction. The radius of the ionic pair is determined by the Bjerrum length

$$l_B = z_A z_B e^2 / (2\varepsilon k_B T),$$

where z_A, z_B are the charge numbers of ions A^+, B^-. For monoions $z_A = z_B = 1$. In the cases of water and benzene is $l_B = 0.35$ nm and 12 nm respectively.

Ionic pairs possess a comparatively large dipole momentum p_0. For example, at concentration 10^{-5} - 10^{-3} mole/l of the salt $R_4N^+X^-$ the measured magnitudes of p_0 consisted of 7 - 20 D [8] (D denotes the Debye unit: 1D = 1 proton charge multiplied by 0.1 nm).

With the increase of the salt concentration there is complex ion formation, for example, ionic triplets $A^+B^-A^+$, $B^-A^+B^-$, which are formed in non-polar liquids at salt concentrations > 10^{-3} mole/l [8].

Ionic triplets are an example of ion-dipole clusters. It can be shown that due to ion-dipole interaction the ion-dipole clusters can form the structure sketched in Fig. 6.1.

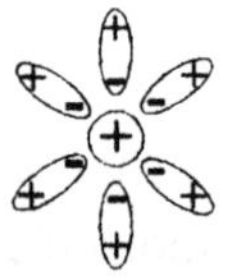

Fig. 6.1, The structure of ion-dipole cluster

On the basis of the statistical method it can be obtained that the radius of a ion-dipole cluster is expressed as

$$l_{id} = (q p_0 / 4\pi\varepsilon k_B T)^{1/2}$$

where q is the charge value of the cluster and p_0 is the value of the dipole momentum of polar molecules which interact with ions. The ion-dipole clusters are expected to be observed in solutions of non-polar DL with polar liquid admixtures, for example, transformer oil with acetone or butyl alcohol etc.

Ion-dipole clusters formed by oxygen in "impure" non-polar liquids were observed in the experiments [9, 10], and in solutions of non-polar liquids with polar admixtures in the work [11]. In non-polar liquids the number, m, of molecules in clusters can reach about

5 and more. For example, in a solution of (iso-$C_5H_{11})_4N^+SCN^-$ in benzene it is m = 5.47 at a concentration $3.8 \cdot 10^{-3}$ mole/l [8]. Equilibrium constants for ionic pairs and ionic triplets are about 10 l/mole and 3000 l/mole respectively [12].

6.3 Conduction at low field

The general shape of a volt-ampere characteristics (VAC), which is usually obtained for gases and DL, can be roughly divided into three regions (Fig. 6.2).

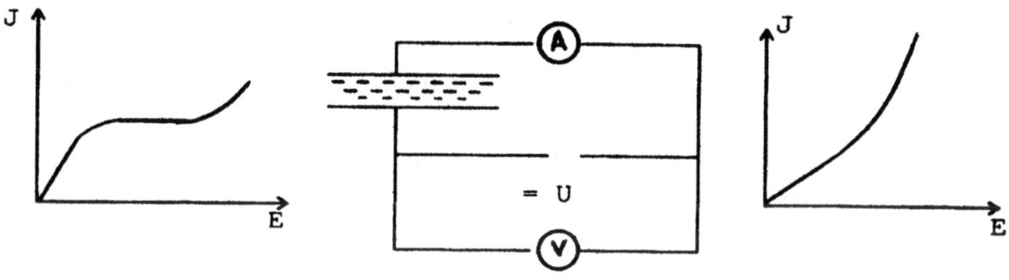

Fig.6.2, VAC for gases (a) and DL (b) for a plane electrode geometry

At low linear electric fields E = V /d < 1 kV/cm the current appears to be a linear function of the applied voltage V, and to obey Ohm's law until about E ~1 kV/cm. At intermediate fields up to about 10 kV/cm, there is a sub-ohmic region where the current tends towards saturation. The saturation region is always present in gases, and is ill-defined or may not be present at all in DL. At high fields the current increases rapidly and, at 40-100 kV/cm, breakdown can occur.

For a continuous current to flow, there must be a source of charge. In the ohmic and sub-ohmic region the natural radiation, dissociation of ionic impurities and small impurity particles could contribute to the current. The influence of natural radiation on the current magnitude is negligibly small (much less than 10 ions per cm^3; this magnitude is the concentration of ions in air under normal condition due to natural ionization).

The conductivity of DL at low fields E ≤ 1 kV/cm, is attributable to impurity ions (so-called uncontrolled chemical impurities) and to ions formed as a result of the self-dissociation of the carrier liquid molecules. The simplest model of low field conductivity is based on the assumption that the chemical impurity is in the form of ion pairs A^+B^- which dissociates with the formation of free impurity ions A^+ and B^- according to the reactions:

$$A^+B^- \underset{\alpha_r}{\overset{\alpha_d}{=}} A^+ + B^-, \tag{6.1}$$

where α_d is the rate of dissociation and α_r is the coefficient of recombination.

The free impurity ions A^+, B^-, which escape recombination, will drift to the electrodes and raise a flow of current. This mechanism of conductivity is usually called the *dissociation mechanism*.

In non-polar DL the ion pairs are usually formed due to an uncontrolled chemical admixture (for example water). In highly purified polar DL the molecules of the parent liquids are the ion pairs A^+B^-.

The dissociation conductivity can be controlled if in a highly purified non-polar DL we add a special ionizing admixture. For example, in the solution of an electron acceptor molecules X in DL the ionizing process can be represented by the reactions

$$X + M = X{\cdot}M = X^-M^+ = X^- + M^+, \tag{6.2}$$

where M is the molecule of parent liquid, X·M is the molecule complex and $X^- M^+$ is the ion pair.

Examples of electron acceptors include iodine J_2, oxygen O_2, water, quinones, sulphur dioxide, mineral salts. So in reaction (6.1) ion B^- is X^- and ion A^+ is M^+.

6.4 Conductivity at high fields

When the field applied to an insulating liquid is increased beyond 1 kV/cm the current tends to rise usually proportionally $J \sim V^2$ and, beyond 40 - 100 kV/cm, the current increases sharply for comparatively small increments of the field. Finally, at a sufficiently high magnitude of electrical strength, breakdown occurs.

6.4.1 Conductivity due to dissociation

There are several alternative theories of conductivity at high fields. Onsager's theory [13] is based on the concept that the dissociation rate of ion pairs A^+B^- is increased increasing the electric intensity E:

$$\alpha_d = \alpha_{d0} F(E), \tag{6.3}$$

where α_{d0} depends only on temperature and F(E) is the factor depending on electric intensity E. In strong fields it takes place

$$F(E) \sim \exp\left[\frac{2e(eE)^{1/2}}{(4\pi\varepsilon)^{1/2} k_B T}\right] \tag{6.4}$$

where e denotes the proton charge, ε the dielectric permittivity and k_B the Boltzmann's constant.

If we assume that the reaction (6.1) is close to equilibrium then the concentration of ions A^+ and B^- is expressed as

$$n_0 = [A^+] = [B^-] = (\alpha_d N / \alpha_r)^{1/2}, \tag{6.5}$$

where $N = [A^+B^-]$ is the concentration of ion pairs. In a plane electrode geometry (see Fig.6.1) direct current j_+, (j_-) of ions $A^+(B^-)$ equals

$$j_+ = e\mu_+ nE, \quad j_- = e\mu_- nE,$$

where μ_+ (μ_-) is the mobility of ion $A^+(B^-)$. So for the total current we have

$$j = j_+ + j_- = \sigma E, \tag{6.6}$$

$$\sigma = e(\mu_+ + \mu_-)n = \sigma_0 [F(E)]^{1/2}, \qquad \sigma_0 = e(\mu_+ + \mu_-)[\alpha_{d0} N / \alpha_r]^{1/2}. \tag{6.7}$$

Onsager's theory of dissociation adequately accounts for the results in solutions of iodine and $SbCl_5$ in Pyralene 1499, but is not correlated for the current due to oxygen, SF_6 and anthraquinones (see [2, 14] and Fig. 6.3).

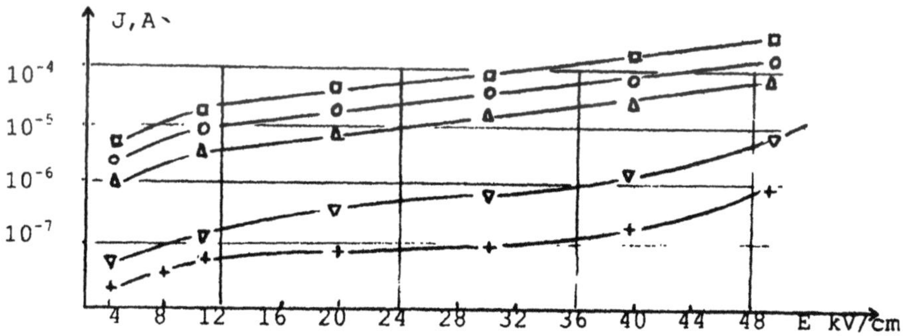

Fig. 6.3, VAC in solutions [14]: + - pure liquid (chlorobiphenyl, trade name Pyralene 1499), Δ - 1 p.p.m. of iodine, o - 2 p.p.m. of iodine, , - 3 p.p.m. of iodine, ∇ - saturated with oxygen.

6.4.2. Electrochemical injection

Simultaneous to a deviation from Ohm's law at high field there is an intense electrohydrodynamical (EHD) motion in DL. Numerous experiments have shown that Onsager's theory did not correlate with data on the EHD motion. Then investigators paid attention to the injection processes on the electrode-liquid interface. It has been shown that a neutral electron acceptor molecule X can participate in an electrochemical ox-reaction

on the cathode: $$X + e^- \underset{\alpha_2}{\overset{\alpha_1}{=}} X^-, \tag{6.8}$$

where α_1 (α_2) is the rate of forward (reverse) electrochemical reaction, e^- stands for electron.

Let us note that the ion A^+ must be discharged on the cathode. Assume that this is the red-reaction

on the cathode: $$A^+ + e^- \underset{k_2}{\overset{k_1}{=}} A\,. \tag{6.9}$$

The reactions (6.8), (6.9) are not independent one from another. Indeed, in the case of a small discharge rate k_1 a thin diffusion layer is formed near the electrode surface. The electric intensity on the electrode surface E (r_0) inside this layer can be much larger than outside layer $E_v = E(\xi_{el})$, where r_0 is the radius of ion A^+ and ξ_{el} is the size of the diffusion layer. The qualitative structure of the diffusion layer is shown in Fig.6.4.

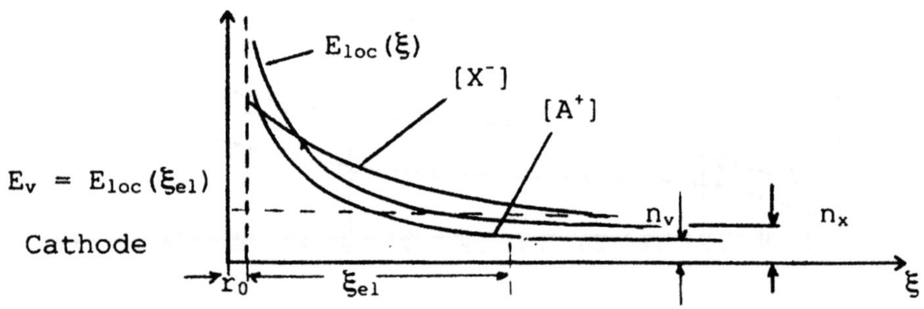

Fig. 6.4, The structure of diffusion electrical layer

To show these distributions we assume the following balance equations for the ionic components $n_1 = [A^+]$, $n_2 = [X^-]$:

$$\varepsilon \frac{dE_{loc}}{d\xi} = e\,(n_1 - n_2)\,; \tag{6.10}$$

$$-D_1 \frac{dn_1}{d\xi} + \mu_1 n_1 E_{loc} - (\mu_1/e)\,n_1 \frac{d\Pi_1}{d\xi} = -\,i_1\,;$$

$$-D_2 \frac{dn_2}{d\xi} - \mu_2 n_2 E_{loc} - (\mu_2/e)\,n_2 \frac{d\Pi_2}{d\xi} = i_2\,; \tag{6.11}$$

with boundary conditions:

$$\xi = r_0: \qquad i_1 = k_1\, n_1(r_0) - k_2\, c_A\,(r_0), i_2 = \alpha_1\, c_x\,(r_0) - \alpha_2\, n_2\,(r_0);$$

$$\xi \to \infty: \qquad E_{loc} \to E_{v}, \qquad n_1 \to n_v; \tag{6.12}$$

where μ_1, μ_2 are the mobilities of A^+, X^-; $r_0 = (\mathbf{r}_{A^+}, r_A, \mathbf{r}_{X^-}, r_X)$ are the radii of A^+, A, X^-, X; $\Pi_1(\xi)$, $\Pi_2(\xi_+)$ are potentials of short range forces acting on the ions A^+, X^-; c_A, c_x are the concentrations of the neutral components A, X; $E_{loc}\,(\xi)$ is the distribution of the local electric intensity in the real electrode-liquid interface.

From the diffusion equations (6.11) it is seen that there are two characteristic lengths. The balance of the diffusion current and the current due to short range forces gives the length of the physically absorbed layer. For example, if $\Pi_i\,(\xi)$ is determined by the image forces

$$\Pi_i\,(\xi) = -\,e^2 / (16\pi\varepsilon\xi), \tag{6.13}$$

then the absorption length is expressed as

$$\xi_{ad} = e\mu_1 / (16\pi\varepsilon D_1) = e^2 / (16\pi\varepsilon k_B T), \tag{6.14}$$

where the following known relationship has been used

$$D = \phi_0 \mu, \qquad \phi_0 = k_B T / e.$$

The balance of diffusion and migration currents gives the length of the diffusion part of the double electric layer in a high external electric field

$$\xi_{el} = \phi_0 / E_v, \qquad \phi_0 = k_B T / e. \tag{6.15}$$

This result is quite different from the case of aqueous solutions where the diffusion length of the double electrical layer is determined by the Debye distance. Putting $\varepsilon = 2\varepsilon_0$, T

= 300°K (room temperature), E_v = 1 kV/cm we obtain $\xi_{ad} = 6.8$ nm, $\xi_{el} = 290$ nm, that is at low fields $\xi_{ad} < \xi_{el}$.

At high fields the diffusion electric layer disappears because of $\xi_{ad} \geq \xi_{el}$.

The case $n_2 >> n_1$ was studied in the work by Felici and Gosse [15] which we will discuss later on. Let us consider another limiting case $n_1 >> n_2$ when there is an increment of the electric intensity inside the diffusion layer by the heterocharges A^+. Using the fact that outside the diffusion layer there is only a migration current, that is $i_1 = n_v E_v$, and taking into account the boundary condition (6.12) we get

$$n_1(r_0) = [\mu_1 n_v E_v + k_2 c_A(r_0)]/k_1. \tag{6.16}$$

This expression shows that the concentration of discharged ions A^+ is proportional to that of the bulk n_v and increases with the decreasing of discharge rate constant k_1. That leads to an increment of the electric intensity inside the diffusion layer.

Electrical rate constants depend on the electric intensity on the electrode surface $E_{loc}(r_0)$:

$$k_1 = k_{10} \exp\left[\frac{eE_{loc}(r_{A^+})r_{A^+}}{k_B T}\right], \quad k_2 = k_{20} \exp\left[\frac{-eE_{loc}(r_{A^+})r_{A^+}}{k_B T}\right]. \tag{6.17}$$

So to determinate the electrochemical rate constants it is necessary to find the electric intensity $E_{loc}(r_0)$.

The electric intensity is usually considered to be constant inside the absorption layer $r_0 < \xi < \xi_{ad}$ [6, 15]. It means

$$E_{loc}(r_0) >> e\bar{n}_1 \xi_{ad}/\varepsilon = E_{ind}, \tag{6.18}$$

where $\bar{n}_1$ is the mean concentration of ions A^+ inside the absorption layer and E_{ind} is the induced field. Under this condition the electric strength $E_{loc}(r_0)$ is induced by the external field E_v and the ions A^- accumulated inside the electrical diffusion layer $\xi_{ad} < \xi < \xi_{el}$. Here we will also assume the condition (6.18).

Let us calculate $E_{loc}(r_0)$. Inside the electric diffusion layer short range forces are negligibly small so the distribution of ions A^+ is governed by the first Eq. in (6.11) at $\Pi_1 = 0$. Simple integration of this equation yields

$$n_1(\xi) - (\varepsilon/2k_B T)E_{loc}^2(\xi) = (i_1/D_1)\xi + C_1.$$

Putting $\xi = \xi_{el}$ we find C_1 and at $\xi = \xi_{ad}$ it follows

$$n_1(\xi_{ad}) - (\varepsilon/2k_B T)(E_{loc}^2(\xi_{ad}) - E_v^2) = (i_1/D_1)(\xi_{ad} - \xi_{el}) + n_v. \tag{6.19}$$

At low field the short range forces will be dominant inside the absorption layer, and this leads to the following charge distribution

$$n_1(\xi) = n_1(\xi_{ad}) \exp(-\Pi_1(\xi)/k_BT), \qquad \text{at } r_0 < \xi < \xi_{ad}. \tag{6.20}$$

Putting $\xi = r_0$ we obtain

$$n_1(\xi_{ad}) = n_1(r_0)/K_{ad}, \quad K_{ad} = \exp(-\Pi_1(r_0)/k_BT), \tag{6.21}$$

where K_{ad} is the absorption constant. Under condition (6.18) it is $E_{loc}(\xi_{ad}) = E_{loc}(r_0)$ and from (6.19), (6.21) we obtain

$$E_{loc}(r_0) = \beta_{el} E_v, \tag{6.22}$$

$$\beta_{el}^2 = 1 + 2k_BT\,(n_1(r_0)/K_{ad} - n_v + i_1(\xi_{el} - \xi_{ad})/D_1)/(\varepsilon E_v^2),$$

where $n_1(r_0)$ is determined by (6.16) and β_{el} is the amplification factor due to the electrical diffusion layer.

We recall that this simple theory is valid at low field and small discharge rate when

$$\xi_{el} > \xi_{ad}, \qquad n_1(r_0) > K_{ad} n_v. \tag{6.23}$$

With an increase of the external field E_v both magnitudes ξ_{el} and $n_1(r_0)$ decrease so at high enough intensity it is always $\beta_{el} = 1$. This conclusion is valid under condition (6.18). Using (6.20) the magnitude $\bar{n}_1$ can be estimated, for example, at $r_0 << \xi_{ad}$:

$$\bar{n}_1 = n_1(r_0)\,(r_0/\xi_{ad})^2.$$

So to check up the conditions (6.18), (6.23) it is necessary to know the magnitude of the discharge rate constant k_1. In classical electrochemistry the value of k_1 is considered to have very high levels due to strong intensity inside the double electric layer about 1 MV/cm and at $E_{loc}(r_0) << 1$ MV/cm the ions A^+ can not be discharged on the cathode. Bearing in mind that at high field $\beta_{el} = 1$ a question arises about the mechanism of the high electric strength generation.

One of the ways to explain this phenomenon is to take into account the real structure of the electrode surface, which microscopically is very rough: on highly polished surfaces (such as steel, aluminium, etc.) the number of microasperities per cm^2 is of the order 10^8, their characteristic dimensions are $\delta_r < 2$ μm with an amplification factor of about 100 - 1000 [16]. The real electrode surface and charge distribution in electrode vicinity are schematically shown in Fig. 6.5. Later on, we will show that outside the diffusion layer there is the non-equilibrium layer where the dissociation-recombination reactions are not in equilibrium. Its characteristic length is

$$\xi_d = \varepsilon E_0 / (en_0),$$

where E_0 is the electric intensity outside the non-equilibrium layer in the liquid bulk and n_0 is the concentration of A^+, B^- at equilibrium state (see Fig. 6.5). The term en_0 is expressed via the ohmic conduction coefficient σ as $en_0 = \sigma/(\mu_1 + \mu_2)$, thus

$$\xi_d = \varepsilon (\mu_1 + \mu_2) E_0/\sigma.$$

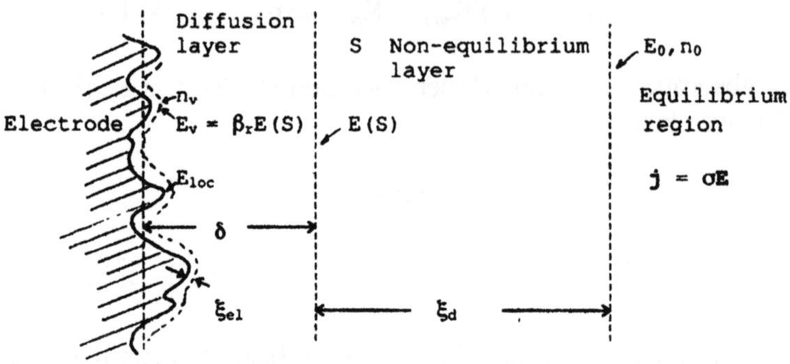

Fig. 6.5 The structure of real electrode surface and distribution of the electric characteristics at low and intermediate field.

At some distance δ from the rough surface we can introduce a "smooth" surface S where the boundary conditions of average characteristics (velocity, electric intensity, etc.) are stated. Note that the distance δ is of the same order as δ_r. If we denote E (S) as the electric intensity on S and E_v the one on the outside of the diffusion layer near the microasperity tip then

$$E_v = \beta_r E(S)$$

where β_r is the amplification factor due to irregularities. Using (6.22) we get

$$E_{loc}(r_0) = \beta E(S), \qquad \beta = \beta_r \beta_{el}, \tag{6.24}$$

where β is the total amplification factor.

We point out that the intensity $E_{loc}(r_0)$ is the active electric strength which causes both discharge and injection reactions. Let us turn now to the discussion of the injection process caused by the reaction (6.8). The solution of the second Eq. in (6.11) under condition (6.18) is expressed as

$$n_2(\xi) = \frac{i_2}{D_2}\int_{\xi}^{\infty} g(\xi)\,/\,g(r_0)\,d\xi, \qquad g(\xi) = \exp(-\Pi_2(\xi)),$$

$$\Pi(\xi) = [eE_{loc}\xi - \Pi_2(\xi)]\,/\,k_BT, \qquad E_{loc} = E_{loc}(r_0).$$

Putting $\xi = r_0$ we can obtain the expression for $n_2(r_0)$ and E_{loc} then assuming $\xi \to \infty$ we obtain the following expression for the density of the injection ions outside the diffusion layer

$$n_2 = \frac{\alpha_1 c_x(r_0)}{\mu_2 E_{loc} + \alpha_2 K(E_{loc})} = n_x(E, T, n_0), \tag{6.25}$$

$$K(E_{loc}) = \frac{eE_{loc}}{k_BT}\int_{r_0}^{\infty} g(\xi)\,/\,g(r_0)\,d\xi, \qquad E = E(S).$$

The formulae (6.24), (6.25) state the following remarkable result: the density of the injection ions at any point on the "smooth" electrode surface S is a function of the electric intensity E(S), temperature T and ohmic conductivity σ, defined at that point. We will call this function $n_x = n_x(E, T, n_0)$ the *injection function*.

If $n_x >> n_0$ then the current is due only to *injection of ions X*. This mode of conductivity is called injectional. If $n_0 >> n_x$ the injection current of ions X is negligibly small and there is only *the dissociation conductivity*. The dissociation - injection model of conductivity gives good qualitative and, in many cases, quantitative correlations between electrical and hydrodynamical experimental data and theoretical calculations.
If we assume

$$\mu_2 E_{loc} << \alpha_2 K(E_{loc}), \tag{6.26}$$

then from (6.25) it follows the result by Felici and Gosse [15]:

$$n_i = n_x(E) = n_A / K(E_{loc}), \tag{6.27}$$

$$n_A = (\alpha_1/\alpha_2)\,c_x.$$

Felici and Gosse have stated that an injection current from metallic electrodes was analogous to electron emission. So n_A in (6.27) is the concentration of absorbed ions X on the metallic surface.

The function $[K(E_{loc})]^{-1}$ at T= 300 °K (room temperature) and E_{loc}< 100 kV/cm can be approximated as [6]:

$$[K(E_{loc})]^{-1} = 1 + r_1 E_{loc} + r_2 E_{loc}^2,$$

where the coefficients r_1, r_2 depend on the radius r_x and dielectric constant.

At $\varepsilon = 2\varepsilon_0$ and $r_x \leq 1$ nm it is: $r_1 = 0.07$ cm/kV, $r_2 = 3.5\cdot10^{-4}$ (cm/kV)^2; and at $\varepsilon = 37\varepsilon_0$ (for nitrobenzene), $r_x < 1$ nm: $r_1 = 0.0046$ kV/cm, $r_2 = 0$. The coefficients r_1, r_2, quickly increase with an increase of the ion radius. These estimates show that the injection function in polar DL may be independent of the electric strength. This case is usually termed an *autonomous injection level.*

For ions with radius $r_x > r_A = [e/(4\varepsilon E_{loc})]^{1/2}$ the function $[K(E_{loc})]^{-1}$ is an exponential one:

$$[K(E_{loc})]^{-1} = \exp(E_{loc} / E_A),$$

$$E_A = \phi_0 / r_x = k_B T / (e r_x)$$

6.4.3 Examples of red-ox reactions

1. It is known that crystal iodine is solved in hydrocarbons in the form of molecules J_2 and molecule complexes $(J_2)_n = J_2 \cdot J_2 \cdot J_2 \ldots J_2$, n= 1,2,3,... Being an electron acceptor iodine participates in ox-reaction on the cathode $(J_2)_n + me^- = (J_2)_n^{-m}$ and charged complexes $(J_2)_n^{-m}$ discharge on the anode $(J_2)_n^{-m} - me^- = (J_2)_n$. Thus solutions of iodine in hydrocarbons are an example of liquids where an unipolar injection of negative ions from a cathode can be produced. Note that iodine possesses a high chemical activity, and because of this the electrode material must be indifferent such as titanium (copper electrodes are not suitable because on its surface copper iodide is formed).
2. In solutions of quinone $C_6H_4O_2$ and hydroquinone $C_6H_4(OH)_2$ in a polar liquid dielectric there are the following red - ox reactions

$$\text{on the anode:} \quad C_6H_4(OH)_2 = C_6H_4O_2 + 2H_+ + 2e^-,$$

$$\text{on the cathode:} \quad C_6H_4O_2 + 2H^+, 2e^- = C_6H_4(OH)_2.$$

This is an example of solutions where an unipolar electrochemical injection of positive ions H from an anode can be formed [17].
3. Denat, Gosse and Gosse reported that in solutions of hydrocarbons with the salt TiAP (triisoamylammonium picrate) with copper electrodes there is an injection of negative charge from the cathode [18].
4. In polar liquid dielectrics such as aromatic hydrocarbons (nitrobenzene, iodobenzene, bromobenzene, spirits, etc.) there are bipolar injections [19, 20]. Generally, the cathode injection is more intensive than the anode one in these cases.

6.4.4 Electron cold emission

Since a long time the electron emission from the cathode has been the most popular explanation for the source of current at high fields. The emission current was assumed to follow the law given by Fowler and Nordheim [21]. It arises due to cold field tunnelling process and it is expressed as

$$j = AE^2 \exp(-B/E); \qquad (6.28)$$

where A, B are constants depending on temperature and liquid-metal interface properties (dielectric permittivity, electron work function, etc.).

For high purified DL high field conduction measurements show little dependence of current on the cathode work function [2]. On the other hand in solutions of non-polar liquids with electronacceptors or polar admixture (iodine, butyl alcohol, etc.) the experiments have shown a strong dependence on the work function [7]. Another striking fact is the low values of electric intensity needed to create a cold emission current, in comparison with the theoretical ones. But this is not surprising, however, because of the microasperities as it was mentioned above. Furthermore, thin oxide films, such as Al_2O_3, BO, MgO with thicknesses about 0.01 - 0.2 μm, having absorbed ions can produce electron emission without an external electric field [22]. These experimental data show that polar and electronacceptive admixtures, having been absorbed on the metallic surface can reduce the work function to zero. This means that it could be $B/E \ll 1$ in expression (6.28), and hence the current dependence is a quadratic function of the electrical strength, $j = AE^2$. Noting that impurities trap electrons emitted from cathode we can assume $j = en_x\mu E$ and we obtain the *linear law of injection*

$$n_x = \eta_c E, \qquad \eta_c = A/e\mu. \qquad (6.29)$$

We can also add that impurity molecules appear to be absorbed by the microasperities which are the centre of the electron emission. Hence the constant A in (6.28), (6.29) must be proportional to the impurity concentration A ~ [X], that is the conductivity is the linear function on the concentration of the electronacceptor:

$$\sigma = en_x\,\mu \sim [X]. \qquad (6.30)$$

Both trends (6.29), (6.30) are generally observed in conduction experiments with solution of non-polar liquids with iodine, butyl alcohol, etc. [14, 7].

6.5 Conduction due to small particles

In DL the charge transport can be only due to ions, but the small particles, which are named as moleons, can also contribute to the current. The charge transport due to moleons is called as *moleon conductivity* (MC). Convinced experimental data on MC had been presented by various authors (see foregoing review by Gallagher [2]. It has been shown that the removal of particles by filtering reduced the conduction current (Fig.6.4).

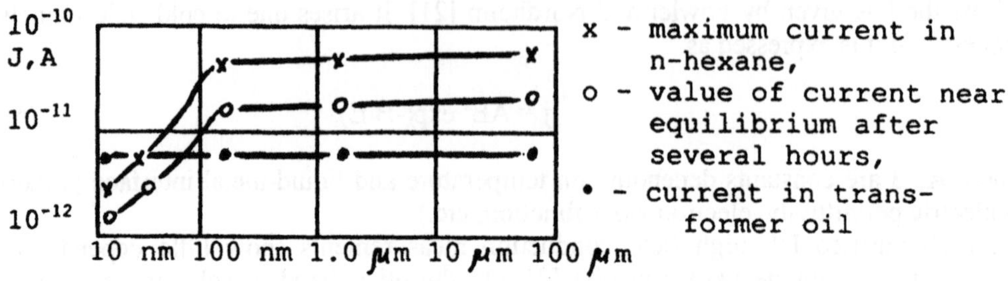

Fig.6.6, Effect of filtering n-hexane (x), (o) and transformer oil (•) on conduction current [2]

Note that small particles may be responsible for the slow decay of current with time, which is generally observed in conduction measurements. The main source of particles can be fresh electrodes introduced into the cell, and even vigorous flushing failed to dislodge them.

The theory of charge transport by small particles was developed by many authors [2]. The simplest theory is based on the equation of particle motion into the form

$$m\, du/dt = qE - 6\pi R\eta u, \tag{6.31}$$

where m, R, q are the mass, radius and charge of the particle, u is the velocity, η is the dynamic viscosity of the parent liquid. The value for q is given by the formula

$$q = (2/3)\, \pi^3 \varepsilon R^2 E. \tag{6.32}$$

Putting u = 0 at t = 0, from (6.31) we get

$$u = E\,(1 - \exp(-t/\tau))\,, \quad \tau = m/\,(6\pi\eta R), \tag{6.33}$$

$$\mu = q/\,(5\pi\eta R) = (\pi\varepsilon R/9\eta)\, E. \tag{6.34}$$

For particles with R = 0.1 μm and mass density 2 g/cm^2, $\varepsilon = 2\varepsilon_0$, $\eta = 0.2$ g/(cm·s) (viscosity of transformer oil) the expressions (6.33), (6.34) yield

$$\tau = 10^{-8}\text{s}, \qquad \mu = 3.3\cdot 10^{-8} \frac{E}{V/cm} \frac{cm^2}{V\cdot s}.$$

The small value of τ shows that the velocity may be considered as constant

$$u = \mu E = bE^2, \qquad b = \pi^2 \varepsilon R/\,(9\eta).$$

Assuming E = 10 kV/cm we have for the velocity v = 3.3 cm/s. The current density is expressed as

$$j = qun_0 = \sigma_p E^3, \qquad \sigma_p = 2\pi^5\varepsilon^2 R^3 n_0 / (27\eta).$$

This theory is not always correlated with experimental data, particularly with electrohydrodynamical characteristics. In real experimental situations the behaviour of small particles is very complicated [7, 23], that indicates the complex character of conductivity. There are many evidences that the main inclusion in current due to ionic impurities and a moleon conductivity is negligibly small [2].

6.6 General multi-ion conduction model in dielectric liquids

The dissociation and the injection are the most general classification of charge generation in DL. More precise graduation takes into consideration the detail characteristics of these processes. For example, the dissociation processes can be caused by gaseous impurities, ion-dipole clusters, ion complexes (ion pairs A^+B^-, ion triplets $A^+B^-A^+$, $B^-A^+B^-$, etc.).
On the other hand, ion injection from electrodes can arise due to electron emission from surface irregularities, electrochemical red-ox reaction and other mechanisms. A remarkable achievement is the discovery of unipolar ion injection from ion-exchange membranes [24]. This injection is reproducible and high injection level is observed only in polar DL (such as nitrobenzene, Pyralene 1499, etc.). These numerous ion generations from the mathematical point of view are described by similar parameters: the rates of dissociation, recombination and injection processes. The characteristics of these processes will be reflected only on the functional type of foregoing coefficients. That is why it is enough to consider the general three ion, dissociation - injection model of conductivity, which is based on the following red-ox reactions:

$$\text{on the cathode:} \quad X + e^- \underset{\alpha_2}{\overset{\alpha_1}{=}} X^-, \quad A^+ + e^- \underset{k_2}{\overset{k_1}{=}} A, \quad A^+B^- + e^- \xrightarrow{k^-_{AB}} A + B^- \qquad (6.35)$$

$$\text{on the anode:} \quad X^- \underset{k_{2x}}{\overset{k_{1x}}{=}} X + e^-, \quad B \underset{k_{2B}}{\overset{k_{1B}}{=}} B + e^-, \quad A^+B^- \xrightarrow{k^+_{AB}} A^+ + B + e^- \qquad (6.36)$$

$$\text{in liquid bulk:} \quad A^+B^- \underset{\alpha_r}{\overset{\alpha_d}{=}} A + B^-, \quad A^+ + X^- \xrightarrow{\alpha_{rx}} A + X \qquad (6.37)$$

For example A^+, B^- can be the molecules of water or parent liquid and X can be the molecule of iodine or other electron acceptors. Discharge reactions of ions X^- in (6.36) are of the regenerative type, which accounts for the stationary direct current and reproducible EHD effects. Usually EHD effects in an industrial grade DL are not reproducible because of the changes in the chemical composition in the injection electrochemical reactions. For example, molecules of water can participate in the electrode processes on the cathode $2H_2O + 2e^- = H_2 + 2OH^-$ and on the anode $2H_2O = 4e^- + 4H^+ + O_2$. In both cases there is not any regeneration of molecules of H_2O. Regenerative red-ox reaction can be realized under the following conditions:

1) the neutral compositions in the reagents,

2) the electrode material must be chemically neutral to reagents (indifferent electrodes).

3) red-ox reactions must not change the chemical compositions in reagents.

So to use EHD effects in practice it is necessary to select carefully both DL and ionizing admixture and the electrode material. The complex liquid + admixture + electrode with regenerative red-ox reaction is called the *ideal red-ox system* [6].

6.7 Conduction model in polar liquids

Polar liquids are presented as aromatic hydrocarbons (nitrobenzene, chlorobenzene, etc.), spirits, acetone, liquid crystals, etc. Polar liquids dissolve and dissociate impurities so efficiently that it is exceedingly difficult to purify them to a state where they exhibit high levels of resistivity. Traditionally, therefore, polar liquids have not been classified as insulators. The interest in their electrical properties was revived when methods of purification, using ion-exchange membranes and electrodialysis, were developed which provided them with resistivities greater than 102 ohm·m [4, 24]. However, it is a most formidable task to maintain this ultra pure state, since even glass can release enough impurity to nearly halve the resistivity. The high chemical activity of polar DL is caused by chemically active groups or atoms in liquid molecules. Usually these groups are strongly oxidizing agents such as NO_2, J, Br, Cl, OH, etc. Chemical formulae of polar dielectric liquids can be represented as R-X where X is the foregoing chemical active group and R- is some radical such as the benzene ring or the group CH3 - CH2- ... in spirits. As mentioned above the dissociation conductivity in high purified polar DL is due to the dissociation of the parent liquid molecules. This ionizing process can be represented as

$$\text{R-X} + \text{R-X} = \text{R-X}^- + \text{X-R}^+ . \tag{6.38}$$

Polar molecules R-X, due to its high chemical activity, can also participate in red-ox reactions on the electrodes

$$\text{on the cathode:} \quad \text{R-X} + e^- = \text{R-X}^-, \quad \text{X-R}^+ + e^- = \text{X-R} ; \tag{6.39}$$

$$\text{on the anode:} \quad \text{X-R} = e^- + \text{X-R}^+, \quad \text{R-X}^- = e^- + \text{R-X} . \tag{6.40}$$

Electrochemical reactions of this type have been observed in liquid crystals [17, 25] and in aromatic hydrocarbons (nitrobenzene [19, 20], chlorobenzene and iodobenzene [20]). Usually cathode electrode red-reaction (6.39) has a higher level than the anode one, therefore EHD motion near the cathode is generally more intensive than near the anode [20]. This remarkable property is used in practice to intensify EHD motion in non-polar liquids adding polar liquid admixture such as butyl alcohol, acetone, spirits, etc. [7]. Specialists point out that the presence of water in polar DL can essentially change the form of ionizing reactions (6.39), (6.40) leading to the destruction of liquid molecules R-X. To avoid this unfavourable phenomenon a special admixture should be added to the liquid. For example, to protect the liquid crystals from electrochemical destruction quinone and hydroquinone were used [17].
Thus the conductivity of high purified polar DL can be described by the bi-ion model represented by reactions (6.38) - (6.40).

6.8 Conduction models in non-polar liquids

Resistivities of about 10^{16} ohm cm can be attained relatively easily in many non-polar DL. The most intensive EHD motion is generally observed at resistivities of about 10 ohm cm, therefore we can take into account the conductivity due to impurities only. The ionizing admixtures used to enhance an electrode charge injection and, as a result EHD motion, were listed in sections 6.3, 6.5. If an admixture exists in the liquid in the form of ion pairs A^+B^- and of a neutral electron acceptor X, we can use the general three-ion model presented by the electrochemical reactions (6.35) - (6.37).
We point out that the foregoing model is the first step to understand the complex conduction phenomenon in DL. Thus at high concentration of ion pairs in non-polar DL the formation of complex triplets is possible:

$$A^+B^- + A^+ = A^+B^-A^+, \qquad A^+B^- + B^- = B^-A^+B^-.$$

Ionic triplets can essentially change the character of conductivity due to small mobilities, non-homogeneous distributions in concentrations and unknown mechanisms of electrode electrochemical reactions.
Model conductivity with complex-ion formation was discussed in chapter 5 where the conditions of negligibly small complex-ion formation were obtained (see inequality (5.30)).

References.

1. Thomson, J.J. and Thomson, G.P.: Conduction of electricity through gases, 3rd edition Vol. 1 pp.193, Cambridge University Press, 1928.
2. Gallagher, T.J.: Simple dielectric liquids. Mobility, conduction and breakdown, Clarendon Press, Oxford 1975.

3. Adamczweski, I.: Ionization, conductivity and breakdown in dielectric liquids, Taylor and Francis, London 1969.
4. Felici, N.J.: - D. C. Conduction in liquid dielectrics (Part 1), Direct Current, 2 (1971 a), 147. - D. C. Conduction in liquid dielectrics (Part 2), Direct Current, 2 (1971 b), 90.
5. Melcher, J.R.: Continuum Electromechanics, M. I. Press, Cambridge, Mass. 1981.
6. Zhakin A.I.: Theory of electrical and electrohydrodynamical effects in low conductive media, Thesis, Kharkov University Press 1990 (in Russian).
7. Stishkov, J.K., Ostapenko A.A.: Electrohydrodynamic flows in liquid dielectrics. Printed in Leningrad University 1989 (in Russian).
8. Copenhafer, D.T., Kraus C.A.: J. Am. Chem. Soc. 73 (1951), 4557.
9. De Groot, K., Gary, L.P. and Jarnagin, R.C.: J. Chem. Phys., 47 (1967), 3084.
10. Secker, P.E.: J. Phys. D, Appl. Phys., 3 (1970), 1073.
11. Kropotcheva, L.V., Rychkov, J.M., Kovalevskaja, T.I.: On the molecular structure of isolating liquids in electric fields, Electr. Obrab. Mater., Mold. Akad. Nauk, Kishinev, 5 (1992), 33-36 (in Russian).
12. Goldman, S.: Cave G. C. B., Can. J. Chem. 49 (1971), 1726.
13. Onsager, L.: Deviation from Ohm's law in Weak Electrolytes, J. Chem. Phys. 2 (1934), 599-615.
14. Durant, P. and Fournic, R.: Proceeding of the Lancaster Conference 1970, 142.
15. Felici, N., Gosse, J.P.: Injection d'ions par des electrodes metalliques dans les hydrocarbures liquides de resistivite elevee, Rever. Phys. Appl., 14 (1979), 629-633.
16. Little, R.P. and Whitney, W.T.: Electron emission preceding electrical breakdown in vacuum, J. Appl. Phys., 34 (1963), 2430.
17. Zhdanov, S.I., Gratchova, I.P.: Electrochemistry of liquid crystals, in: Liquid crystals (Ed. S.I. Zhdanov), Chemistry Press, Moscow 1979 (in Russian).
18. Denat, A., Gosse B. And Gosse, J.-P.: Ion injection in hydrocarbons, J. Electrostatics, 7 (1979), 205-225.
19. Zann, M., Chatelon, H: Charge injection between concentric cylindric electrodes, J. Appl. Phys., 48, N5, (1977), 1797-1805.
20. Zhakin, A.I., Tarapov, I.E., Fedonenko, A.I.: Experimental investigation of conduction mechanisms in polar dielectric liquids, Electronnaja Obrabotka Materialov, Moldovia Akad. Nauk, Kishinev, 5 (1983), 37-41 (in Russian).
21. Fowler, R.H. and Nordheim, L.W.: proc. R. Soc. A119 (1928), 173.
22. Elinson, M.I., Vasiljev G.E.: Autoelectron emission. Phys.-Mat. State. Press, Moscow 1958 (in Russian).
23. Ostroumov, G.A.: Interaction between electrical and hydrodynamical fields, Nauka Press, Moscow 1979 (in Russian).
24. Felici, N.J. and Tobazeon, R.E.: Charge carrier elimination and production by electrodialytic polymers in contact with dielectric liquids, J. Electrostatics, 11 (1981), 135-161.
25. Voinov, M., Dunnett, J.S.: Electrochemistry of nematic liquid crystals, J. Electrochem. Soc., 120 (1973), 922-924.

Chapter 7

GENERAL INVESTIGATION OF CONDUCTIVITY PROCESSES IN DIELECTRIC LIQUIDS

7.1 Introduction

The basic concepts of Electrohydrodynamics have been considered by many authors. In the first four chapters of this book there is a comprehensive review on EHD foundations. Numerous new results have been obtained recently concerning the Electrohydrodynamics on the basis of the dissociation-injection model of conductivity, and in this section we shall discuss the major conclusions of these investigations.

7.2 Electrohydrodynamics' approach

The low conductivity of liquid dielectrics (less than 10^{-10} S/cm) leads to small magnitudes of magnetic energy in comparison with the electrical one:

$$\mu H^2 \ll \varepsilon E^2 \tag{7.1}$$

(here and later we use S.I. units).
Indeed, from the Maxwell equation

$$\nabla\times \mathbf{H} = \mathbf{j} + \varepsilon\, d\mathbf{E}/dt,$$

we have $H = \max(d{\cdot}j,\ d{\cdot}\omega\varepsilon E)$, where d, ω are the characteristic scales of length and frequency of the electric field. Putting $j = \sigma E$ we get the following conditions:

$$\mu H^2/\varepsilon E^2 = \max(\mu\sigma^2 d^2/\varepsilon,\ \mu\omega^2 d^2\varepsilon) \ll 1. \tag{7.2}$$

These inequalities are valid until $\sigma \le 10^{-5}$ S/cm, $\omega \le 10^6\ s^{-1}$ at $d \sim 1$ cm, $\varepsilon \sim 2\varepsilon_0$, $\mu \sim \mu_0$.
Under conditions (7.2) an electric field is a potential one:

$$\mathbf{E} = - \nabla\phi, \tag{7.3}$$

and Maxwell stress tensor is expressed as

$$T_{ik} = - (p_{str} + \varepsilon/2\, E^2)\delta_{ik} + \varepsilon E_i E_k, \tag{7.4}$$

$$p_{str} = - \rho\, (d\varepsilon/d\rho)E^2/2, \tag{7.5}$$

where ϕ is the electrical potential, δ_{ik} is the symmetric unit tensor, p_{str} is the electrostriction pressure. Here and later on we assume that the dielectric permittivity does not depend on the electric intensity E.

7.3 Basic system of EHD equations

Dielectric liquids, being weak electrolytes, are described by the model of homogeneous viscous multicomponent fluid medium. Basic system of EHD equations hold [1, 2]:
Conservation of mass

$$d\rho/dt + \rho u_{k,\,k} = 0; \tag{7.6}$$

Conservation of momentum

$$du_i/dt = (p_{ik} + T_{ik})_{,k}; \tag{7.7}$$

Conservation of energy

$$\frac{d}{dt}\,(u^2/2 + U_p) + \frac{\partial}{\partial t}\,E^2/2 = -(q_{Tk} + \sum_{i=1}^{n} U_{pi}\, I_{(i)\,k} - p'u_k + S_k - p_{kj}\, u_j)_{,\,k}\,; \tag{7.8}$$

Balance of concentrations

$$dc_i/dt = -I_{(i)\,k,\,k} + m_i\,\dot{\xi}_i\;; \tag{7.9}$$

Electrostatic equations

$$\nabla\cdot \mathbf{D} = q, \qquad \mathbf{E} = - \nabla\phi, \qquad \mathbf{D} = \varepsilon\mathbf{E}. \tag{7.10}$$

In this system of equations the comma before the subscript denotes the covariant derivative, the summation rule over repeated indexes is meant. Here i-th components are introduced by the following notations: ρ_i is the mass density, $\mathbf{u}_i$ is the velocity, $\mathbf{I}_{(i)} = \rho_i(\mathbf{u}_i\text{-}\mathbf{u})$ is the diffusion mass flux, $c_i = \rho_i/\rho$ is the concentration, U_{pi} is the internal energy, $\dot{\xi}_i$ is the mass sources, determined by electrochemical reactions (see (6.24). Solution's properties "on the whole" are introduced: $\rho = \sum_i \rho_i$ is the mass density, $\mathbf{u} = \sum_i c_i \mathbf{u}_i$ is the velocity, p_{ik} is the symmetrical stress tensor, $\mathbf{q}_T$ is the vector of heat flux density, $U_p = \sum_i c_i U_{pi}$ is the internal energy.

Electric and magnetic magnitudes are the following: **E**, **H** are the electric and magnetic intensities, $\mathbf{D} = \varepsilon\mathbf{E}$, $\mathbf{P} = \mathbf{D} - \varepsilon_0\mathbf{E}$, $\mathbf{S} = \mathbf{E}\times\mathbf{H}$ are the electrical induction, polarization and density of electromagnetic energy flux, q is the charge density and $p' = p_{str} - \mathbf{P}\cdot\mathbf{E}/2$ is the polarization pressure.

At low concentrations in ionic and neutral components the basic system of equations (7.6) - (7.10) can be reduced to the form

$$\rho\,(\partial\mathbf{u}/\partial t + (\mathbf{u}\cdot\nabla)\,\mathbf{u}) = -\nabla\tilde{p} + \tau_{ik,\,k} - (E^2/2)\nabla\varepsilon + q\mathbf{E} + \rho\mathbf{g}\,, \qquad 7.11)$$

$$\partial\rho/\partial t + (\rho u_k)_{,k} = 0, \qquad \partial n_j/\partial t + (i_{(j)k} + n_j u_k)_{,k} = \dot{\xi}_i\,, \qquad (7.12)$$

$$\rho c_p\,(\partial T/\partial t + u_k T_{,k}) = (kT_{,k})_{,k} - p u_{k,\,k} + \dot{\xi}_T\,, \qquad (7.13)$$

$$(\varepsilon E_k)_{,k} = q\,, \quad \mathbf{E} = -\nabla\phi\,, \qquad q = \sum_i e_i n_i\,, \qquad (7.14)$$

$$\tilde{p} = p + p_{str}\,, \qquad (7.15)$$

$$\tau_{ij} = 2\eta\,(V_{ij} - (1/3)\,u_{k,\,k}\delta_{ij}) + \eta_v u_{k,\,k}\delta_{ik}\,, \qquad (7.16)$$

$$V_{ij} = (1/2)\,(u_{i,\,j} + u_{j,\,i}),$$

$$\mathbf{i}_{(j)} = (\mathrm{sign}\ e_j)\,\mu_j n_j \mathbf{E}\ \text{(for charged components)}, \qquad (7.17)$$

$$\mathbf{i} = -D_j \nabla n_j\ \text{(for neutral components)}.$$

Here τ_{ij} is the viscous stress tensor, **g** is the free falling acceleration, e_i is the ion charge, ($e_i = 0$ if component is neutral), sign(e_i)= 1 if $e_i > 0$, and sign(e_i)= -1 if $e_i < 0$, k is the coefficient of thermal energy flux, $\dot{\xi}_T$ is the thermal source determined by viscous and Joule dissipations:

$$\dot{\xi}_T = \tau_{ij} u_{i,\,j} + \mathbf{E}\cdot\mathbf{j}^*, \qquad (7.18)$$

$$\mathbf{j}^* = \sigma\mathbf{E}\,, \qquad \sigma = \sum_i |e_i|\mu_i n_i\,,$$

where μ_i is the mobility of ion in i-th component and σ is the coefficient of conductivity.
The sum of the electrical terms in Eq. (7.11) is the so-called electrical force

$$\mathbf{f}_E = q\mathbf{E} - (1/2)\, E^2 \nabla\varepsilon - \nabla p_e . \tag{7.19}$$

The first term is the force per unit volume on a medium containing free electric charge of density q. It is the strongest EHD force term and usually dominates when DC electric field are present in conducting dielectric liquids. The second term is the force exerted on a dielectric liquid by a non uniform electric field. It is usually weaker than the free charge force term and only dominates when an AC electric field is imposed on an insulating DL. The third term is the electrostriction term, and it can be combined with the pressure p as it has been done in Eq. (7.11).

7.4 Three-ion EHD model

Let us consider a three-ion EHD model presented by the electrochemical reactions (6.35) - (6.37). Noting $n_1 = [A^+]$, $n_2 = [B^-]$, $n_4 = [X^-]$, $n_3 = N = [A^+B^-]$, $c_x = [X]$, we get five components, but it is necessary to take into account only four components because generally $c_x = \text{const}$. So in this case we have

$$q = e\,(n_1 - n_2 - n_4)\,; \tag{7.20}$$

$$\mathbf{i}_j^* = (-1)^{j-1} \mu_j n_j \mathbf{E}\,, \qquad j = 1, 2, 4\,; \qquad \mathbf{i}_4^* = \mathbf{i}_N = - D\nabla N;$$

$$\dot{\xi}_1 = \dot{\xi}_2 - \alpha_1 n_1 n_4\,, \quad \dot{\xi}_2 = -\dot{\xi}_3 = \alpha_d N - \alpha_r n_1 n_2\,, \quad \dot{\xi}_4 = - \alpha_1 n_1 n_4 .$$

These expressions allow us to close the basic system of EHD equations (7.11) - (7.17). We term this case the *three-ion* EHD *model.*

7.5 Boundary conditions

Boundary conditions on the metallic electrodes are determined by the expression of the injection function (6.25), the current balance of ion pairs A^+B^- and impurity ions A^+, B^-, constant values of electric potential on the electrodes and general hydrodynamics correlations:

on the anode: $\quad n_1 = n_A(E), \;\; \mathbf{i}_N \cdot \mathbf{n} + \mu_1 n_A(E)\, E = 0, \;\; \phi = V, \quad$ (7.21)

$$\mathbf{u} = 0, \;\; T = T_1; \tag{7.22}$$

on the cathode: $n_2 = n_B(E),\ n_4 = n_x(E),$ (7.23)

$$\mathbf{i}_N \cdot \mathbf{n} + n_B(E)\,E = 0,\ \ \phi = 0, \tag{7.24}$$

$$\mathbf{u} = 0,\ \ T = T_2. \tag{7.25}$$

Here $\mathbf{n}$ is *the normal to electrode surface vector* directed towards the liquid; $n_j(E)$, $j = A, B, X$ are the injection function of ions A^+, B^-, X^-, determined accordingly to formula (6.25) as

$$n_A(E) = k^+_{AB}\,N, \qquad n_B(E) = k^-_{AB}\,N, \tag{7.26}$$

$$n_x(E) = \alpha_1\, c_x \,/\, (\mu_4 E_{loc} + \alpha_2 K(E_{loc})\,), \tag{7.27}$$

$$K(E_{loc}) = (E_{loc}/\phi_0) \int_{r_x}^{\infty} \exp\left(\,(\,\Pi(\xi) - \Pi(r_x)\,)\,/\,k_BT\,\right) d\xi\,, \tag{7.28}$$

$$\Pi(\xi) = eE_{loc}\xi - \Pi_x(\xi)\,, \quad E_{loc} = \beta E\,, \quad \phi_0 = k_BT/e\,, \tag{7.29}$$

where r_x is the radius of X, and β is the amplification factor.

7.6 Non-dimensional parameters and general analysis of three-ion EHD model at low and high electric fields

Let us consider the three-ion EHD model for incompressible liquids and the isothermal case. Introducing the characteristic scales of length d, electrical potential V, concentration of impurity ions A^+, B^-, $n_0 = \sqrt{\alpha_{d0} N / \alpha_r}$; injection ions X^-, $n_x(E_0)$, $E_0 = V/d$; ion pairs A^+B^-, N_0 and defining the scales for velocity, $u_0 = \nu/d$, time d^2/ν, $\nu = \eta/\rho$ and pressure, $\nu\eta/d^2$, we can write the basic system of EHD equations in the following dimensionless form:

$$\partial \mathbf{u}/\partial t + (\mathbf{u}\cdot\nabla)\,\mathbf{u} = -\nabla \tilde{p} + \nabla^2 \mathbf{u} + G\,n\,\mathbf{E}\,, \qquad \nabla\cdot \mathbf{u} = 0\,; \tag{7.30}$$

$$\nabla\cdot\mathbf{E} = C\,n\,, \qquad \mathbf{E} = -\nabla\phi\,, \qquad n = n_1 - n_2 - \sigma_c n_4\,; \tag{7.31}$$

$$P_\sigma C\, \partial n_j/\partial t + \nabla\cdot \left[\,(-1)^{j-1} \beta_j\, n_j\, \mathbf{E} + P_\sigma C\, n_j\, \mathbf{u}\right] = C\dot{\xi}_j\,, \qquad j = 1, 2, 4\,; \tag{7.32}$$

$$\partial N/\partial t + \nabla\cdot \left[\,-(\,1/P_d\,)\,\nabla N + N\,\mathbf{u}\right] = -(1/P_{ch}\,)\,\dot{\xi}_N\,; \tag{7.33}$$

$$\dot{\xi}_1 = \dot{\xi}_2 - \Gamma_1\, \sigma_c n_1 n_4\,, \qquad \dot{\xi}_2 = \Gamma_p \dot{\xi}_N\,; \tag{7.34}$$

$$\dot{\xi}_N = F(E)\,N - n_1 n_2, \qquad \dot{\xi}_4 = -\Gamma_1 n_1 n_4; \tag{7.35}$$

on the anode: $n_1 = f_A(E), \quad \mathbf{n}\cdot\nabla N = (\beta_1 P_d\, \alpha E / P_\sigma C)\, f_A(E)$; (7.36)

$$\phi = 0, \; \mathbf{u} = 0. \tag{7.37}$$

on the cathode: $n_2 = f_B(E), \quad \mathbf{n}\cdot\nabla N = (P_d\, \alpha E / P_\sigma C)\, f_A(E)$; (7.38)

$$n_4 = f_x(E), \; \phi = 0, \; \mathbf{u} = 0. \tag{7.39}$$

Here the following set of dimensionless numbers appear:

$$G = M^2/(P_\sigma^{\,2}\, C) = e n_0 E_0 d^3 / (\rho \nu^2), \tag{7.40}$$

is the ratio of Coulomb and viscous forces. At high fields, when injection processes are dominant, the important parameter is

$$G_c = \sigma_c\, G = e\, n_x(E_0)\, E_0\, d^3 / (\rho \nu^2), \quad \sigma_c = n_x(E_0)/n_0, \tag{7.41}$$

where the number σ_c is the criterion of dissociation ($\sigma_c << 1$) and injection ($\sigma_c >> 1$) regime of conductivity,

$$C = d/\xi_d, \qquad \xi_d = \varepsilon E_0 / (e n_0); \tag{7.42}$$

is the measure of non-equilibrium in the dissociation-recombination reaction (6.1). At $C >> 1$ the reaction (6.1) is close to equilibrium. This condition is carried out at low fields. At $C \sim 1$, $\sigma_c \sim 1$ there is a dissociation-injection conductivity. This is the region of intermediate field. At $\sigma_c >> 1$ and $C \le 1$ the injection conductivity is dominant. The characteristic length ξ_d was for the first time introduced by Thompson and Thompson [3]. This parameter determines the size of the layers near the electrodes or interfaces where the reaction (6.1) is not in equilibrium. This boundary layer we will call the *non-equilibrium layer*.
The dimensionless parameter

$$M = (\varepsilon/\rho)^{1/2} / \mu_2; \tag{7.43}$$

which is the ratio of the so-called electrohydrodynamical mobility $(\varepsilon/\rho)^{1/2}$ and the true ion mobility, was introduced by Ostroumov [4]. The parameters

$$P_\sigma = 2\varepsilon\nu / (d^2 \sigma_2), \quad P_d = \nu/D, \quad P_{ch} = \nu / (\alpha_{d0} d^2); \; \sigma_2 = 2e\mu_2 n_0 \tag{7.44}$$

are the electrical, diffusion and chemical Prandtl numbers,

$$\alpha = n_0 / N_0, \quad \beta_i = \mu_i / \mu_2, \quad i = 1, 2, 4; \tag{7.45}$$

are the dissociation's degree of ion pairs and the ratios of mobilities. The non-dimensional injection functions are defined as

$$f_x(E) = n_x(E)/n_x(E_0), \quad f_j(E) = n_j(E) / n_0, \quad j = A, B. \tag{7.46}$$

Finally, the parameters Γ_p, Γ_1 are expressed as

$$\Gamma_p = 1 + \beta_1, \quad \Gamma_1 = \beta_1 + \beta_3. \tag{7.47}$$

Dimensionless parameters were defined using the correlations

$$\alpha_r = e\,(\mu_1 + \mu_2)/\varepsilon, \quad \alpha_1 = e\,(\mu_1 + \mu_4)/\varepsilon, \quad \alpha_{d0}N = \alpha_r\, n_0^2. \tag{7.48}$$

The first two formulae are known as Langevin correlations.

7.7 Ohmic model of conductivity and the structure of non-equilibrium layers.

The Ohmic EHD model for incompressible, viscous, homogeneous DL, assumes Ohm's law of conductivity:

$$\mathbf{j} = \sigma \mathbf{E}, \tag{7.49}$$

where the conductivity coefficient is a known function of temperature and electric intensity

$$\sigma = \sigma\,(T, E). \tag{7.50}$$

The governing EHD equations for an ohmic model of conductivity are the following:

$$\nabla\cdot \mathbf{u} = 0, \tag{7.51}$$

$$\rho\,(\partial\mathbf{u}/\partial t + (\mathbf{u}\nabla)\,\mathbf{u} = -\nabla\,\tilde{p} + \eta\nabla^2\mathbf{u} + q\mathbf{E} - (E^2/2)\nabla\varepsilon + \rho\mathbf{g}; \tag{7.52}$$

$$\nabla\cdot(\varepsilon\mathbf{E}) = q, \quad \nabla\times\mathbf{E} = 0; \tag{7.53}$$

$$\partial q/\partial t + \nabla\cdot(\sigma\mathbf{E} + q\mathbf{u}) = 0; \tag{7.54}$$

$$\rho c_p\,(\partial T/\partial t + \mathbf{u}\cdot\nabla T) = k\cdot\nabla^2 T. \tag{7.55}$$

These equations traditionally have been used in the early investigations on EHD effects [1].

It is interesting to trace out how the general system (7.30) - (7.33) is transformed into the Ohmic EHD model (7.51) - (7.55). This interest is caused by various reasons, but the most important is to quantify the error made in the approximation. To simplify the problem we will consider the isothermal case. As it is known, an ohmic conductivity is realized at low fields, at $C \gg 1$; $\sigma_c \leq 1$; $f_j(E) \leq 1$, $j = A, B$; $f_x \leq 1$. Taking into account that generally $P_\sigma \leq 1$, $P_{ch} \gg 1$, $\beta_j \sim 1$, $j = 1, 2, 4$, $\alpha \ll 1$, we will seek the solution into the form of asymptotic series on the small parameter $m = C^{-1}$:

$$\mathbf{u} = \mathbf{u}_0 + m\mathbf{u}_1 + m^2\mathbf{u}_2 + \ldots; \qquad \tilde{p} = p + mp_1 + m^2p_2 + \ldots; \qquad n_4 = 0;$$

$$n_j = n_{j0} + m\, n_{j1} + m^2 n_{j2} + \ldots, j = 1, 2; \; N = N_{(0)} + m\, N_1 + m^2 N_2 + \ldots;$$

$$\mathbf{E} = \mathbf{E}_0 + m\mathbf{E}_1 + m^2\mathbf{E}_2 + \ldots, \qquad \mathbf{E}_i = -\nabla\phi_i, \qquad i = 0, 1, 2, \ldots$$

The zeroth-order approximation yields:

$$N_{(0)} = 1, \qquad n_{10} = n_{20} = \sqrt{F_0}, \qquad F_0 = F(E_0).$$

In the first-order approximation we have

$$\partial\mathbf{u}_0/\partial t + (\mathbf{u}_0\cdot\nabla)\mathbf{u}_0 = -\nabla p + \nabla^2\mathbf{u}_0 + (G/C)\mathbf{E}_0\, q_1, \qquad \nabla\cdot\mathbf{u}_0 = 0; \tag{7.56}$$

$$\nabla\cdot\mathbf{E}_0 = q_1, \qquad \mathbf{E}_0 = -\nabla\phi_0, \qquad q_1 = n_{11} - n_{21}; \tag{7.57}$$

$$\begin{aligned} &P_\sigma\, \partial n_{j1}/\partial t + \nabla\cdot [\, (-1)^{j-1}\, \beta_j \sqrt{F_0}\, \mathbf{E} + (P_e\sqrt{F_0} + n_{j1}\, P_\sigma)\, \mathbf{u}_0] = \\ &\Gamma_p\, [\partial F_0/\partial E\; \mathbf{E}_0\cdot\mathbf{E}_1/E_0 - \sqrt{F_0}\;(n_{11} + n_{21})\,], \quad j = 1, 2; \end{aligned} \tag{7.58}$$

where $P_e = P_\sigma C \gg 1$. So we suggest that

$$C\, \partial F_0/\partial E < 1 \;\text{ or }\; \mathbf{u}_0 = 0. \tag{7.59}$$

Taking the difference of Eq. (7.58), we get

$$P_\sigma \partial q_1/\partial t + \nabla\cdot [\, (\beta_1 + \beta_2)\sqrt{F_0}\, \mathbf{E}_0 + P_\sigma\, q_1\, \mathbf{u}_0] = 0. \tag{7.60}$$

The equations (7.56), (7.57) and (7.60) define an ohmic EHD model and coincide with (7.51) - (7.54) in the isothermal case. The solåtions of an ohmic EHD model are valid outside the non-equilibrium layers. So an Ohmic model is valid 1) at ξ_d

<< d, 2) at low field when $F(E) \sim 1$, $dF/dE << 1$. It is necessary to note that Onsager's dissociation theory of conductivity based on an ohmic model (see (6.6), (6.7)) does not satisfy both conditions at high field.

Let us consider the charge and electric intensity distributions inside a non-equilibrium layer under steady conditions.

Introducing the normal co-ordinate x to cathode surface (in Fig. 6.5 cathode surface is that of S), from (7.31) - (7.32) we can get the following equations describing the foregoing distributions:

$$\frac{dE}{d\xi} = n_2 - n_1 + \sigma_c n_4, \quad b_2 \frac{d(n_2 E)}{d\xi} = (1 + b_2)(1 - n_1 n_2), \tag{7.61}$$

$$-\frac{d(n_1 E)}{d\xi} = (1 + b_2)(1 - n_1 n_2) - (1 + b_4)\sigma_c n_1 n_4, \tag{7.62}$$

$$b_4 \frac{d(n_4 E)}{d\xi} = -(1 + b_4)\sigma_4 n_1 n_4, \qquad E = |E_x| \tag{7.63}$$

where $\xi = xC$, $b_2 = \mu_2/\mu_1$, $b_4 = \mu_4/\mu_1$, E is the absolute value of the normal intensity component and other parameters are defined in section 7.6. These equations are obtained under the condition that all normal derivatives are much greater than the ones along the cathode surface.

Accordingly to (7.38), (7.39) boundary conditions have the form

$$\xi = 0: \quad n_2 = f_B, \quad n_4 = f_x; \tag{7.64}$$

$$\xi \to \infty: \qquad n_1 \to 1, \qquad E \to 1.$$

Results of numerical calculations of the system boundary value problem (7.61) - (7.64) are presented in Fig. 7.1, Fig. 7.2, Fig. 7.3. Numerical analysis allows to conclude that inside a non-equilibrium layer three cases of field distribution may be observed: 1) electric intensity E increases towards the electrode (low level of injection $n_x(E) < n_0$, see Fig. 7.1), 2) E decreases towards the electrode (high level of injection $n_x(E) > n_0$ and $\mu_1 \sim \mu_2 \sim \mu_4$, see Fig. 7.2), 3) at high injection level and low mobility of injection ions: $n_x(E) > n_0$, $\mu_4 << \mu_1 \sim \mu_2$ the bipolar charge structure can be observed, see Fig. 7.3.

These non-equilibrium charged layers are often observed in the experimental works: Rychkov and Stishkov [5] using a probe method have shown that in a transformer oil saturated with iodine there are bi-polar structures (Fig. 7.3).

Those structures have been found in solutions of transformer oil with polar admixtures such as butyl alcohol [6]. Thin charged layers at the electrode surfaces were observed in non-polar hydrocarbon by Kosvinzev [7] who had used a Schlieren optical system.

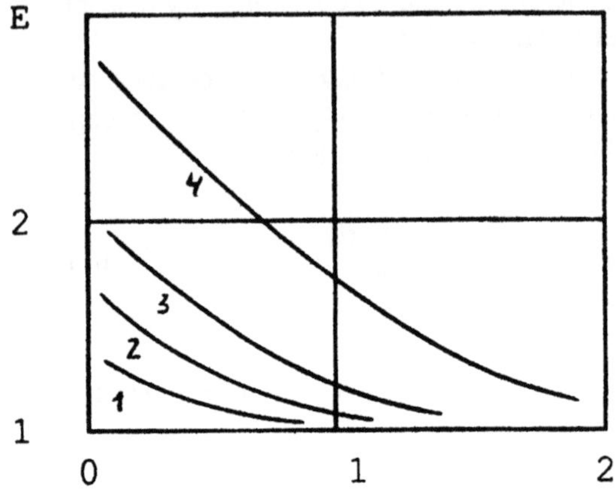

Fig. 7.1, Distribution of the electric intensity in non equilibrium layer near cathode in absence of the charge injection $\sigma_c = 0$ and various mobilities of homoions (μ_2) and heteroions (μ_1). Curves 1 - 4 are corresponded to $b_2 = \mu_2/\mu_1 = 0.5$; 1; 2; 5 accordingly.

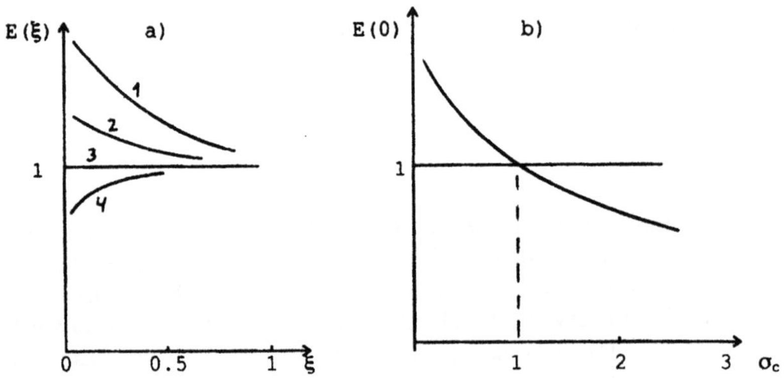

Fig. 7.2, a) Distribution of the electric intensity in non-equilibrium layer at various levels in charge injection and equal mobilities $\mu_1 = \mu_2 = \mu_4$. Curves 1 - 4 are corresponded to σ_c= 0; 0.5; 1; 2. b) Dependence of the electric intensity on the cathode surface versus injection parameter sc at equal mobilities.

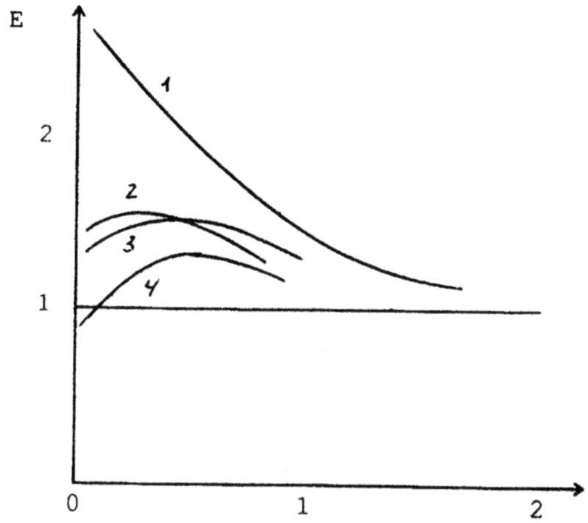

Fig. 7.3, Distribution of the electric intensity in non equilibrium layer near cathode at small mobility of injection ions $\mu_4 / \mu_1 = 0.5$ and high mobility of homoions $\mu_2 / \mu_1 = 4$; 1- σ_c = 0, 2- = 3, 3- = 5, 4- = 10.

7.7. Investigation of bi-ion model.

Detail investigations of bi-ion model in the case of plane and cylindrical geometries were performed by F. Pontiga & A. Castellanos [8-9]. They took into account Onsager dissociation enhancement and Schottky charge injection of positive ions from an anode and have used the following system of equations

$$\mathrm{div}\ \varepsilon\ \mathbf{E} = e\,(n_1 - n_2), \quad \mathbf{E} = -\ \nabla\Phi,$$

$$(-1)^{j-1}\,\mathrm{div}\,(\mu_j\, n_j\, \mathbf{E}) = \alpha_d c - \alpha_r n_1 n_2\,,\ j = 1,2; \tag{7.45}$$

with the boundary conditions

$$\text{on the cathode: } \Phi = 0,\ n_2 = 0\,; \qquad \text{on the anode: } \Phi = V\,,\ n_1 = n_i \tag{7.46}$$

Dissociation rate α_d and injection function n_i were accordingly to (5.78) and (6.25) taken into the forms

$$\alpha_d = \alpha_{d0}\, F(p), \quad n_i = n_{i0} / K(E), \tag{7.47}$$

The results of numerical calculations [1] were compared with experimental data obtained by Denat, Gosse and Gosse [10] who measured the current-voltage characteristic in cyclohexane doped with triisoanylammonium picrate (TIAP) at intense electric field. The conductivity of the solution was $3.3 \cdot 10^{-12}\ \Omega^{-1}m^{-1}$ and the electrode separation 0.21 mm. Fig. 7.4. shows the experimental (dots) and theoretical current - voltage characteristics (solid lines) and the dissociation current (dashed line). From Fig. 7.4. it is seen that at high field experimental values of current density are higher that the values predicted with theory. The authors discuss the different possibilities to account for the discrepancy observed at high fields above $2 \cdot 10^7$ V/m.

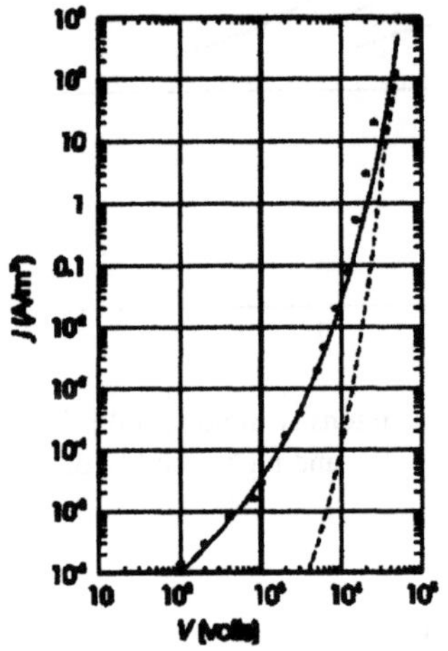

Fig. 7.4. Experimental current - voltage characteristic corresponding to solution of TIAP in cyclohexane (dots) (from Denat et al [10]) and theoretical current density (solid line) and dissociation current (dashed line) (from Pontiga and Castellanos [8]).

References.

1. Melcher, J.R.: Continuum Electromechanics, M. I. Press, Cambridge, Mass. 1981.
2. Zhakin A.I.: Theory of electrical and electrohydrodynamical effects in low conductive media, Thesis, Kharkov University Press 1990 (in Russian).
3. Thomson, J.J. and Thomson, G.P.: Conduction of electricity through gases, 3rd edition Vol. 1 pp.193, Cambridge University Press, 1928.
4. Ostroumov, G.A.: Interaction between electrical and hydrodynamical fields, Nauka Press, Moscow 1979 (in Russian).
5. Stishkov, J.K., Ostapenko A.A.: Electrohydrodynamic flows in liquid dielectrics. Printed in Leningrad University 1989 (in Russian).
6. Kropotcheva, L.V., Rychkov, J.M., Kovalevskaja, T.I.: On the molecular structure of isolating liquids in electric fields, Electr. Obrab. Mater., Mold. Akad. Nauk, Kishinev, 5 (1992), 33-36 (in Russian).
7. Kosvinzev, S.R.: Experimental investigation of EHD stability in non-homogenously heated weak-conductive liquids, Thesis, Perm State University, Perm 1993 (in Russian).

8. Pontiga F. and Castellanos A. Electrical Conduction of Electrolyte Solutions in Nonpolar Liquids, IEEE Trans. Of Indust. Appl. 32, N4 (1996), 816-824.
9. Pontiga F. and Castellanos A. A dissociation-injection Model for non-polar Liquid Conduction and Wire-cylinder Geometry, IEEE Trans. on Diel. and Wire-cylinder Geometry, IEEE Trans. on Diel. and Elect. Insul. 4, N2 (1997), 224-237.
10. Denat A., Gosse B. and Gosse J.P. High field dc and ac conductivity of electrolyte solutions in hydrocarbons. J. Electrostat. 11 (1982), 179-194.

Chapter 8

SURFACE WAVES AND INSTABILITIES IN PLANE CHARGED INTERFACES

8.1 Introduction

Besides the Coulomb force, exerted by an electric field on free charge injected into liquid dielectrics, there is another mechanism of EHD motion caused by surface interface electric tensions. So this type of electroconvection arises when a liquid is in contact with air or with other inmiscible liquids.

A knowledge of the behaviour of liquids with air-liquid or liquid-liquid interfaces in a strong external electric field is important for various applications (liquid displacement, liquid dispersion into small drops and so on). Surface interface electric forces can be of two types. The first is due to polarization of liquid dielectric and the second is caused by the action of the electric field on the free surface charges. We will consider the role of free surface forces in EHD and in particular we will investigate the major linear EHD instabilities in plane charged interface.

8.2 Boundary conditions on the free surface

Let us assume that the equation of free surface is determined as

$$F(t, x^1, x^2, x^3) = 0, \tag{8.1}$$

where (x^i) is some curve co-ordinate system. Boundary conditions on the free surface include [1-3] :

kinematic

$$\partial F/\partial t + \nabla F \cdot \mathbf{u} = 0\,; \tag{8.2}$$

dynamic:

$$(\langle p_{ik} \rangle + \langle T_{ik} \rangle)\, n^k = 2\gamma K n^i\,; \tag{8.3}$$

and electrostatic, which in the simplest form can be expressed as

$$q_s = \langle \varepsilon E_n \rangle, \quad \langle \phi \rangle = 0; \tag{8.4}$$

$$\partial q/\partial t + \nabla_s \cdot (\mathbf{j}_s^* + q_s \mathbf{u}_t) + 2Kq_s v_n + \langle j_n^* \rangle = \xi_s. \tag{8.5}$$

Here the sign $\langle \dots \rangle$ denotes the jump of the quantity through the discontinuity surface, $\mathbf{n} = (n^i)$ is the normal vector to surface, $v_n = \mathbf{u} \cdot \mathbf{n}$, $E_n = \mathbf{E} \cdot \mathbf{n}$; q_s, $\mathbf{j}_s$ are the surface charge and current, $\mathbf{u}_t$ is the tangent to surface component of velocity, $\nabla_s \cdot (\dots)$ denotes the operation of surface divergence, γ is the surface tension coefficient, K is the average curvature, $j_n^* = \mathbf{j}^* \cdot \mathbf{n}$, where $\mathbf{j}^*$ is the current density in the bulk, ξ_s is the surface source which is determined by electrochemical reactions on a free surface. The stars in $\mathbf{j}_s^*, \mathbf{j}^*$ denote the drift part of the total currents, that is $\mathbf{j} = \mathbf{j}^* + q\mathbf{u}$, $\mathbf{j}_s = \mathbf{j}_s^* + q_s\mathbf{u}_t$. (detail derivation of the boundary conditions (8.5) see, for example, in the paper of A. Castellanos & A. Gonzalez [3]).

In the Ohmic conductivity model it takes place

$$\mathbf{j}^* = \sigma \mathbf{E}, \qquad \mathbf{j}_s^* = \sigma_s \mathbf{E}_t, \tag{8.6}$$

where σ_s is the coefficient of surface conductivity. In a general case, when different ions contribute to the free surface conductivity, the balance equation of surface charge (8.5) must be written for each ionic species.

8.3 Field-coupled linear surface waves and instabilities in plane charged interface

The first exploration on the linear stability of the horizontal interface between conducting and non conducting fluids under the influence of an initially uniform vertical electric field (Fig. 8.1) was performed by Taylor and McEwan [4]. Non-linear analysis in long-wave approximation was studied by Zhakin [5]. Another interesting dynamic effects of electrical stress acting at electrified interface are discussed, for example, in the papers [1-3]. In this section we will discuss the major conclusions of investigations [4,5].

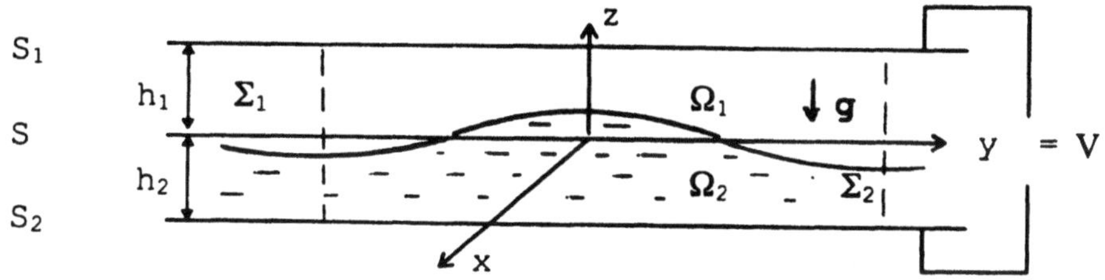

Fig. 8.1. Instabilities and surface waves in an initially plane interface

8.3.1 Formulation of the problem

The problem will be considered at isothermal conditions assuming that the interface contacts with a non-conductive ideal liquid (gas) and the ohmic liquid conductivity is constant. The basic boundary value problem, describing the non-linear motion and the non-linear shapes of the interface, follows from equations (7.56) - (7.58) where $q = 0$, $\varepsilon =$ const. The boundary conditions are determined on the interface by (8.1) - (8.5) and on the electrodes S_1, S_2 by the conditions (7.22), (7.25) for the velocity and electrical potential.

Let us assume that at equilibrium state the interface is plane and coincides with the Cartesian x,y - plane (Fig. 8.1). So in a disturbed state the interface equation is

$$z = f(t,x,y)$$

At an equilibrium state we have

$$f_0 = 0,\ \mathbf{E}_0 = E_0\,\mathbf{e}_z,\quad p_{10} = p_0 - \rho_1 g\,(z - h_1)\quad \text{for } 0 < z < h_1;$$

$$p_{20} = p_{20} + \rho_1 gh - \rho_2 gz - (1/2)\,E_0^2,\quad \mathbf{E}_0 = 0\quad \text{for } -h_2 < z < 0;$$

$$q_{s0} = \varepsilon_1 E_0,\quad E_0 = V/h_1,\quad \phi_0 = -E_0\,(z - h_0).$$

Here zero in the lower indexes notes the magnitudes at equilibrium state, p_{10} (p_{20}) is the pressure in the upper (bottom) liquid. Introducing the disturbed state into the form $f_0 + f'$, $\phi_0 + \phi'$, $\mathbf{u}_i$, $p_{i0} + p_i'$, $q_s + q_s'$ for small disturbances we get the following set of equations and boundary conditions (upper indexes prime are omitted)

$$(0 < z < h_1):\quad \nabla^2\phi_1 = 0,\quad \rho_1\,\partial\mathbf{u}_1/\partial t = -\nabla p_1,\ \nabla\cdot\mathbf{u}_1 = 0; \tag{8.7}$$

$$(-h_2 < z < 0):\quad \nabla^2\phi_2 = 0,\quad \rho_2\,\partial\mathbf{u}_2/\partial t = -\nabla p_2 + \eta\nabla^2\mathbf{u}_2,\ \nabla\cdot\mathbf{u}_2 = 0; \tag{8.8}$$

S (z=0):

$$\frac{\partial f}{\partial t} = u_{1z} = u_{2z},\quad p_2 + <\frac{dp_0}{dz}> f - 2\eta\frac{\partial u_{2z}}{\partial z} - \varepsilon E_0\frac{\partial\phi_1}{\partial z} = -\gamma\,\nabla_1^2 f;$$

$$\eta\,(\frac{\partial u_{2z}}{\partial x} + \frac{\partial u_{2x}}{\partial z}) = q_{s0}\frac{\partial\phi_2}{\partial x},\quad \eta\,(\frac{\partial u_{2x}}{\partial y} + \frac{\partial u_{2y}}{\partial z}) = q_{s0}\frac{\partial\phi_2}{\partial y};$$

$$q_s = -\,(\varepsilon_1\frac{\partial\phi_1}{\partial z} - \varepsilon_2\frac{\partial\phi_2}{\partial z}),\quad \phi_1 + \frac{\partial\phi_0}{\partial z}\,f = 0;$$

$$\partial q_s/\partial t + \nabla_1\cdot(-\sigma_s\,\nabla_1\,\phi_2 + q_{s0}\,\mathbf{u}_{1t}) + \sigma\,\partial\phi_2/\partial z = 0; \tag{8.9}$$

$$S_1\,(z=h_1):\ \phi_1=0,\ u_{1z}=0,\qquad S_2\,(z=-h_2):\ \phi_2=0,\ \mathbf{u}=0. \tag{8.10}$$

Here the lower index 1(2) denotes the variables for upper (bottom) liquid and ∇_1, ∇_1^2, $\nabla_1\cdot$ are the differential operations on the variables x,y.

Let us consider the important case of a perfectly conductive liquid ($\sigma \to \infty$). From the physical point of view the limiting case means that the time of charge relaxation $\tau_e = \varepsilon/\sigma$ is much less than the period of interface oscillations $T = 2\pi/\omega$ where ω is the frequency of oscillations.

In this case from (8.9) we get $\phi_2 = 0$, which means that the free surface is equipotential. This approach is usually called as metallic. So instead of the last five conditions in (8.9) we will have

$$z=0:\quad \partial u_{2z}/\partial x + \partial u_{2x}/\partial z = \partial u_{2z}/\partial y + \partial u_{2y}/\partial z = 0,\quad \phi_1 = E_0 f. \tag{8.11}$$

8.3.2 General investigation of smal perturbations

The general property of small oscillations traditionally is studied by the method of the quadratic forms. To simplify the problem we assume that the upper liquid is a gas and disturbances are periodic in x,y directions. So the regions Ω_1, Ω_2 have the periodic boundary conditions on Σ_1, Σ_2 (Fig. 8.1).

Taking the scalar product of Eq. (8.8) and the complex conjugate function $\mathbf{u}^*$ and then integrating over the region Ω_2 we get (to simplify the notations the lower indexes are omitted and the asterisk in the upper index designates the complex conjugate):

$$\rho\int_V \partial\mathbf{u}/\partial t\cdot\mathbf{u}^* dV = -\,\eta\int_V |\nabla\times\mathbf{u}|^2\, dV + \int_S [\eta\,(\partial u_z/\partial x + \partial u_x/\partial z)\, u_x^* +$$
$$\eta\,(\partial u_z/\partial y + \partial u_y/\partial z)\, u_y^* - (p - 2\eta\partial u_z/\partial z)\, u_z^*\,]\, dS.$$

Using boundary conditions (8.9), (8.11) allow to reduce this equality to the form

$$\rho\int_{\Omega} \partial\mathbf{u}/\partial t\cdot\mathbf{u}^* dV = -\eta\int_{\Omega} |\nabla\times\mathbf{u}|^2\, dV - \int_S [\,-\gamma\nabla_1^2 f + \rho g f + \varepsilon E_0\partial\phi/\partial z]\, \partial f^*/\partial t\ \ dS.$$

analogously, from Eq. (8.7) we have

$$\int_{\Omega} \nabla\phi\cdot\nabla(\partial\phi^*/\partial t)\, dV = -E_0\int_S \partial\phi/\partial z\ \ \partial f^*/\partial t\ \ dS.$$

Summarizing the last two integrals, they can be reduced to the equality:

$$\int_{\Omega} \rho\, \partial\mathbf{u}/\partial t\cdot\mathbf{u}^*\, dV + \int_S\ [\gamma\nabla_1 f\cdot\nabla_1(\partial f^*/\partial t) + \rho g f\, \partial f^*/\partial t]\, dS\ -$$

$$\varepsilon_1 \int_{\Omega} \nabla\phi\cdot\nabla(\partial\phi^*/\partial t)\, dV = \eta \int_{\Omega} |\nabla\times\mathbf{u}|^2\, dV. \tag{8.12}$$

For small oscillations, when the disturbances are proportional to $e^{\lambda t}$:

$$(\mathbf{u}, f, \phi) = (\mathbf{U}\,(r), F\,(r), \Phi\,(r)\,)\, e^{\lambda t},$$

from (8.12) it follows

$$(\lambda + \lambda^*)\, I_0 = -2I_1, \qquad I_1 = \eta \int_{\Omega} |\nabla\times\mathbf{U}|^2\, dV; \tag{8.13}$$

$$I_0 = \int_{\Omega} \rho|\mathbf{U}|^2\, dV + \int_{S} (\gamma|\nabla_1 F|^2 + \rho g|F|^2\,)\, dS - \varepsilon_1 \int_{\Omega} |\nabla\phi|^2\, dV.$$

The criterion of instability is defined by the condition $\Re(\lambda) = (\lambda + \lambda^*)/2 = 0$. So from (8.13) it is seen that at critical values of parameters it follows $I_1 = 0$, that is, the liquid motion is potential $\nabla\times\mathbf{u} = 0$. Thus, the threshold of instability can be found in a model of ideal liquid and the viscosity of liquid does not influence upon the critical values of parameters.

8.3.3 Small oscillations in charged interface

Introducing the normal mode decomposition

$$(\mathbf{u}, f, \phi) = (\mathbf{U}(z), F, \Phi\,(z)\,) \exp\,[i\,(\omega t - k_1\, x - k_2\, y)\,].$$

and using the standard procedure, for small oscillations of the plane charged interface it can be found

$$\omega^2 = [\gamma k^2 + \langle\rho\rangle g - \varepsilon_1 E_0^2\; k\; \coth(kh_1)]\; k/\, (\rho_1\; \coth(kh_1) + \rho_2\; \coth(kh_2)\,),$$

$$k^2 = k_1^2 + k_2^2, \langle\rho\rangle = \rho_2 - \rho_1.$$

The instability takes place if

$$E_0^2 > E_*^2 = \min[\varepsilon_1^{-1} \tanh(kh_1)(\gamma k^2 + \langle\rho\rangle\, g)\,/k]\,. \tag{8.14}$$

Let us rewrite (8.14) in the following dimensionless form

$$W_* = \varepsilon_1\, E_*^2\, /\, (gh_1\, (\rho_2 - \rho_1)\,), \qquad \chi = kh_1, \qquad G = l_c^2/h_1^2,$$

$$l_c = [\gamma\, /\, g\, (\rho_2 - \rho_1)\,]^{1/2},$$

$$W_* = \min[W(G,\chi)], \qquad W(G,\chi) = (1 + G\chi^2)\tanh(\chi)/\chi.$$

where l_c is the capillary length and χ is the non-dimensional wave number. At small χ we have

$$dW/d\chi = 2\chi\,(G - 1/3) + O(\chi), \qquad \chi << 1.$$

since at $G > 1/3$ the critical disturbances are long waves. Under condition $G > 1/3$ the critical value $W_*=1$. In other words, if the distance between the upper electrode and the interface is small

$$h_1 < \sqrt{3}\; l_c = h_*, \tag{8.15}$$

then the critical magnitude of the electric field intensity is expressed as

$$E_* = \sqrt{(\rho_2 - \rho_1) g h_1 / \varepsilon_1}\;. \tag{8.16}$$

In the limiting case $h_1 >> l_c$ the critical electric intensity E and critical wavelength $l_* = 2\pi / k_*$ are expressed as

$$E_* = \sqrt{2(\rho_2 - \rho_1) g l_c / \varepsilon_1}\;, \qquad l_* = 2\pi l_c, \qquad l_c = [\gamma / (g(\rho_2 - \rho_1))]^{1/2}. \tag{8.17}$$

Thus, if the upper electrode is far from the charged interface the capillary forces play the major role determining both E_* and l_*.

8.4 Non-linear analysis of free surface shapes and non-linear waves in an initially plane charged interface

The linear analysis does not permit the determination of the structure and amplitude of curved interface which has been formed after the development of an instability. The classical bifurcation theory predicts that in the supercritical region $E > E_*$ at small supercritical values $(E - E_*) / E_* << 1$ the initially plane charged interface can be transformed into roll, hexagon and quadrate cells (see, for example, the review in the monograph by Rosensweig R.E. [6]). The evolution of the interface structure with the increase of an electrical intensity is the following (Fig. 8.2): at $E = E_*$ the hexagon cells are arisen by a jump (or "hard") way; at $E = E_q > E_*$ the hexagon structure is changed into quadrate cells. Experiments have shown that at $E >> E_q$ spikes are developed and breakdown can occur.

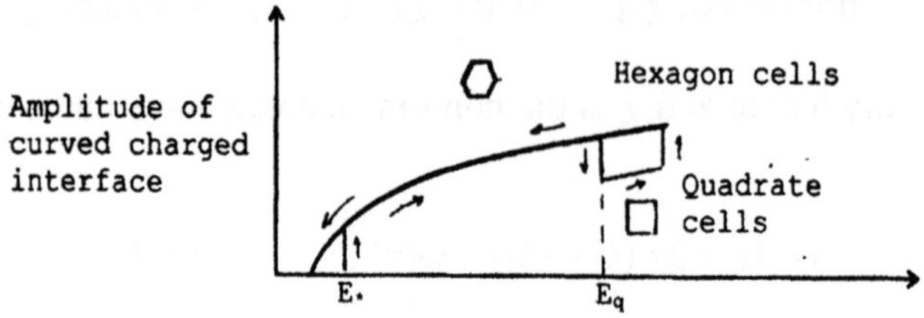

Fig. 8.2, The evolution of the curved charged interface with increasing of electrical intensity

The conclusions of the bifurcation theory accounted for the critical intensity E but did not correlate with the surface shapes and their evolutions completely. The experiments have shown [5] that there are single surface shapes like dimples, dumps, non-linear rolls etc. The single surface shapes can be moved along the interface like particles. At small electric field they disappear and at high field they are transformed into spikes.

In this section we present the new method of investigation of the temporal evolution of the small disturbances on the initially plane charged interface [5], and then an analysis of the behaviour of non-linear disturbances in a long wave approach.

8.4.1 Method of non-linear analysis

Non-linear analysis will be carried out under the following assumptions: both liquids are ideal and the motion of the liquid is potential $\mathbf{u} = \nabla\Psi_i$, where Ψ_i is the velocity potential of i-th liquid, i = 1,2. Let us introduce the following set of dimensionless quantities (denoted with asterisks):

$$x^* = x/\lambda, \qquad y^* = y/\lambda, \qquad z^* = z/h_2, \qquad t^* = t/t_0, \qquad t_0 = \lambda/c_0,$$

$$c_0 = \sqrt{gh_2}\,, \qquad \Psi_1{}^* = \Psi_1/(\lambda c_0), \qquad \phi^* = \phi/V, \qquad f^* = f/a.$$

Here a, λ are the characteristic scales of the disturbance's amplitude and the horizontal length. Both a and λ must be determined after solving the problem. The parameter c_0 is the phase velocity of long waves in the absence of an electric field. Thus, the boundary value problem in dimensionless form is

$$\Psi_{zz} + \mu^2\nabla_1{}^2\phi = 0 \qquad (\Omega_1), \tag{8.18}$$

$$\Psi_{izz} + \mu^2\nabla_1{}^2\Psi_1 = 0 \qquad (\Omega_1\,, i = 1, 2)\,; \tag{8.19}$$

at z = εf, we have

$$\varepsilon f_t = \mu^{-2}\Psi_{1z} - \varepsilon\nabla_1 f\cdot\nabla_1\Psi_1 = \mu^{-2}\Psi_{2zz} - \varepsilon\nabla_1 f\cdot\nabla_1\Psi_2, \tag{8.20}$$

$$\Psi_{2t} - \rho\Psi_{1t} + (1/2\,\mu^2)\,[\Psi_{2z}{}^2 + \mu^2|\nabla_1\Psi_2|^2 - \rho\,(\Psi_{1z}{}^2 + \mu^2|\nabla_1\Psi_1|^2)] +$$

$$+\varepsilon<\rho>f - <\rho>/2\,h^3 W\,(\phi_z{}^2 + \mu^2|\nabla_1\phi|^2) =$$

$$= \varepsilon\mu^2 G <\rho> h^2\,\nabla_1\cdot[\nabla_1 f(1 + \varepsilon^2\mu^2|\nabla_1 f|^2)^{-1/2}] + C, \tag{8.21}$$

$$\phi = 1;$$

$$\text{at } z = h: \quad \phi = 0,\ \Psi_{1z} = 0; \qquad z = -1: \quad \Psi_{2z} = 0. \tag{8.22}$$

Here the asterisks have been omitted and the lower letter indexes denote the partial derivatives. The following set of dimensionless quantities appear:

$$W = E^2/E_*{}^2, \quad E = V/h_1, \qquad E_*{}^2 = (\rho_2 - \rho_1)\,gh_1/\varepsilon_1, \tag{8.23}$$

is the ratio of Coulomb and gravity forces,

$$G = l_c^2/h_1^2, \qquad l_c^2 = \gamma/[g\,(\rho_2 - \rho_1)], \quad \varepsilon = a/h_2, \qquad \mu = h_2/\lambda,$$

$$h = h_1/h_2, \qquad \rho = \rho_1/\rho_2, \qquad <\rho> = 1 - \rho.$$

We suppose that the disturbances of the free surface are small $\varepsilon \ll 1$. Then, we will seek the functions Ψ_j, ϕ in the form of asympthotic expansions:

$$\phi = f_0 + \varepsilon\phi_1 + \varepsilon^2\phi_2 + \ldots, \qquad \Psi_j = \varepsilon\Psi_{j1} + \varepsilon\Psi_{j2} + \ldots \tag{8.24}$$

where the functions ϕ_i, Ψ_{ji} also depend on the small parameter ε and are of the order of unity. For example, they can be functions of the form $\sin(x/\varepsilon)$ or $\cosh^{-1}(\varepsilon x)$.

The usual linearization technique allows us to obtain

$$\phi_0 = -z/h + 1, \quad \phi_i = \int_{-\infty}^{\infty} \phi_{i*} G_3 \exp(-i\mathbf{k}\cdot\mathbf{r})\,dk_1\,dk_2, \qquad i = 1, 2;$$

$$\phi_{1*} = f_*/h, \qquad \phi_{2*} = -(f\phi_{1z})_*, \qquad G_3 = \sinh[\mu k(h-z)]/\sinh(\mu kh);$$

$$\Psi_{1i} = \int_{-\infty}^{\infty} \Psi_{1i*} G_1 \exp(-i\mathbf{k}\cdot\mathbf{r})\,dk_1\,dk_2,$$

$$G_1 = \cosh[\mu k\,(z - h)]/\cosh(\mu kh);$$

$$\Psi_{2i} = \int_{-\infty}^{\infty} \Psi_{2i*}\, G_2 \exp(-i\mathbf{k}\cdot\mathbf{r})\, dk_1\, dk_2,$$

$$G_2 = \cosh[\mu k (z+1)]\cosh(\mu kh);$$

$$\Psi_{j1*} = \Psi_{j*}, \qquad \Psi_{j2*} = -(f\Psi_{j1z})_*, \qquad j = 1, 2; \tag{8.25}$$

$$k = (k_1^2 + k_2^2);$$

where the asterisk in lower indexes denotes the Fourier transformations in the plane (x, y) and

$$\Psi_j = \Psi_j(t, x, y) = \Psi_j(t, x, y, z) \qquad \text{at} \quad z = 0.$$

Note that the functions ϕ_1, Ψ_{j1} are expressed via f, Ψ_j which are determined by Eqs. (8.20), (8.21) (some of the boundary conditions have been used). Thus, substituting (8.24) into (8.20), (8.21) we obtain the following integral-differential system for f, Ψ_1, Ψ_2 describing the temporal evolution of non-linear small finite amplitude disturbances.

$$f_t = \mu^{-2}\Psi_{21z} - \varepsilon\nabla_1 f\cdot\nabla_1\Psi_2 + \varepsilon\mu^{-2}(\Psi_{22z} + f\Psi_{21zz}) =$$

$$= \mu^{-2}\Psi_{11z} - \varepsilon\nabla_1 f\cdot\nabla_1\Psi_1 + \varepsilon\mu^{-2}(\Psi_{12z} + f\Psi_{11zz}); \tag{8.26}$$

$$\Psi_{2t} - \rho\Psi_{1t} + \langle\rho\rangle f - \mu^2 G\langle\rho\rangle h^2 \nabla_1^2 f + h^2\langle\rho\rangle W\phi_{1z} +$$

$$+\varepsilon\,[\,(\rho\Psi_{11z} - \Psi_{21z})\, f_t + (1/2\mu^2)(\Psi_{21z}^2 + \mu^2|\nabla_1\Psi_2|^2)$$

$$-\rho(\Psi_{11z}^2 + \mu^2|\nabla_1\Psi_1|^2) + h^2\langle\rho\rangle W(\phi_{1zz} f + \phi_{2z})$$

$$-h\langle\rho\rangle W(h^2\phi_{1z}^2 + \mu^2|\nabla_1 f|^2)/2] = C_1. \tag{8.27}$$

Here the partial derivatives are taken at z = 0 and the magnitudes of higher order are neglected. Let us consider the long wave approach assuming

$$\mu h \ll 1, \qquad \mu \ll 1, \varepsilon \sim \mu^2 h^2. \tag{8.28}$$

Combining (8.28) with (8.15) we conclude that long wave disturbances will be dominant under conditions

$$h_1 < \sqrt{3}\, l_c, \qquad h_2 \ll \lambda, \qquad h_1 \sim h_2, \qquad a \sim h_2^3/\lambda^2. \tag{8.29}$$

The first condition is controlled in experiments and the others are realized by special types of initial disturbances. Under conditions (8.28) the integrals (8.25) could be easily calculated and after some transformations we obtain the following system of equations for the functions f, $\Psi = \Psi_2$:

$$f_t + \nabla_1{}^2\Psi + \varepsilon\, \nabla_1 \cdot (f\, \nabla_1 \Psi) + \mu^2 \nabla_1{}^2(\nabla_1{}^2\Psi)/3 = 0; \qquad (8.30)$$

$$(1 + \rho/h)\, \Psi_t + \langle\rho\rangle\, (1 - W)\, f + \mu^2 \langle\rho\rangle\, h^2\, (W/3 - G)\, \nabla_1{}^2 f +$$

$$+ \mu^2\, (1 - h^3\,) / (3h)\, \nabla_1{}^2\Psi_t + \varepsilon\, [(1 - /h)\, |\nabla_1\Psi|^2 - 3 \langle\rho\rangle\, W\, f^2 /h\,] /2 = C\,. \qquad (8.31)$$

These equations describe a temporal evolution of long wave disturbances and non-linear surface shapes in the long wave approach.

8.5 Non-linear equilibrium shapes of charged interface

At equilibrium state ($f_t = 0$, $\Psi = 0$) from Eq. (8.31) it follows

$$(1 - W)\, f + \mu^2\, h^2\, (W/3 - G)\, \nabla_1{}^2 f - 3\varepsilon\, (W/2h)\, f^2 = C = \text{const}\ ;$$

or in dimensional form

$$\nabla_1{}^2 f = a_1\, f + a_2\, f^2 + C, \qquad \nabla_1{}^2 = \partial^2/\partial x^2 + \partial^2/\partial y^2; \qquad (8.32)$$

$$a_1 = 3\, (E^2 - E_*{}^2\,) / [h_1{}^2\, (E^2 - E_s{}^2\,)\,], \quad a_2 = 9E^2 / [2h_1{}^3\, (E^2 - E_s{}^2\,)\,];$$

where E, E_* are defined in (8.23) and $E_s = (3\gamma /h_1\,)^{1/2}$. Taking the transformation

$$f = a\, \vartheta(x_1, x_2\,), \qquad x_1 = x/\lambda, \qquad x_2 = y/\lambda\,; \qquad (8.33)$$

$$a = |a_1| /a_2, \qquad \lambda = |a_1|^{-1/2} = h_1\, |\, (E^2 - E_s{}^2) /3\, (E^2 - E_*{}^2\,)\, |^{1/2}; \qquad (8.34)$$

from (8.32) we get

$$\nabla^2\vartheta - \text{Sign}\, (a_1\,)\, \vartheta - \vartheta^2 = C = \text{const}, \quad \nabla^2 = \partial^2/\partial x_1{}^2 + \partial^2/\partial x_2{}^2\,. \qquad (8.35)$$

So the solutions of the Eq. (8.35) are determined by the universal function $\vartheta(x_1, x_2)$ which does not depend on the parameters of the initial problem. That is why the expressions (8.34) determine the characteristic scales of the amplitude a and horizontal length λ. Thus our considerations are valid under conditions

$$\varepsilon = |a| /h_1 = (2/3)\, |E^2 - E_s{}^2| /E^2 = (2/3)\, |W - 1| / W \ll 1,$$

$$\mu h = h_1 /\lambda = [3\, |(E^2 - E_*{}^2\,) / (E^2 - E_s{}^2\,)\, |\,] = [\, 3\, |\, (W - 1) / (W - 3\varepsilon_1 G)\, |\,]^{1/2} \ll 1.$$

Two dimensional case

The equation (8.35) in the two dimensional case $\vartheta = \vartheta(x_1)$ at $a_1 > 0$ has the solution in closed form

$$\vartheta = -(3/2)\cosh^{-2}(x_1/2) \tag{8.36}$$

and at $a_1 < 0$ is expressed via the elliptical cosine

$$\vartheta = \vartheta_2 + (\vartheta_1 - \vartheta_2)\,\mathrm{cn}\left(x_1\sqrt{(\vartheta_3 - \vartheta_1)/6m}\right), \tag{8.37}$$

$$m = (\vartheta_2 - \vartheta_1)/(\vartheta_3 - \vartheta_1);$$

where ϑ_1, ϑ_2, ϑ_3 are the roots of the polinome

$$P(\vartheta) = \vartheta^3 - (2/3)\,\vartheta^2 + 3C_1\vartheta + C_2;$$

where C_1, C_2 are arbitrary constants.

Axisymmetrical case

In this case $\nabla^2 = r\, d/dr(r\, d/dr)$, where r is the distance between the axis of symmetry and a point on the free surface. Boundary conditions are the following: $du/dr = 0$ at $z = 0$ and $\vartheta = 0$ at $r \to \infty$. Numerical calculations have shown that at $a_1 > 0$ there is a single solution of a dimple type (Fig. 8.3, curve 1) and at $a_1 < 0$ there is an infinite number of solutions of an "oscillatory" type (Fig. 8.3, curves 2,3).

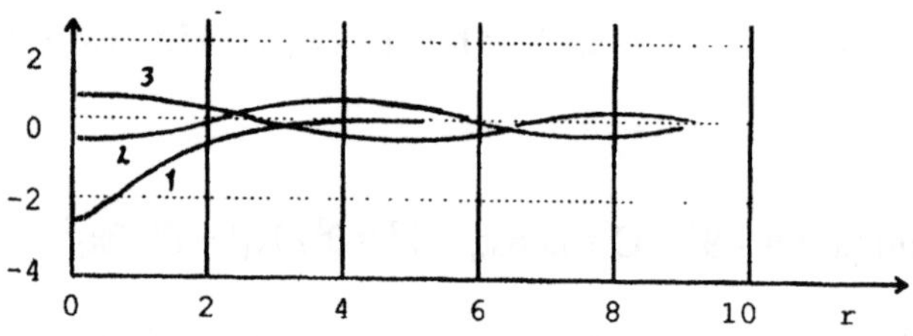

Fig. 8.3, Solution of the equation (8.32): 1 - single solution at $a_1 > 0$: 2,3 - solution of "oscillatory" type at $a_1 < 0$

Let us define the values $W > 1$ as supercritical and consider two limiting cases of high

($E_s >> E$) and small ($E_s << E$) surface tensions,

At $E_s >> E$ in the supercritical region $E > E_*$ ($W > 1$) we have

$$f = -(2h_1/3)(1 - E_*^2/E^2)\,\vartheta(x_1, x_2).$$

For single solutions $\vartheta(x_1,x_2) < 0$ and in that case "bump" forms may be developed.

At $E_s << E$ in the supercritical region it takes place

$$f = (2h_1/3)(1 - E_*^2/E^2)\,\vartheta(x_1, x_2);$$

and this solution describes the single surface forms of a dimple type. Non-linear surface forms have been observed on the interface between glycerine ($\rho_2 = 1.26$ g/cm^3) and solution iodine in transformer oil ($\rho_1 = 0.98$ g/cm^3) [5]. For $\gamma = 2$ dn/cm, $\varepsilon = 2.2\varepsilon_0$ and $h = 0.5$ cm we have $E_* = 8.5$ kV/cm, $E_s = 3.6$ kV/cm. As $E \sim E_*$ then $E_s < E$. That is why in experiments [5] dimples had been observed.

8.6 Non-linear surface waves on a charged interface

The analyses of equations (8.30), (8.31) have shown that there are two types of non-linear long waves on charged interfaces [5]. The first type is hydrodynamic waves subjected to electric forces. These waves are realized under condition $1 - W >> \varepsilon$. The second one appears under strong influence of electric field $1 - W \sim \varepsilon$, when hydrodynamic inertia forces are much smaller than the electrical ones. Let us consider these two wave motions separately.

8.6.1 Hydrodynamic non-linear waves in an electric field

The pair of Eqs. (8.30), (8.31) are of Boussinesq type and describe a long-wave motion in the shallow water. As it had been shown in [7], Boussinesq system can be reduced into KdV equation. For the discussed problem in the two-dimensional case the foregoing KdV equation is expressed as

$$u_t + cu_x + \varepsilon\delta\, u\, u_x + \mu^2\beta u_{xxx} = 0, \qquad u = f = \Psi_x, \tag{8.38}$$

$$c = [\, <\rho> (1 - W)/(1 + \rho/h)\,]^{1/2},$$

$$\delta = \frac{c^2(3h+\rho)}{2(h+\rho)}[1 - \frac{3<\rho>W}{c^4(3h+\rho)}], \qquad \beta = \frac{c^2}{6}\left[1 + \frac{h^3(W-3G)}{c(h+\rho)} + \frac{(1-h^2)}{c(h+\rho)}\right].$$

Let us consider the effect of an electric field on the shape and velocity of hydrodynamic single wave (soliton). The solution describing a soliton has the form

$$u = f = f_0 / \cosh^2(b\xi), \qquad \xi = x - c(1+\zeta)t, \tag{8.39}$$

$$\zeta = \varepsilon\delta f / (3c), \qquad b^2 = \varepsilon\delta f_0 / (12\mu^2\beta).$$

The derivation of KdV equation (8.38) was done supposing that $c >> \varepsilon$ [7]. As $c \sim 1$, $\delta \sim 1$, $\beta \sim 1$, then $\zeta \sim \varepsilon$, $b^2 \sim \varepsilon$, therefore the magnitude c is the dimensionless velocity of soliton. As c is decreased when the field increases, the liquid motion slows down. Suppressions of small oscillations on the charged interface by an electric field have been observed in the experiments [8]. From (8.39) also follows that the parameter b is decreased with the increase of the electric strength, that is, a soliton shape in an electric field becomes more flat.

8.6.2 Electrohydrodynamical solitons on charged interface

The strong influence of the field on the EHD motion of a liquid with a charged interface is defined by condition $1 - W \sim \varepsilon$. As it follows from Eq. (8.30) the partial derivative f_t is of similar order as the function Ψ. On the other hand from (8.31) it is seen that $\Psi_t \sim \varepsilon$, therefore $\Psi_t \cdot (1 - W) \sim \varepsilon$, $f_t \sim \Psi$, $f_{tt} \sim \Psi_t \sim \varepsilon$. These equivalences show that the functions f, Ψ depend on the time as $f \sim F(\sqrt{\varepsilon}\, t)$, $\Psi \sim \sqrt{\varepsilon}\, H(\sqrt{\varepsilon}\, t)$, where F, H have unit order. Summarizing these considerations we can try to seek the solution of the system (8.30), (8.31) in the following asymptotic expansions

$$f = F(\tau, x, y) + \varepsilon_1 F_1(\tau, x, y) + \varepsilon_1^2 F_2(\tau, x, y) + \dots$$

$$\Psi = \varepsilon_1 H(\tau, x, y) + \varepsilon_1^2 (\tau, x, y) + \varepsilon_1^3 (\tau, x, y) + \dots; \tag{8.40}$$

where $\tau = \varepsilon_1 t$ is the "slow" time and $\varepsilon_1 = \sqrt{\varepsilon}$. Substituting (8.40) into (8.30), (8.31) and neglecting the magnitudes of the order ε in (8.30) and ε^2 in (8.31) we obtain

$$F_\tau + \nabla_1^2 H = 0,$$

$$\varepsilon(1 + \rho/h) H_\tau + \langle\rho\rangle (1 - W) F + \langle\rho\rangle \mu^2 h^2 (W/3 - G) \nabla_1^2 F - 3\langle\rho\rangle (W/2h) F^2 = C.$$

Both these equations are equivalent to the following non-linear equation

$$\alpha_0 \partial^2 F/\partial\tau^2 = \nabla_1^2 (\nabla_1^2 F - \alpha_1 F - \alpha_2 F^2), \tag{8.41}$$

$$\alpha_0 = 3\varepsilon(1 + \rho/h)B, \quad \alpha_1 = 3\langle\rho\rangle (W - 1) B,$$

$$\alpha_2 = \varepsilon (9/2) WB/h, \quad B = 1/[\mu^2 h^2 (W - 3G)].$$

The equation (8.41) describes the motion of single non-linear forms (dimples, dumps, etc.). For example, in two-dimensional cases $\nabla_1^2 = \partial^2/\partial x^2$,

$$F = A \cosh^{-2}(b\xi), \qquad \xi = x - U\tau, \tag{8.42}$$

$$A = -(3/2)(\alpha_1 + U\alpha_0)/\alpha_2, \quad b = (\alpha_1 + U^2\alpha_0)^{1/2}/2;$$

where U is the dimensionless soliton velocity. From (8.42) it is seen that the depth of the dimple is increased and its width decreases with the increase of velocity U and at $U = 0$ the solution (8.42) coincides with the solution (8.33), (8.36) describing a motionless dimple. Thus the non-linear equation (8.40) describes the non-linear single surface forms which can move along the interface like particles.

8.7. Another aspect in non-linear electrohydrodynamics of free surface.

1. Let us discuss another limiting case of short wave approach (SWA) when $\lambda \ll h_1$, $\lambda \ll h_2$. Formally SWA equations can be obtained from (8.26), (8.27) at $\mu \gg 1$. Assuming that upper liquid is gas and putting $\psi_2 = \psi$ we obtain the following equations in short wave approach:

$$f_t = \mu^{-1} K[\psi] - \varepsilon(\nabla_1 f \cdot \nabla_1 \psi + f\Delta_1\psi + K[fK[\psi]]), \tag{8.43}$$

$$\psi_t + f - G_\lambda \Delta_1 f - W_\lambda K[f] - \varepsilon\mu W_\lambda (f\Delta_1 f + |\nabla_1 f|^2 + K^2[f] + K[fK[f]]) = C = \text{const}, \tag{8.44}$$

$$G_\lambda = l_c^2/\lambda^2, \qquad W_\lambda = E^2/E_\lambda^2, \qquad E_\lambda^2 = \rho_2 g\lambda/\varepsilon_1,$$

where K[f] is the integral operator defined as

$$K[f] = \iint_{-\infty}^{\infty} f(\mathbf{r}')\, K(\mathbf{r}, \mathbf{r}')\, dx'\, dy',$$

$$K(\mathbf{r}, \mathbf{r}') = \frac{1}{4\pi^2} \iint_{-\infty}^{\infty} k \cos[\mathbf{k}\cdot(\mathbf{r}' - \mathbf{r})]\, dk_1\, dk_2,$$

$$\mathbf{r} = (x, y), \qquad \mathbf{r}' = (x', y'), \qquad k = (k_1^2 + k_2^2)^{1/2}.$$

We recall that equations (8.43), (8.44) are valid under conditions $\varepsilon\mu \ll 1$, $\mu \gg 1$.

With mathematical point of view this limiting case is more complicated then one of SWA becouse equations (8.43), (8.44) are system of integro-differential equations.

2. In section 8.3.2. we have shown that viscosity does not influence upon the critical values of instability parameters. But nonlinear dynamics of liquid layer with free surface can be essentially depended on viscosity due to viscous losses. Taking into account the

finite conductivity of liquid A. Castellanos and A. Gonzalez have shown that viscous losses can not be compensated by electrical energy [3].

References.

1. Melcher J.R., Taylor G.I. Electrohydrodynamics: a review of the role of interfacial shear stress, Annual Review of Fluid Mechanics, ed. by W.R. Sears, M. Van Dyke, v.1, Palo Alto, California (1969), 111-145.
2. Zhakin A.I. The stability of a horizontal interface of weakly conductive liquid in a tangent electric field, Magnitnaja Gidrodinamica, Latvia, Riga, 3 (1981), 74-80 (in Russian).
3. Castellanos A. & González A. Nonlinear electrohydrodynamics of free surfaces, IEEE Trans. on Diel. and Electr. Insul., **5** (1998), 334-343.
4. Taylor, G.I, McEvan, A.D.: The stability of a horizontal fluid interface in vertical electric field, J. Fluid mech., 22 (1965), 1-15.
5. Zhakin A.I.: On non-linear shapes and non-linear waves on the interface perfectly conductive liquids (ferrofluid) in transverse electrical (magnetic) fields, Magnitnaja Gidrodynamica, Latvia, Riga, 4 (1983), 41-48 (in Russian).
6. Rosensweig, R.E.: Ferrohydrodynamics, Cambridge University Press, Cambridge (England) 1985.
7. Non-linear waves, Edited by S. Leibovich and R. Seebass, Carnel Univ. Press, 1974.
8. Rubalko A.S., Borzinetz V.V. and Grigorjev V.V. Small oscillation spectrum of a charged liquid helium interface, Low Temperature Physics, Ukraine, Kharkov, 9, N6 (1983), 570-573 (in Russian).

Chapter 9

EHD PUMPING: ELECTROKINETIC PUMPING BY A HIGH VOLTAGE TRAVELLING WAVE. ION-DRAG PUMPING

9.1 Introduction

There are two methods to pump DL with the Coulomb electric force. The first one is electrokinetic pumping by a high voltage travelling wave on the exciting electrode of the form

$$V(\mathbf{r}, t) = V_1(\mathbf{r})\cos(\omega t) + V_2(\mathbf{r})\sin(\omega t), \tag{9.1}$$

where for a voltage travelling wave it is $V_1 = V_0 \cos(\mathbf{k}\cdot\mathbf{r})$, $V_2 = V_0 \sin(\mathbf{k}\cdot\mathbf{r})$, hence $V = V_0 \cos(\omega t - \mathbf{k}\cdot\mathbf{r})$, and the wave vector $\mathbf{k}$ is determined by an electrode pumping performance. At $V_1 = V_0$, $V_2 = 0$ the boundary condition implies a high frequency harmonic voltage.

At high frequency harmonic voltage the space charge disappears in the liquid bulk and the behaviour of liquid dielectrics in AC electric fields is equivalent to the situation when an ideal liquid dielectric is subjected to a stationary electrical strength. Space charge in the liquid volume can be induced by distributed temperature gradient or charge injection from electrodes. Both mechanisms of space charge generation were used to pump dielectric liquids by a voltage travelling wave [1, 2].

The basic features of pumping by a voltage travelling wave are the following: (i) If the polarity of the charge induced near the exciting electrode is the same (opposite) as that on the exciting electrode it results in a repulsive force causing backwards (forward) pumping. (ii) If the frequency is too high then a very small charge near the exciting electrode is induced, while if the frequency is too low, a phase lag between induced charge and travelling electric intensity does not result in Coulomb pumping force. Thus at very low and very high frequencies the pumping by a voltage travelling wave is not effective.

The second alternative method to pump DL is the use of a DC electric field at working liquids with unipolar injection. In this case the pumping liquid moves from the injector towards the ions collector. This manner to displace liquids is usually called ion-drag pumping.

There are many ion-drag pump concepts and devices (see, for example, [2-7]). Usually pointed emitters are used [4-6] and/or in thin plane grids in both emitter and collector [3, 7]. All types of ion-drag pumps have the same operating characteristics: (i) a small pressure generation per length of the pumping section (up to 5 Pa/cm at applied voltage about 30 kV/cm); (ii) a low efficiency of about 6% at 30 kV/cm.

Low efficiencies are explained by the viscous losses inside the pumping section itself, and are difficult to avoid.

Let us give some typical values on the operating characteristics. The ion-drag pump by Stuetzer [4] created, in a close loop system, a volume flow rate of 0.8 cm/s across 0.6 mm diameter by means of a 1.2 mm long pump stage with kerosene. Sharbaugh and Walker [5] using 84 sharp points on the emitter obtained flow rates as high as 11.5 cm/s at 30 kV DC with negative ion emission.

Electrokinetic pumping by high voltage travelling wave and ion-drag pumping are non-mechanical, they have neither moving mechanical parts or the need of external pressure to operate. Thus, EHD pumping is attractive for industrial applications such as cooling of transformers and high-power underground cables. EHD pumping may have an impact on the design of novel pumping devices for space applications. A remarkable achievement is the creation of a micro-EHD pump [7]. It consists of two parallel silicon grids with dimension 3x3 mm^2, an electrode gap of about 400 μm with porous holes of about 70x70 μm^2. Under an applied voltage of about 700 V, a volume flow rate of about 20 cm^3 per minute can be reached with low viscous polar liquids such as spirits, acetone, etc.

In this section we shall derive a system of average EHD equations for both a harmonic voltage and a travelling wave electric field excitation. The criteria determining the influence of conductivity effects on the behaviour of liquid dielectrics in AC electric fields will be calculated (see inequality (9.4), (9.6)). Then we shall consider the basic features of electrokinetic pumping by a high voltage travelling in a plane tube channel and discuss a simple theory of ion-drag pump.

9.2 Derivation of average EHD equations

At high alternating fields the degree of injected charge and the depth to which it penetrates into the liquid volume are related by the balance of ion migration and convection. The migration depth is estimated by the distance to which an injected charge migrates during the period $2\pi/\omega$:

$$\delta = 2\mu E/\omega , \tag{9.2}$$

with μ the mobility of injection ions. If the length δ has the order of the electrode gap d:

$$\delta = 2\mu E/\omega \sim d, \tag{9.3}$$

then the conductivity effects due to injection phenomena can be significant, while under condition

$$\delta = 2\mu E/\omega << d \tag{9.4}$$

the influence of injection charges on the liquid motion is negligibly small.

To determine the criterion for the influence of space charge in ohmic liquids is more complicated. We shall get this criterion by the averaging method of differential equations [8]. Let us assume a two-ion model suggesting that both species of ions A^+, B^- are generated due to dissociation of ion pairs A^+B^- and injection from electrodes with recombination in the liquid volume. Take the characteristic scales for the variables as

$$t_0 = 1/\omega, \qquad u_0 = \nu/d, \qquad p_0 = \eta\, u_0/d, \quad E_0 = V/d,$$

$$n_0 = [\alpha_{d0}(T_1)\, N_0/\alpha_r(T_1)]^{1/2},$$

where T_1 is the anode temperature.
Dimensionless EHD equations are the following:

$$s^{-1}\partial \mathbf{u}/\partial t + (\mathbf{u}\cdot\nabla)\,\mathbf{u} = -\nabla\tilde{p} + \nabla^2\mathbf{u} + Gn\mathbf{E} - G_d E^2\nabla\varepsilon\,; \quad \nabla\cdot\mathbf{u} = 0\,; \tag{9.5}$$

$$\nabla\cdot\varepsilon\mathbf{E} = C\,n, \quad n = n_1 - n_2, \qquad \mathbf{E} = -\nabla\phi\,; \tag{9.6}$$

$$(Pe/S)\,\partial n_j/\partial t + \nabla\cdot[\,(-1)^{j-1}\beta_j\, n_j\,\mathbf{E} + Pe\, n_j\,\mathbf{u}\,] = \Gamma_p\, C\xi_N\,; \tag{9.7}$$

$$S^{-1}\,\partial T/\partial t + \mathbf{u}\cdot\nabla T = P^{-1}\nabla^2 T\,, \qquad \xi_N = F_1 - \alpha n_1\, n_2\,; \tag{9.8}$$

$$\beta_j = \mu_j(T)/\mu_2(T_1), \quad \alpha = \alpha_r(T)/\alpha_r(T_1);$$

$$F_1 = \alpha_{d0}(T)\,F(E,T)/\alpha_{d0}(T_1), \qquad Pe = P_\sigma C.$$

Here $S = u_0/(\omega d) = \nu/(\omega d^2)$, $G_d = E_0^2\, d/(2\eta u_0)$ and other dimensionless numbers are given in section 7.6. Dimensionless temperature is defined as $\theta = (T - T_1)/\Delta T$, $\Delta T = T_2 - T_1$, where T_1, T_2 are respectively the temperatures of the electrodes and T is the dimensional temperature. At small temperature difference ΔT, we have

$$\varepsilon = \varepsilon_1(1 - \beta_\varepsilon\,\theta)\,, \qquad \mu_j = \mu_j(T_1)[1 + \beta_{\sigma j}\,\theta], \qquad \alpha_{d0} = \alpha_{d0}(T_1)[1 + \beta_d\,\theta],$$

$$\beta_\varepsilon << \beta_{\sigma j} << 1, \qquad \varepsilon_1 = \varepsilon(T_1)\,. \tag{9.9}$$

At high frequency the parameter S is small. We will consider that the frequency is high enough so that

$$SG << 1, \qquad SG_d << 1, \qquad S/Pe << 1, \qquad S/P_\sigma << 1\,. \tag{9.10}$$

The first two inequalities mean that the Coulomb and polarization forces are much smaller than the inertial force. Taking into account that $S/Pe = \mu_2(T_1)E_0/(\omega d) \sim \delta/d$, we see that the third inequality implies a small penetration of injected charge into the liquid bulk. Finally, the last inequality $S/P_\sigma \sim \sigma/\varepsilon\omega << 1$ is the condition that the period of the electric field oscillation is much less than the time of charge relaxation. The conditions (9.10) are very strong. Later on we will address more "gentle" ones (see (9.46)). Accordingly to the known method of averaging [8], we will seek the solution of Eqs. (9.5) - (9.8) into the form

$$\mathbf{u} = \mathbf{u}_a(\tau, \mathbf{r}) + \mathbf{u}'(t, \mathbf{r}), \qquad \tilde{p} = p(\tau, \mathbf{r}) + p'(t, \mathbf{r});$$

$$n_j = n_{ja}(\tau, \mathbf{r}) + n_j'(t, \mathbf{r}), \qquad T = \theta_a(\tau, \mathbf{r}) + T'(t, \mathbf{r});$$

$$\mathbf{E} = \mathbf{E}_a(\tau, \mathbf{r}) + \mathbf{E}'(t, \mathbf{r}); \tag{9.11}$$

where $\tau = St$ is the "slow" time and the upper index prime denotes the pulsation variables. The first terms respond to the time average ones and the averaging operation is defined as

$$f_a = \langle f \rangle = t_0^{-1} \int_0^{t_0} f(t + t_1)\, dt_1 \quad (t_0 >> 1). \tag{9.12}$$

the magnitude $\langle f \rangle$ is the integral mean of the function f(t) at time range $(0, \bar{t}_0)$ where $\bar{t}_0 >> \omega^{-1}$. In the expression (9.12) $t_0 = \bar{t}_0\omega$, hence $t_0 >> 1$.

Substituting (9.11) into Eqs. (9.5) - (9.8) and averaging each equation, we get the following system of average equations:

$$\partial \mathbf{u}_a/\partial\tau + (\mathbf{u}_a \nabla)\mathbf{u}_a = -\nabla p_a + \nabla^2 \mathbf{u}_a - \langle \mathbf{u}'\nabla)\mathbf{u}' \rangle + G[q_a \mathbf{E}_a + \langle q'\mathbf{E}' \rangle] + G_d \langle |\mathbf{E}_a + \mathbf{E}'|^2 \nabla\varepsilon \rangle; \tag{9.13}$$

$$\nabla \cdot \mathbf{u}_a = 0, \qquad q_a = n_{1a} - n_{2a}, \qquad q' = n_1' - n_2'; \tag{9.14}$$

$$\nabla \cdot (\langle\varepsilon\rangle \mathbf{E}_a + \langle \varepsilon \mathbf{E}' \rangle) = C q_a, \qquad \mathbf{E}_a = -\nabla\phi_a; \tag{9.15}$$

$$Pe\, \partial n_{ja}/\partial\tau + \nabla \cdot [(-1)^{j-1}(n_{ja}\mathbf{E}_a \langle\beta_j\rangle + \langle \beta_j n_j' \rangle \mathbf{E}_a + n_{ja}\langle \beta_j \mathbf{E}_a \rangle + \langle \beta_j n_j' \mathbf{E}' \rangle) + Pe(n_{ja}\mathbf{u}_a + \langle n_j \mathbf{u}' \rangle)] = \Gamma_p C \langle \xi_N \rangle; \tag{9.16}$$

$$\partial\theta_a/\partial\tau + \mathbf{u}_a \cdot \nabla\theta_a + \langle \mathbf{u}' \cdot \nabla T' \rangle = P^{-1}\nabla^2\theta_a. \tag{9.17}$$

For pulsation variables with closing relations (9.13) - (9.17) we obtain

$$\partial \mathbf{u}'/\partial t = S\ \{ < (\mathbf{u}'\cdot\nabla)\ \mathbf{u}' > - (\mathbf{u}'\cdot\nabla)\ \mathbf{u}' - (\mathbf{u}'\cdot\nabla)\ \mathbf{u}_a - (\mathbf{u}_a\cdot\nabla)\ \mathbf{u}' - \nabla p' +$$

$$+G\ [q'\mathbf{E}_a + q_a\mathbf{E}' + q'\mathbf{E}' - <q'\mathbf{E}'>] - G_d\ [\ |\mathbf{E}_a + \mathbf{E}'|^2\ \nabla\varepsilon - < |\mathbf{E}_a + \mathbf{E}'|^2\ \nabla\varepsilon>]\}\ ; \quad (9.18)$$

$$\nabla\cdot\mathbf{u}' = 0, \qquad \nabla\cdot\ [\mathbf{E}_a\ (\varepsilon - <\varepsilon>) + \varepsilon\mathbf{E}' - <\varepsilon\mathbf{E}'>] = Cq'\ ; \qquad \mathbf{E}' = -\nabla\phi'\ ; \quad (9.19)$$

$$\mathrm{Pe}\ \partial n'/\partial t = S\ \{ -\nabla\cdot\ [\ (-1)^{j-1}\ (n_{ja}\ \mathbf{E}_a\ (\beta_j - <\beta_j>) + \mathbf{E}_a\ (\beta_j\ n_j' - <\beta_j\ n_j'>) +$$

$$+\ n_{ja}\ (\beta_j\ \mathbf{E}' - <\beta_j\ \mathbf{E}'>) + \beta_j\ n_j'\mathbf{E}' - <\beta_j\ n_j'\mathbf{E}'>) +$$

$$\mathrm{Pe}\ (n_j'\ \mathbf{u}_a + n_{ja}\ \mathbf{u}' + n'\mathbf{u}' - <n'\mathbf{u}'>)\ ] + \Gamma_p\ C\ (\xi_N - <\xi_N>)\ \}\ ; \quad (9.20)$$

$$\partial T'/\partial t = S\ \{ <\mathbf{u}'\ \nabla T'> - \mathbf{u}'\cdot\nabla T' - \mathbf{u}'\cdot\nabla\theta_a - \mathbf{u}_a\cdot\nabla T' + P^{-1}\ \nabla^2 T'\ \}. \quad (9.21)$$

The average boundary conditions (7.41) - (7.44) have the form

$$S_1 : \mathbf{u}_a = 0, \quad \theta = 0, \quad \phi_a = V, \quad n_{1a} = <f_A>;$$

$$S_2 : \mathbf{u}_a = 0, \quad \theta = 1, \quad \phi_a = 0, \quad n_{2a} = <f_B>\ ; \quad (9.22)$$

where we introduce the voltage V_a due to which the injected charge penetrates into the liquid volume and S_1 (S_2) denotes the anode (cathode) surface.

For the pulsation variables it follows

$$S_1 : \mathbf{u}' = 0, \quad T' = 0, \quad \phi' = V_{10}\cos t + V_{20}\sin t, \quad n_1' = f_A - <f_A>,$$

$$S_2 : \mathbf{u}' = 0, \quad T' = 0, \quad \phi' = 0, \quad n_2' = f_B - <f_B>\ . \quad (9.23)$$

To satisfy the boundary conditions for pulsation potential ϕ' in (9.23) it is convenient to take the transformation

$$\mathbf{E}' = \mathbf{E}_{10}\ (\mathbf{r})\cos t + \mathbf{E}_{20}\ (\mathbf{r})\sin t + \mathbf{E}_{in}'\ (t, \mathbf{r})\ , \quad (9.24)$$

where $\mathbf{E}_{j0}\ (\mathbf{r})$, j = 1,2 are determined by the problems

$$\nabla\cdot\ (<\varepsilon>\ \mathbf{E}_{j0}) = 0, \qquad \mathbf{E}_{j0} = -\nabla\phi_{j0}$$

$$S_1 : \phi_{j0} = V_{j0}, \quad S_2 : \phi_{20} = 0, \quad j = 1, 2\ . \quad (9.25)$$

The physical meaning of the decomposition (9.24) is that the pulsation $\mathbf{E}'$ is expanded into the sum of the main harmonics plus high order spatial harmonics.

So, to determine the average Eqs. (9.13) - (9.17) in the closed form it is necessary to find the pulsation variables solving the system (9.18) - (9.21). Taking into account the inequality (9.10) we can consider the right-hand terms of (9.18), (9.20), (9.21) as small perturbations and search for the solution of (9.18) - (9.21) in the form of asymptotic expansions:

$$\mathbf{u}' = S\mathbf{u}_1 + S^2\mathbf{u}_2 + \dots; \qquad p' = p_0' + Sp_1 + S^2p_2 + \dots;$$

$$n_j' = Sn_{j1} + S^2n_{j2} + \dots; \qquad \mathbf{E}_{in}' = S\mathbf{E}_1 + S^2\mathbf{E}_2 + \dots;$$

$$T' = ST_1 + S^2T_2 + \dots; \tag{9.26}$$

For the terms of the first-order approximation it follows

$$\partial\mathbf{u}_1/\partial t = -\nabla\tilde{p}_0' + Gq\,(\mathbf{E}_{10}\cos t + \mathbf{E}_{20}\sin t) - 0.5\,G_d\,[\,(E_{10}^2 - E_{20}^2)\cos 2t +$$

$$+2\mathbf{E}_{10}\cdot\mathbf{E}_{20}\sin 2t + 4\mathbf{E}_a\cdot(\mathbf{E}_{10}\cos t + \mathbf{E}_{20}\sin t)\,]\,\nabla\varepsilon(\theta_a); \tag{9.27}$$

$$\nabla\cdot\mathbf{u}_1 = 0, \qquad \nabla\cdot[\,\varepsilon(\theta_a)\,\mathbf{E}_1\,] = Cq_1, \qquad q_1 = n_{11} - n_{21}; \tag{9.28}$$

$$Pe\,\partial n_j/\partial t = -\nabla\cdot[\,(-1)^{j-1}\,n_{ja}\,\beta_j(\theta_a)\,(\mathbf{E}_{10}\cos t + \mathbf{E}_{20}\sin t)\,] +$$

$$+\Gamma_p\,C\,(F_1 - \langle F_1\rangle), \tag{9.29}$$

$$F_1 = F_1(\,|\mathbf{E}_{10}\cos t + \mathbf{E}_{20}\sin t|\,,\theta_a), \quad \theta_1 = 0. \tag{9.30}$$

So, up to first-order the temperature pulsation is zero and other pulsation functions are expressed via average terms q_a, $\mathbf{E}_a$, the equations of which follow from Eqs. (9.15), (9.16):

$$\nabla\cdot[\varepsilon(\theta_a)\,\mathbf{E}_a\,] = C\,q_a,\ q_a = n_{1a} - n_{2a},\quad \mathbf{E}_a = -\nabla\phi_a;$$

$$Pe\,\partial n_{ja}/\partial\tau + \nabla\cdot[\,(-1)^{j-1}\,\beta_j(\theta_a)\,n_{ja}\,\mathbf{E}_a + Pe\,n_{ja}\,\mathbf{u}_a\,] = \Gamma_p C\,\langle\xi_N\rangle;$$

$$\langle\xi_N\rangle = \langle F_1\rangle - \alpha(\theta_a)\,n_{1a}\,n_{2a},\ j = 1, 2. \tag{9.31}$$

These equations must be solved with boundary conditions (9.22) for ϕ_a and n_{1a}, n_{2a}.

It should be pointed out that there are two interesting cases:

1. Charges are generated only due to injection: $V_a \neq 0$, $F_1 = 0$, $f_A = 0$ or $f_B = 0$. As seen from (9.31) in this case $q_a \neq 0$ and injected charges can penetrate into a liquid volume due to charge migration.

2. There is no injection $f_A = f_B = 0$ and charge generation is caused by dissociation $F_1 \neq 0$. At small size of non-equilibrium layers: $C = d/\xi_d >> 1$ (see section 7.7) it is

$$q_a = 0,\ n_{1a} = n_{2a} = [F_1/\alpha(\theta_a)]^{1/2}, \qquad C >> 1. \tag{9.32}$$

At high fields or small conductivity when $C < 1$ the space charge can change its polarity in the liquid volume and has a homogeneous distribution near the electrodes (see Fig. 7.2). In the following paragraph we will consider two major cases of electric field travelling wave excitation $\mathbf{E}_{10} \neq 0$, $\mathbf{E}_{20} \neq 0$ and harmonic voltage $\mathbf{E}_{10} \neq 0$, $\mathbf{E}_{20} = 0$.

9.3 Average EHD equations at voltage travelling wave

To simplify the analysis we will consider the case of small temperature difference when inequalities (9.9) take place. The average momentum equation (9.13), including the Coulomb force to first-order, is expressed as

$$\partial \mathbf{u}_a/\partial \tau + (\mathbf{u}_a \cdot \nabla)\mathbf{u}_a = -\nabla p + \nabla^2 \mathbf{u}_a + G\,\mathbf{f}_E - 0.5\,G_d\,(E_{10}^2 + E_{20}^2)\,\nabla\varepsilon(\theta_a)\,; \tag{9.33}$$

$$\mathbf{f}_E = q_a\,\mathbf{E}_a + \mathbf{f}_{E1}, \qquad \mathbf{f}_{E1} = S < q_1\,(\mathbf{E}_{10}\cos t + \mathbf{E}_{20}\sin t) >. \tag{9.34}$$

Taking the difference of Eqs. (9.29) and integrating yields

$$q_1 = Q_1 \sin t + Q_2 \cos t\,, \qquad Q_j = -\nabla\cdot\,[n_{ja}\,\beta_j(\theta_a)\mathbf{E}_{j0}\,], \tag{9.35}$$

therefore

$$\mathbf{f}_{E1} = 0.5\,(Q_2\,\mathbf{E}_{10} + Q_1\,\mathbf{E}_{20}\,)\,S/Pe. \tag{9.36}$$

Let us rewrite the average EHD equation for the discussed case in dimensional form

$$\rho\,(\partial \mathbf{u}_a/\partial t + (\mathbf{u}_a \cdot \nabla)\,\mathbf{u}_a) = -\nabla\,\tilde{p} + \eta\nabla^2\,\mathbf{u}_a + \mathbf{f}_E - 0.5\,(E_{10}^2 + E_{20}^2)\,\nabla\varepsilon, \tag{9.37}$$

$$\nabla\cdot\,\mathbf{u}_a = 0, \qquad \mathbf{f}_E = q\mathbf{E} - 0.5\,(\varepsilon/\omega)\,[(\nabla\tau_2\cdot\,\mathbf{E}_{20})\,\mathbf{E}_{10} + (\nabla\tau_1\cdot\,\mathbf{E}_{10})\,\mathbf{E}_{20}\,],$$

$$q = e\,(n_1 - n_2), \qquad \tau_j = e\,n_j\,\mu_j/\varepsilon\,, \qquad j = 1, 2. \tag{9.38}$$

where the functions n_j, $\mathbf{E}$, $\mathbf{E}_{j0}$, $j = 1, 2$ are determined by the problems

$$\nabla\cdot\,\varepsilon\mathbf{E} = q, \qquad \mathbf{E} = -\,\nabla\phi, \qquad \nabla\cdot\,\varepsilon\mathbf{E}_{j0} = 0, \qquad \mathbf{E}_{j0} = -\,\nabla\phi_{j0}\,; \tag{9.39}$$

$$\partial n_j/\partial t + \nabla\cdot\,[\,(-1)^{j-1}\,\mu_j\,n_j\,\mathbf{E} + n_j\,\mathbf{u}_a\,] = F\,N_0 - \alpha_r n_1\,n_2\,, \qquad j=1,2 \tag{9.40}$$

$$q = e\,(n_1 - n_2)\,.$$

Boundary conditions for the dimensional variables are the following

$$S_1 : \mathbf{u}_a = 0, \quad \phi = V_a, \quad \phi_{j0} = V_{j0}, \quad n_1 = <n_A>, \quad j=1, 2,$$

$$S_2 : \mathbf{u}_a = 0, \quad \phi = 0, \quad \phi_{j0} = 0, \quad n_2 = <n_B>. \qquad (9.41)$$

This boundary value problem describes the electrokinetic pumping of dielectric liquids by a high voltage travelling wave (for a specific choice of V_{10}, V_{20} functions) with injection or dissociation charge generations both in thermal or isothermal cases. These equations will be used later on to study the pumping of dielectric liquid in a planar tube channel (see section 9.5).

9.4 Average EHD equations at alternating harmonic voltage

Consider the case of harmonic alternating voltage $V_{10} \neq 0$, $V_{20} = 0$ and high conductivity when $C >> 1$. In this case, from (9.31) it follows

$$q_a = 0, \quad \mathbf{E}_a = 0, \quad n_{1a} = n_{2a} = n_a = [<F_1>/\alpha]^{1/2}, \quad \mathbf{f}_{E1} = 0 .$$

Hence to calculate the average electric force it is necessary to take into account the second-order terms. Let us write the final result in dimensionless form

$$\mathbf{f}_E = S^2G/(2Pe\, P_\sigma)\, [(\varepsilon\, \nabla h_e \cdot \mathbf{E}_1 + h_e q + P_\sigma\, \nabla q \cdot \mathbf{u}_a)\, \mathbf{E}_0 + q\mathbf{E}_1] ;$$

$$q = - \varepsilon\, \nabla h_e \cdot \mathbf{E}_0 , \qquad h_e = n_a(\beta_1 + \beta_2)/\varepsilon ; \qquad (9.42)$$

where electric intensities $\mathbf{E}_0$, $\mathbf{E}_1$ are determined by the problems:

$$\nabla \cdot (\varepsilon(\theta)\mathbf{E}_0) = 0, \quad \nabla \cdot (\varepsilon(\theta)\mathbf{E}_1) = q, \quad \mathbf{E}_j = - \nabla \phi_j , \quad j=0,1 ;$$

$$S_1 : \quad \phi_0 = V_{10}, \quad \phi_1 = 0, \qquad S_2 : \quad \phi_0 = \phi_1 = 0. \qquad (9.43)$$

From (9.42) it follows that the high voltage approximation is valid under condition

$$qS^2G/(2PeP_\sigma) \sim d^2\sigma^2E_0^2\beta\Delta T / (4\rho\omega^2v^2\varepsilon) << 1 ; \qquad (9.44)$$

$$\sigma = en_0(\mu_1 + \mu_2) , \qquad \beta = \sigma^{-1}d\sigma/dT, \qquad \Delta T = T_2 - T_1 .$$

Note that this inequality is carried out at much smaller frequencies than the one in (9.10), hence the condition (9.44) is a more "gentle" one.

At small temperature difference it follows

$$q \sim \beta_\sigma, \; E \sim \beta_\sigma, \; h_e \sim 1, \; |\nabla h_e \cdot \mathbf{E}_1| \sim \beta_\sigma^2, \; qE_1 \sim \beta_\sigma^2, \text{ while } h_e q \sim |\nabla q \cdot \mathbf{u}_a| \sim \beta_\sigma.$$

Therefore the average electric force (9.42) can be represented as

$$\mathbf{f}_E = W\,(h_e q + P_\sigma\, \nabla q \cdot \mathbf{u}_a)\, \mathbf{E}_0, \quad W = \pi\,(\sigma_2/\varepsilon\omega)^2\, \varepsilon E_0^2\,/8p_0,$$

$$p_0 = \rho u_0^2; \qquad \sigma_2 = 2e\mu_2\,(T_1)\, n_0\,.$$

Let us introduce the dimensional average EHD systems at high frequency harmonic voltages in an ohmic model of conductivity:

$$\rho\,(\partial \mathbf{u}_a/\partial t + (\mathbf{u}_a\cdot\nabla)\,\mathbf{u}_a) = -\nabla p + \eta\nabla^2\mathbf{u}_a - (1/4)\,E_0^2\nabla\varepsilon +$$

$$-\,0.5\,\omega^{-2}\,[\sigma\,(\nabla\sigma\cdot\mathbf{E}_0)\,/\,\varepsilon + \nabla(\nabla\sigma\cdot\mathbf{E}_0)\cdot\mathbf{u}_a]\cdot\mathbf{E}_0\,; \tag{9.45}$$

$$\nabla\cdot\mathbf{u}_a = 0, \qquad \nabla\cdot(\varepsilon\mathbf{E}_0) = 0, \qquad \mathbf{E}_0 = -\nabla\phi_0\,; \tag{9.46}$$

$$\rho c_p\,(\partial T/\partial t + \mathbf{u}_a\cdot\nabla T) = k\nabla^2 T, \sigma = e\,(\mu_1 + \mu_2)\, n_0\,; \tag{9.47}$$

$$S_1 : \mathbf{u}_a = 0,\; T = T_1,\; \phi_0 = V;\; S_2 : \mathbf{u}_a = 0,\; T = T_2,\; \phi_0 = 0\,. \tag{9.48}$$

where V is the amplitude of harmonic voltage.

Note that this system responds to the situation of a small temperature difference because it had to take into consideration the linear terms on the small parameters β_σ, β_ε (see (9.9)). In the general case, the average electric force is in the form (9.42).

9.5 Electrokinetic pumping by a voltage travelling wave

Consider a steady flow of dielectric liquid in a planar tube channel due to a voltage travelling wave, the arrangement is sketched in Fig. 9.1.

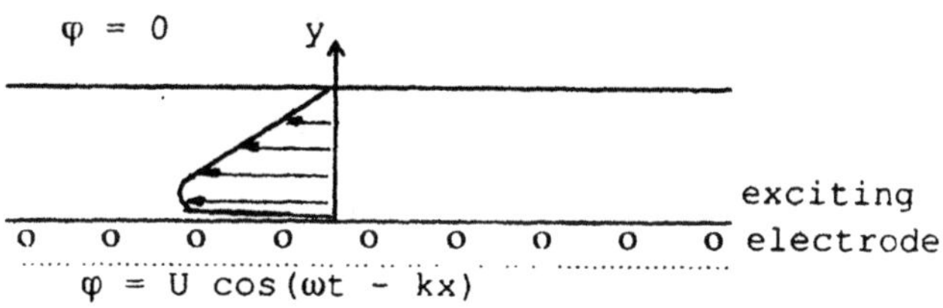

Fig. 9.1, Pumping of DL by voltage travelling wave. The velocity profile answers to injection from exciting electrode

To simplify the analysis assume a linear problem (Reynolds number Re << 1), isothermal conditions and unipolar charge generation of positive charges from the anode. For this problem the equations of motion follow from (9.37) - (9.41):

$$-\partial p/\partial x + \eta \nabla^2 u_{ax} + f_{Ex} = 0; \qquad -\partial p/\partial y + \eta \nabla^2 u_{ay} + f_{Ey} = 0;$$

$$\partial u_{ax}/\partial x + \partial u_{ay}/\partial y = 0. \tag{9.49}$$

$$f_{Ex} = -0.5\,\omega^{-1}\,(d\sigma/dy\ \partial\phi_1/\partial y)\,\partial\phi_2/\partial x, \qquad \sigma = q\mu\,;$$

$$f_{Ey} = qE - 0.5\,\omega^{-1}\,(d\sigma/dy\ \partial\phi_1/\partial y)\,\partial\phi_2/\partial y\,;$$

$$\phi_1 = V\frac{\sinh[k(d-y)]}{\sinh(kd)}\cos(kx), \qquad \phi_2 = V\frac{\sinh[k(d-y)]}{\sinh(kd)}\sin(kx).$$

The space charge distribution and the electric intensity $\mathbf{E} = (0, E)$ are given by

$$q = q_A E(0)/E(y), \qquad E(y) = [\,(2\,q_A/\varepsilon)\,E(0)\,y + E^2(0)\,] \tag{9.50}$$

where $E(0)$ is the electric intensity on the injector at $y = 0$. In the case of high level of injection, $C = 2q_A d/(\varepsilon E(0)) >> 1$, it can be obtained:

$$E(0) = (9/8)\,\varepsilon V^2/(q_A d^3)\,;$$

while at weak injection $C << 1$:

$$q = q_A\,(1 - 0.5\,C\,y/d), \qquad E(y) = E_a = V_a/d = \text{const}.$$

The system (9.49) is reduced into the form

$$-\partial p/\partial x + \eta\nabla^2 u_{ax} - Aq'\sinh 2k(y-d)\,(1+\cos 2kx)/\sinh^2 kd = 0\,;$$

$$-\partial p/\partial y + \eta\nabla^2 u_{ay} - Aq'\,(1+\cosh 2k(y-d))\sin 2kx/\sinh^2 kd = 0\,;$$

$$\partial u_{ax}/\partial x + \partial u_{ay}/\partial y = 0, \quad A = e\mu k^2 V^2/(8\omega), \qquad q' = dq/dy\,.$$

It is seen that the velocity components have the following structure

$$u_{ax} = U(y) + w_x, \qquad u_{ay} = w_y,$$

$$w_x = f_1(y)\cos(2kx), \quad w_y = f_2(y)\sin(2kx)\,;$$

where U(y) is the velocity component responding to the pumping along the tube channel and w_x, w_y correspond to an "agitation" motion inside the channel. For U(y) we obtain

$$\eta d^2U(y)/dy^2 = Aq'(y)\sinh[2k(y-d)].$$

Integrating this equation with the zero velocity conditions at the wall yields

$$U(y) = -U_0\,[s - F(s)/F(1)], \quad s = y/d; \tag{9.51}$$

$$U_0 = (\sigma_A/\omega)\,\chi^2 V^2 F(1)/(\eta d), \qquad \sigma_A = q_A\eta, \qquad \chi = kd.$$

$$F(s) = 0.5\,C\int_0^s\Big[\int_0^s (Cs+1)^{-3/2}\sinh[2\chi(1-s)]\,ds\Big]\,ds/\sinh^2[\chi].$$

At weak injection $C \ll 1$ it can be obtained

$$U(y) = -U_0\,[1 - s - \sinh(2\chi(1-s))/\sinh(2\chi)]; \tag{9.52}$$

$$U_0 = (\sigma_A/\omega)\,V^2 C\,[\sinh(2\chi)/\sinh^2\chi]/(\eta d);$$

and at high injection level for parameter U and the function F(s) in the formula (9.51), it follows

$$U_0 = (\varepsilon V_0/\omega)\,\chi^2 V^2 F(1)/(4\eta d^3); \tag{9.53}$$

$$F(s) = \int_0^s s^{-1}\,\mathrm{sh}[2\chi(s-1)]\,ds - 2\chi\int_0^s\Big[\int_0^s t^{-1/2}\cosh(2\chi(t-1))\,dt\Big]\,ds/\sinh^2\chi.$$

The analysis of the latter expressions allows to conclude:

1. The pumping velocity is increased with the increasing of parameter k (wavelength of voltage travelling wave), injection level (charge q_A) and amplitude of voltage travelling wave;
2. The pumping direction depends on the position of the injecting electrode: if the exciting electrode is an injector then the pumping velocity is opposite to the direction of the voltage travelling wave motion. The velocities of pumping and voltage travelling wave are in the same direction if the excitation and injection electrodes are not the same.
3. Maximum pumping velocity is reached at smallest frequencies.

9.6 Simple theory of ion-drag pumping

Let us consider a stationary EHD flow between two parallel vertical permeable porous electrodes (grids). To avoid a hydrodynamical instability leading to a vortex formation and viscous losses we suppose that there are thin horizontal dielectric plates disposed at the same distance h from one another (Fig. 9.2).

The simplest theory is based on the following electrostatic and conservation laws which we present in the integral form:

$$(\rho/h)\int_0^h\int_0^L \partial^2u_x/\partial x^2 dydx = -\Delta p + (\eta/h)\int_0^h\int_0^L \partial u_x/\partial y\, dydx +$$

$$+(e/h)\int_0^h\int_0^L n\,\partial\phi/\partial x\, dydx + (1/h)\int_0^h \Delta p_{str}\, dy\,; \qquad (9.54)$$

$$\varepsilon\,(\partial^2\phi/\partial x^2 + \partial^2\phi/\partial y^2) = en\,; \qquad (9.55)$$

$$e\int_0^h (n\mu\,\partial\phi/\partial x + nu_x)\, dy = J, \qquad \int_0^h u_x\, dy = Q; \qquad (9.56)$$

$$\Delta p = \int_0^h p\Big|_{x=0}^{x=L} dy/h\,, \qquad \Delta p_{str} = \rho\,(\partial\varepsilon/\partial\rho)\, E^2/2\Big|_{x=0}^{x=L}.$$

Here we assume that the left electrode is the cathode, generating negative charges, Q, J denote volume velocity rate and total current per unit length in transversal direction. We also suppose that pressure difference Δp, injection law and applied voltage are known:

$$x = 0 : \phi = 0, \quad n = n_i\,(E), \qquad x = L : \phi = V\,.$$

The problem is to determine the dependence of Q with Δp and V. Let us assume the parabolic approximation

$$u_x = 6U_0\,(y/h)\,(1 - y/h), \qquad U_0 = Q/h\,;$$

where U_0 is the average velocity. Then for n(x), $\phi'(x)$ we obtain $n(x) = n_i(E)\, A/(R_E + f(x))$, $E = |\phi'(0)|$, $R_E = Q/(\mu E_0 h)$;

$$f(x) = (A^2 + AC\, x/L)^{1/2} - R_E, \qquad \phi'(x) = E_0 f(x)\,;$$

$$E_0 = V/L, \qquad C = 2\, e\, n_i(E)\, L/(\varepsilon E_0)\,.$$

The unknown constant A is determined by the condition that the electrical potential difference must be equal to V, which can be presented as

$$(A^2 + AC)^{3/2} - A^3 - 1.5\, AC\,(1 + R_E) = 0.$$

Finally from the momentum equation (9.54) we find the following relationships among Q, p, V:

$$Q = h^3 / (12\eta L)(p_e - \Delta p) ; \tag{9.57}$$

$$p_e = q_i E_0 LA [1 - 2 (R_E / CA)(\sqrt{CA + A^2} - A)] + \Delta p_{str}, \qquad q = en(E) ;$$

$$\Delta p_{str} = \rho\, \partial\varepsilon/\partial\rho\, (E_0^2/2) [(\sqrt{CA + A^2} - R_E)^2 - (A - R_E)^2] .$$

Let us consider two limiting cases:
At weak injection level we obtain

$$p_e = q_i E_0 L [1 + \rho (\partial\varepsilon/\partial\rho)/\varepsilon] \qquad \text{at } C \ll 1 . \tag{9.58}$$

At high injection level it follows

$$p_e = p_+ (1 + R_E)(3 - R_E) \qquad \text{at } C \gg 1 ; \tag{9.59}$$

$$p_+ = (3/8) [\varepsilon + \rho (\partial\varepsilon/\partial\rho)] E_0^2 .$$

Thus in the last case the electrical pressure depends on the Reynolds electrical number. Substituting (9.59) into (9.57) we can determine the flow rate dependence from electrical parameters and pressure difference Δp into the form

$$R_E = 1 - D/P_E + [(1 - D/P_E)^2 + 3 - 1/P_E]^{1/2} ; \tag{9.60}$$

$$P_E = p_+ / \Delta p, \quad D = 6\eta\mu U / (h^2 \Delta p) .$$

At electrical stress $P_E \gg 1$ from (5.60) it follows $R_E = 3 - 2\, D/P_E$. So the flow rate is restricted by the following maximum

$$U_0 < U_{max} = \mu E_0 (3 - D/P_E) \sim 3\mu E_0 ;$$

and the electrical pressure is determined as

$$p_e = 48\eta\mu V / h^2 . \tag{9.61}$$

Formulae (9.60), (9.61) show that with the increasing of liquid viscosity the flow rate will decrease while the electric pressure will increase. It is also seen that the flow rate may be significant (about $U \sim 3\mu E \sim 30$ cm/s at $E \sim 10$ kV/cm) while the electrical pressure for typical liquid parameters is very small.

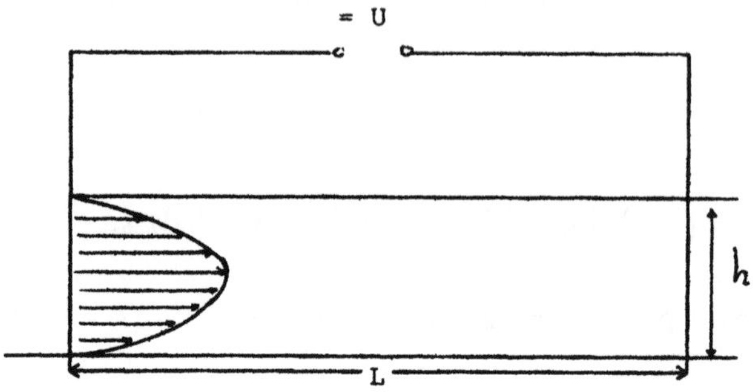

Fig. 9.2. The scheme of ion-drag pumping

References.

1. Melcher, J.R.: Continuum Electromechanics, M. I. Press, Cambridge, Mass. 1981.
2. Washabaugh, A.P., Zann, M., Melcher, J.R.: Electrokinetic pumping using travelling wave high voltages. Proceeding of the IEEE Industry Applications Society Annual Meeting, Pittsburg, Pennsylvania 1988.
3. Zhakin A.I.: Theory of electrical and electrohydrodynamical effects in low conductive media, Thesis, Kharkov University Press 1990 (in Russian).
4. Stutzer, O.M.: Ion-drag pumps, J. Appl. Phys., 31 (1960), 136.-146.
5. Sharbaugh, A.M., Walker, G.W.: The design and evaluation of an ion-drag dielectric pump to enhance cooling in a small oil-filled transformer, IEEE Trans. Ind. Appl. 21 (1983), 950-955.
6. Bryan, J.E., Seyed-Yagoobi, J.: Experimental study of ion-drag pumping using various working fluids, IEEE Trans. Electr. Ins. 26, N4 (1991), 647-655.
7. Richter, A.: EUSA, 4 (1990), 37-38.
8. Bogolubov, N.M., Mitropolsky, Yu. A.: Asymptotic methods in the theory of non-linear oscillations, Nauka Press, Moscow 1974 (in Russian).

PART III

CONDUCTION, INSTABILITIES AND BREAKDOWN IN LIQUID DIELECTRICS

P.K. Watson
Xerox Corporation, Webster, NY, USA

Abstract

Many of the liquids that are used in EHD processes are relatively impure, or doped materials, in which dissociable species give rise to mobile ions, and hence determine the electrical conductivity. Several factors play a role in the ion dissociation process, including the dielectric constant of the liquid, the concentration of the dissociable species, the electric field, ion recombination, and space charge effects. These various factors are analyzed in chapter 10 and their relative importance is examined. The majority of conduction experiments in liquid dielectrics are based on the use of metal electrodes, but such currents are emission limited, and are therefore more characteristic of the electrode than of the liquid. There are, however, methods by which space charge limited (SCL) currents can be generated in liquids, such as via ion-emitting electrodes. Another method, involving the use of an electron injection technique, is described in this lecture. Steady state and transient SCL currents can be generated by this technique; these are analyzed in chapter 11 and the results are used to interpret a range of experiments. Space charge limited currents can be generated in dielectric liquids by means of controlled injection processes, and above a critical level of injection the conduction process becomes electrohydrodynamically unstable. In chapter 12 we show that electrically driven instabilities have marked similarities with thermally driven convection processes, and there is an interesting relationship between the Rayleigh criterion for thermal instability and the corresponding EHD instability threshold. The growth rate of an EHD instability of an advancing space charge is derived and experimental results that illustrate this process are presented. In chapter 13 we analyse the role of EHD processes in the electrical breakdown of liquid dielectrics. Optical studies of pre-breakdown processes in insulating liquids enable one to identify several stages in the breakdown process. In the case of negative point/plane gaps the process starts with the creation of a rapidly expanding vapor cavity adjacent to the point electrode. As it expands, the surface of this cavity becomes unstable and one observes the growth of an EHD instability. The runaway growth of this instability leads to streamers that bridge the gap and cause the actual breakdown. By combining simple electrostatic and hydrodynamic concepts we can derive equations for cavity growth, the instability growth rate, and streamer velocity and these results are in reasonable agreement with experiment. Finally, in chapter 14 we consider electrospray and atomization. Electrospray ionization is an important technique for producing macroions in vacuo from large, complex molecular species in solution. The technique has extended the usefulness of mass spectrometric analysis to the large, fragile, polar molecules that play a vital role in biological systems. Among the distinguishing features of electrospray ions are their multiple charges, which result in coherent sequences of peaks in the mass spectrometer output, the peaks differing from one another by unit charge. The EHD aspects of ion production include droplet emission from the tip of a Taylor cone, and the evaporation of solvent from these droplets, leading to the Rayleigh instability of the droplets and their fission.

Chapter 10

A SURVEY OF CONDUCTION PHENOMENA IN LIQUID DIELECTRICS

10.1 Introduction

The purpose of this chapter is to survey the experimental data on electrical conduction in the types of insulating liquids that are encountered in EHD studies, and to introduce the underlying theory of conduction in weakly ionized dielectric media (sometimes referred to as "leaky dielectrics.")

It is probably a fair generalization to claim that almost all experimental work on electrohydrodynamics has been carried out on relatively impure liquids such as transformer oil, corn oil, unpurified silicone fluids, doped freon, etc. Indeed, many EHD processes rely on injected ionic species, or impurity ions, or ionized dopants. Our main concern in this lecture is therefore the study of conduction processes in impure, practical dielectric liquids. However, it is appropriate to acknowledge the extensive, elegant work that has been carried out on electronic and ionic conduction processes in pure liquids; these studies have been discussed previously in this volume.

10.2 The Relationship between Conductivity and Dielectric Constant

It is well known that the conductivity of dielectric liquids increases with increasing dielectric constant. This dependence arises because the electrostatic attraction between ion pairs decreases as dielectric constant increases -- behavior that was postulated in the 19th century in qualitative form as the Nernst-Thomson rule. Ions in liquids play a dominant role in EHD phenomena and it is therefore useful to begin this review of conduction processes in liquids with the relationship between dielectric constant and ion concentration in organic liquids.

An extensive survey was made by Sharbaugh about 35 years ago, in which he collected published data on dc resistivity and dielectric constant for about 70 nominally pure organic liquids from a variety of reference sources including the International Critical Tables, Weissberger and Proskauer's "Organic Solvents," Landölt and Bornstein Zahlenwerte, and the French Tables Annuelles Intern. de Constantes. This survey was later published by Sharbaugh and Barker [1]. No attempt was made to assess the reliability of the data, which is shown in Fig. 1. In spite of the non-uniformity of the test conditions implied by the diversity of the sources, there is a clear trend in resistivity versus dielectric constant.

As one would expect, the high dielectric constant, self-ionized liquids such as the alcohols and acids exhibit high conductivity; the aldehydes, ketones and ethers are more resistive, and the esters, hydrocarbons and halogen substituted hydrocarbons are good insulators. Low dielectric constant liquids do not self-ionize, so the source of their conductivity must be an extrinsic one. Sharbaugh argued that water is probably the main source of ions in liquids of low dielectric constant, ($\varepsilon'<10$), and he cited his own experimental work on hexane as evidence: when dried over sodium and measured in an enclosed system hexane has a resistivity $\rho \sim 10^{19}$ Ωcm; at the other extreme, hexane saturated with water has a resistivity, $\rho \sim 10^{14}\Omega$cm (the solubility of water in hexane is about 10^{-3} moles/liter). A reasonable assumption is, therefore, that practical grade liquids contain some water in solution. The published data are consistent with this assumption as may be seen from Fig. 1, which shows a theoretical curve, based on a hypothetical water concentration of 10^{-4} moles/liter.

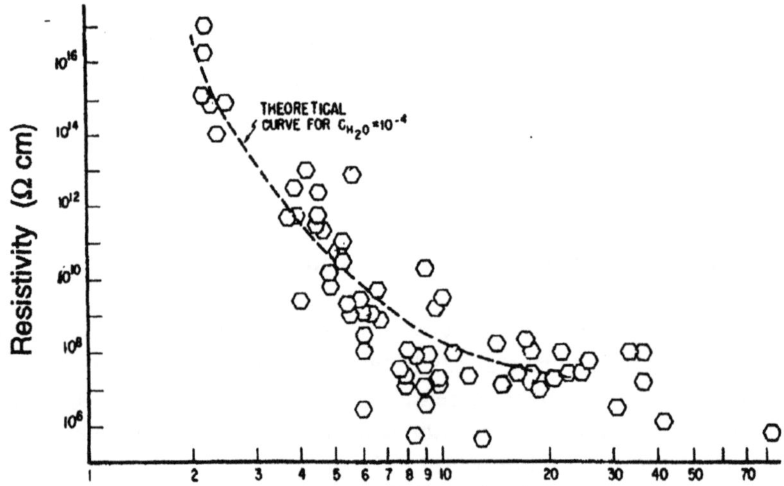

Fig. 1, Published values for dc volume resistivity vs dielectric constant. The dashed line shows the theoretical trend for ρ versus ε', assuming that water is a common contaminant at a level of 10^{-4} moles/liter (Sharbaugh and Barker [1]).

As further confirmation of the role of impurities in the conductivity of liquids we cite the work of Brière, Felici et al. [2] in which various liquids, including acetone, ethyl alcohol and nitrobenzene, were purified by ion-exchange and electro-dialysis; the resulting ultra-pure liquids exhibited resistivities some five or six decades higher than when purified by conventional techniques.

The relationship between resistivity and dielectric constant of impure or doped liquids may be seen most easily by applying the law of mass action to the equilibrium between the dissolved impurity and the dissociated ions,

$$\frac{\alpha_d^2}{(1-\alpha_d)}=\left(\frac{A}{N}\right)\exp\left(-\frac{U}{kT}\right) \tag{10.1}$$

where α_d is the degree of dissociation of the impurities into free ions, A is approximately constant, N is the number of impurity molecules per unit volume, U is the energy required to separate the ions, k is Boltzmann's constant and T is absolute temperature.

The ion generation process involves the separation of positive and negative ions in their Coulomb field in the medium of dielectric constant ε, so that $U = U_o/\varepsilon$, where U_o is the energy required to separate the charges in vacuo. For low values of ε (less than 20) it is known that $\alpha_d \ll 1$, so that $\alpha_d^2/(1-\alpha_d)\approx\alpha_d^2$.

$$\text{Therefore } \alpha_d \approx \sqrt{(A/N)}\exp(-U_o/2\varepsilon kT) \tag{10.2}$$

$$\text{Hence, } \log\rho \approx \left(a/\varepsilon\right)+b \tag{10.3}$$

where a and b are constants. The calculated line on Fig. 1 is of this form.

10.3 Field Dependence of the Ion Generation Process

The Sharbaugh-Barker model of the relationship between conductivity and dielectric constant does not address the details of the ion generation mechanism, which is best seen as a two-step process: (i) an initial, thermally induced ion pair separation, in which the ions, separated by a short distance r_o, are held together by their mutual attraction (these ions constitute a so-called geminate ion pair); this step is followed by (ii) a competition between complete separation of the geminate ion pair in the applied field versus the recombination of the pair due to their mutual Coulomb attraction.

This process was analyzed by Onsager in a classic paper [3] in which he showed that the problem of geminate ion separation vs recombination can be reduced to that of Brownian motion of the charge carriers in the presence of the applied field and the Coulomb field of the ion pair. After the initial separation step the geminate pair is thermalized while still bound by its Coulomb field; there is then a finite probability that this pair will separate into free carriers by diffusion, subject to the combined effects of the two fields. Onsager showed that in the limit of zero applied field the escape probability is

$$P(0) = \exp(-r_c/r_o) \tag{10.4}$$

where r_c is a separation distance at which the energy of Coulomb attraction of an ion pair is equal to kT, given by

$$r_c = e^2/4\pi\varepsilon'\varepsilon_o kT \tag{10.5}$$

Onsager's general solution to the carrier generation problem involves an infinite, double series in r_o, E and an angle θ expressing the orientation of the ion pair to the external field. In the low field limit the general equation simplifies to

$$P(E) = P(0)\left[1 + (e/kT)r_c\, E/2 + \ldots\right] \tag{10.6}$$

Equation 6 and the full Onsager equation have been extensively tested for both thermally generated and radiation generated carriers, and for both ionic and electronic carriers. Measurements of the field dependence and temperature dependence of carrier production and of conductivity in a variety of liquid dielectrics, molecular crystals, and amorphous photoconductors have confirmed the validity of the theory.

Details of the theory has been discussed by Zakhin in chapter 5. For the present, however, first-order theory will suffice to illustrate this important, field-dependent carrier-generation mechanism.

10.4 Equilibrium Concentration of Ions at Low Fields

The equilibrium concentration of ions depends on both the generation rate and the rate at which carriers are removed from the liquid. Three processes contribute to the removal of ions: their drift to the electrodes, their recombination in the bulk, and their diffusion to the walls and the electrodes.

In the low field limit the loss of ions due to recombination dominates over drift and diffusion, so, ignoring the loss to the electrodes and the walls, we can equate the generation and recombination rates. Writing the concentrations of positive and negative ions as n_+ and n_-

respectively, q, the generation rate and α_r the recombination rate constant we have, $q = \alpha_r n_+ n_-$. Now at low fields the concentrations of positive and negative ions must be equal, hence

$$n_+ = n_- = [q/\alpha_r]^{1/2} \tag{10.7}$$

There are two limiting forms to the recombination rate constant, that of Langevin, for liquids and high pressure gases, and that due to Thomson for low pressure gases. The Langevin form of the rate constant, assuming univalent ions, is

$$\alpha_r = e(\mu_+ + \mu_-)/\varepsilon \tag{10.8}$$

where μ_+ and μ_- are the positive and negative ion mobilities. These ions are usually of molecular dimensions, but there are exceptions, to be discussed in Section 5 below.

When a field is applied to the ionized liquid the ions drift towards the electrodes and space charge accumulates in the liquid adjacent to the electrodes. The analysis of this problem will be considered in the next section. If we restrict ourselves to fields that are so low that ion drift and space charge accumulation can be neglected, and with electrode and wall spacing large enough that diffusion effects are negligible, then current density in the univalent case is simply

$$J = \sum enE\mu = e(q/\alpha_r)^{1/2} E(\mu_+ + \mu_-) \tag{10.9}$$

This relationship between current and the square root of carrier generation rate is well known for weak electrolytes. A clear demonstration in the low dielectric constant limit is provided by the work of Hummel and Allen on hexane irradiated with 1.5 MeV x rays [4]. Their results, shown in Fig. 3, follow the expected square root relationship very closely, and it is interesting to note the authors' comments regarding the experimental conditions that were necessary to achieve this, "To avoid losing a considerable fraction of ions as measuring current and by diffusion to the electrodes one has to work at high radiation intensities, large plate separations and relatively low measuring voltages." This is consistent with the limits implied by the low-field model.

The decay of current following the removal of the source of ionization offers a good experimental method of measuring the ratio of recombination coefficient to mobility. From the definition of recombination coefficient,

$$dn/dt = -\alpha_r n^2 \tag{10.10}$$

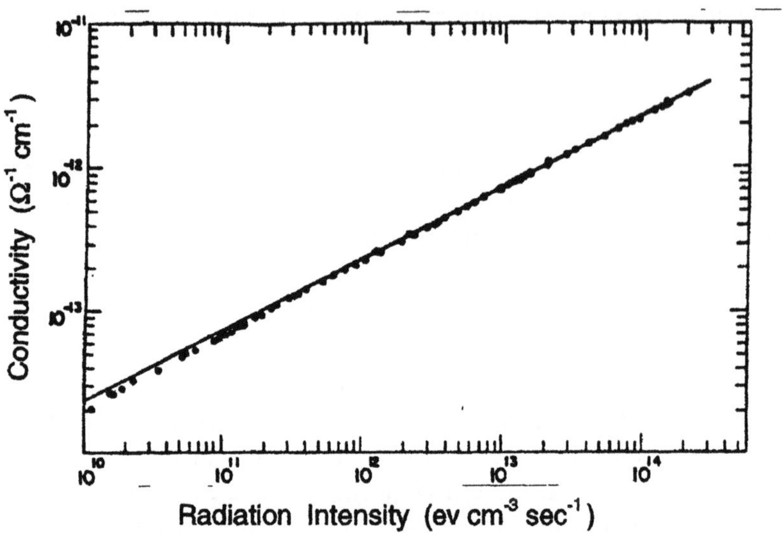

Fig. 2, Conductivity of n-hexane vs dose rate.
Sample irradiated with 1.5 MeV x rays (Hummel and Allen [4]).

$$\therefore 1/n - 1/n_o = \alpha_r t \tag{10.11}$$

where n_o is the number density of ions of each sign in steady state. Now, current density is $ne\mu$, where μ is the sum of the mobilities, μ_+ and μ_-. Hence

$$e\mu E\left[1/I(t) - 1/I_o\right] = \alpha_r t \tag{10.12}$$

Using this relationship to analyze their current decay results Hummel and Allen concluded that the ratio α_T/μ for n hexane is in the range 0.93 to 1.02×10^{-6} v.cm. This compares favorably with the theoretical value, $e/\varepsilon = 0.96\times10^{-6}$v.cm.

10.5 The Stability of Ions in Liquid Dielectrics

Some impure liquid dielectrics exhibit an anomalously high level of conductivity. Indeed, part of the motivation for the study of adventitious ions in non-aqueous media came from the electrical power industry where, many years ago, there was a problem with high dielectric loss in aging transformer oil. It was found that the source of the problem was the build up of metal soaps caused by corrosion in the transformers, and that these soaps were in a colloidal state; the question was, how could these colloidal structures be related to the increase in conductivity? The answer, which is related to the ion recombination process, has wide applicability, as will appear. In the Langevin model it is assumed that when oppositely charged ions drift together, they recombine. However, recombination will not occur if the energy of Coulomb attraction is less than kT at the point of closest approach of the ions. Thus the problem is related to the Onsager model in which the probability of escape of an ion pair, separation r_0, scales with exp $(-r_c/r_0)$, where r_c is the critical distance at which the electrical potential energy between the ions is equal to kT.

We take two extreme cases as examples of this relationship: in water at room temperature the critical distance r_c is about 0.7 nm, and ions that are stable in aqueous solution must be held apart at a greater distance than this; but, aqueous ions are surrounded by a shell of water molecules held in place by polarization forces, and it is this shell that stabilizes the ions against recombination. In contrast, the critical distance r_c in hydrocarbons is about 28 nm, and polarization forces are not normally capable of organizing parent molecules in structures of this size. As the ions are smaller than the critical size, the equilibrium discussed in Section 2 is set up between the ions and the undissociated species.

If now a long chain, polar solute such as a soap is added to the insulating liquid, these polar molecules will associate to form structures known as micelles, which may be disc shaped, or spherical, with the polar groups facing inwards and the hydrocarbon tails facing outwards. Because of their highly polar interior these micelles often contain polar impurities such as water, and this provides an environment in which ionization can occur; the micelle is then effective in stabilizing ions against recombination; thus micelles promote conductivity of what would otherwise be a highly insulating medium. Clearly this has important implications for the control of conductivity in insulating liquids.

Details of this subject are the province of the physical chemist, and a recent paper by Morrison [5] provides a comprehensive survey of the topic. For present purposes this brief survey will suffice to explain the remarkable stability of some types of ions in insulating liquids, and the ability of certain additives to enhance the conductivity of liquids of low dielectric constant.

10.6 Space Charge Effects in Conduction

So far we have restricted ourselves to relatively low fields, where the balance between carrier generation and recombination dominates over carrier drift. If however higher fields are applied to the liquid, so that an increasing number of ions escape recombination in the region adjacent to the electrodes, then space charge will accumulate in those regions and the local field will be enhanced. This type of problem was first examined at the beginning of this century in the context of conduction in ionized gases and in seeded flames (i.e., flames to which thermally ionizable salts have been added), and J. J. Thomson wrote a valuable treatment of the steady state solutions to many of these problems [6]. This analysis includes the first clear statement of what is now referred to as unipolar space charge limited conduction -- a topic that will appear repeatedly in the course of these lectures.

In the case of a uniformly ionized gas, or liquid, Thomson's equations describe the relationship between current density J, field E, carrier mobilities (μ_+ and μ_-), generation rate q, and recombination rate constant α_r. (The original equations are in cgs units with unit dielectric constant, but it is trivial to convert the equations to a rationalized system and explicitly include the dielectric constant ε.)

In the presence of a relatively strong field, space charge builds up adjacent to the electrodes in an ionized medium; in mid gap, away from the space charges, the field has its minimum value E_o, the region is net neutral ($n_+ = n_-$) and field is related to current density by Eq. (9) (Section 4 above). But this equation cannot hold in the regions adjacent to the electrodes; in the space charge regions, one sign of carrier dominates over the other, and drift dominates over recombination. To estimate the extent of the space charge regions, note that the length of the space charges cannot be less than the distance required to generate the carriers needed to sustain the current.

Thomson's analysis is best illustrated by the limiting case $\mu_+ = \mu_-$, (this is often the case when both negative and positive carriers are molecular ions). Then the thicknesses of the space charge regions are equal, and are of order J/qe (assuming that recombination is negligible in the high field regions); the field at each electrode is, approximately $(J/eq\mu\varepsilon)^{1/2}$ and the potential drop across the each space charge scale with J^2. In contrast, the potential drop across the space-charge-free region is proportional to J. The relationship between current and applied voltage therefore has the form,

$$V_A = AJ^2 + BJ \tag{10.13}$$

In general the term AJ^2 is the more important. Thomson's theory has been demonstrated for many examples of conduction in weakly ionized gases and liquids [7] though there are interesting departures from the theory at high fields.

References

1. Sharbaugh, A. H. and Barker, R. E., Ionic Impurity Conduction in Organic Liquids. Proc. Conf. on Conduction Processes in Dielectric Liquids; Grenoble, 1968. Also, J Polymer Sci C 10 (1965) 139.

2. Brière, G. and Felici, N., Compt. Rend. 261 (1965) 5097; ibid 259 (1964) 3237; ibid 261 (1965) 4400.
3. Onsager, L., Deviations from Ohm's Law in Weak Electrolytes, J Chem Phys 2 (1934) 599-615.
4. Hummel, A. and Allen, A. O., Ionization of Liquids by Radiation, J Chem Phys 44 (1966) 3426-31.
5. Morrison, I. D., Electrical Charges in Nonaqueous Media, Colloids and Surfaces, 71 (1993) 1-37.
6. Thomson, J. J. and Thomson, G. P., Conduction of Electricity through Gases, Cambridge Univ. Press (1928).
7. Denat, A, Gosse, B., and Gosse, J. P., Electrical Conduction of Solutions of an Ionic Surfactant in Hydrocarbons, J Electrostatics 12 (1982) 197-205.

Chapter 11

SPACE CHARGED LIMITED CURRENTS IN LIQUID DIELECTRICS

11.1 Introduction

The majority of conduction experiments in liquid dielectrics are based on the use of metal electrodes, but the metal-to-dielectric interface acts as a blocking contact, or at best a poor injector. Charge injection from metal electrodes does set in at high fields, but it occurs from microscopic asperities on the electrodes, so the emission is nonuniform. For the purpose of studying charge transport phenomena, therefore, experimental techniques are required that provide a uniform, copious source of charge carriers, and several approaches to this have been proposed: for example, Adamczewski [1], Freeman [2], Schmidt [3] Hummel and Allen and others have used x-rays to introduce excess carriers into dielectric liquids; Silver et al. [4] have shown that the tunnel junction can be used as an injecting electrode; members of the CNRS group have exploited the ion-exchange membrane as an injecting contact [5]. LeBlanc [6] used a photocathode to inject electrons into hydrocarbon liquids; Hartmann and Schmidlin [7] used a photoconductive layer as an injecting contact, and Watson, Schneider, and Till [8] have used an electron beam as an ohmic contact.

11.1.1 The Space Charge Limit

Before discussing some of these experimental results it is appropriate to ask, is there a limit to the current that can be carried by a dielectric? In the case of an aqueous electrolyte, for example, this is not a well-posed problem as the space charge is screened, but in a dielectric the field and the current are coupled via the space charge, and it is this coupling that gives rise to an important current-limiting effect.

There are, in fact, two limits to be considered, the emission limit (i.e., the maximum current that can be delivered by an injecting electrode at a given field) and the space charge limit (SCL). Consider the case of unipolar injection from a planar electrode into a dielectric ε, and suppose that injection can occur at low values of interface field, including zero. We then have two equations that describe the steady state, unipolar conduction process in the one-dimensional, planar case, with univalent ions of mobility μ, diffusion coefficient D, number density n(x), drifting in the space-charge field E(x),

$$\text{Poisson's equation,} \quad dE/dx = n(x)e/\varepsilon \tag{11.1}$$

$$\text{Current flow,} \quad J = n(x)e\mu E(x) - De\,dn/dx \tag{11.2}$$

For all cases of interest in EHD, diffusion is negligible, and hence we obtain

$$E(x) = \left[E_o^2 + 2Jx/\mu\varepsilon\right]^{1/2} \tag{11.3}$$

In the limiting case $E_o = 0$ at $x = 0$, the voltage across the dielectric, thickness L, is given by integrating Eq. (3). This gives $V = (2/3)\left[2JL^3/\mu\varepsilon\right]^{1/2}$ and hence we obtain the familiar form

$$J = (9/8)\,\mu\varepsilon V^2/L^3 \tag{11.4}$$

Another frequently occurring SCL situation is that of radial flow from a point contact, which can be conveniently approximated by a concentric geometry, with cathode radius r_c and anode radius r_A. In the usual case, $r_c \ll r_A$, and so the radial flow SCL current takes the form

$$I = (3\pi/2)\mu\varepsilon\left(V_A^2/r_A\right) \tag{11.5}$$

11.1.2 The Emission Limit

The question then is, how closely can currents approach the space charge limit for typical dielectric liquids and for various types of electrodes?

(i) To illustrate the emission limit for planar, metal electrodes consider the case of n-hexane -- a liquid that has been studied more extensively than any other liquid dielectric. Typical values of dc current density at high fields between planar electrodes, obtained by Goodwin and Macfadyen [9] are of order 10^{-9} amp/cm^2 at 220 kV/cm, L = 400 μm, V = 8.8 kV. House [10] reported 10^{-7} amp/cm^2 at 1,000 kV/cm, L~100 μm, V = 10 kV. The corresponding values of SCL current density are $2{\cdot}2\times10^{-4}$ amp/cm^2 and $1{\cdot}8\times10^{-2}$ amp/cm^2 respectively, so in these two examples the currents are a factor of 2×10^5 below the space charge limit.

These low values of current density occur partly because of depletion and polarization effects which are known to affect dc conductivity. Measurements under pulse conditions (ms and μs duration pulses of voltage) by Watson and Sharbaugh [11] indicate that at a field of 1·4 Mv/cm currents within one order of magnitude of the space charge limit can be observed, but at lower fields (8×10^4 v/cm) the current is a factor of 4×10^5 below the space charge limit.

(ii) In striking contrast to the planar electrode case, point emitters are quite capable of sustaining currents at a level close to the space charge limit, as shown by Halpern and Gomer [12] in their studies of field emission from finely etched tungsten emitters into liquid H_2, D_2 H_e, A_r, N_2, O_2 and C_6H_6. In this work they observed high currents that were consistent with Eq. (5), and from this they were able to deduce the mobilities of the charge carriers in these various pure liquids.

(iii) An interesting variant on the planar electrode case has been investigated by Denat, Gosse, and Gosse [13]. They studied the conductivity of hydrocarbons into which electrolytes such as tetramethyl phenylene diamine (TMPD) and tri-isoamylammonium picrate (TIAP) had been dissolved. Dissociation occurs preferentially on the electrode surfaces, and because of their large size the ions are desorbed at relatively low field. Thus ions of the sign corresponding to the electrode are injected, and current transients are observed at a level approaching the space charge limit.

(iv) Ion exchange membranes have been used extensively as injecting electrodes by the team at CNRS, Grenoble and emission currents close to the space charge limit are reported [5].

(v) Watson, Schneider, and Till [8] have used electron beams as ohmic contacts to films of liquid dieletrics. In this experimental technique there is no upper electrode on the sample, but electrons are injected through the free surface of the dielectric, and become trapped to form a reservoir of negative ions that constitutes a virtual cathode. By this means one is able to generate steady state or transient space charge flows in insulating liquids and this method has been used to measure charge carrier mobilities in liquids of low vapor pressure; this same technique has also been used to study the EHD behavior of these fluids [8], and this will be the topic of a subsequent chapter. In these experiments the injected electrons lose their energy in multiple collisions with the electrons of the molecules of the fluid, and after a short distance (0.15 μm for 2.2 keV electrons) they become thermalized; then, depending on the trapping environment the electrons may move a short distance before being deeply trapped to form negative molecular ions. These ions constitute the source of the charge carriers and of the space charge that drives the conduction process.

11.2 Steady State SCLC and the Measurement of Ion Mobilities

The apparatus used in these electron injection experiments is shown in Fig. 1. In the absence of bulk ohmic conductivity the conduction is entirely determined by the injected space charge, and this satisfies one of the conditions of space charge limited conduction. It is also necessary to maintain a condition of zero field at the liquid surface if the resulting conduction process is to be at the space charge limit, and to accomplish this an electrometer is used to monitor the surface potential, and to bias the system to the required zero-field condition. In the apparatus shown in Fig. 1 an injection beam is used to create the space charge, and a second electron beam is used to monitor the surface potential of the liquid film and to provide feedback to the reference grid, which maintains the required zero-field condition at the liquid surface. The thickness of the liquid is measured optically. From these measurements of current density, surface potential and liquid film thickness, one obtains the information that specifies the conductivity of the liquid. The equipment is operated at a vacuum of about 10^{-6} torr, and this restricts the types of fluid to those of low vapor pressure, so the majority of these tests have been conducted on silicone fluids. A convenient, well characterized homologous series of methyl-phenyl siloxanes was available, and in this section measurements of the conductivities of three of these liquids are described. A typical set of data is shown in Fig. 2 in which the values of surface potential and sample thickness are plotted as a function of current density for one of these siloxane fluids.

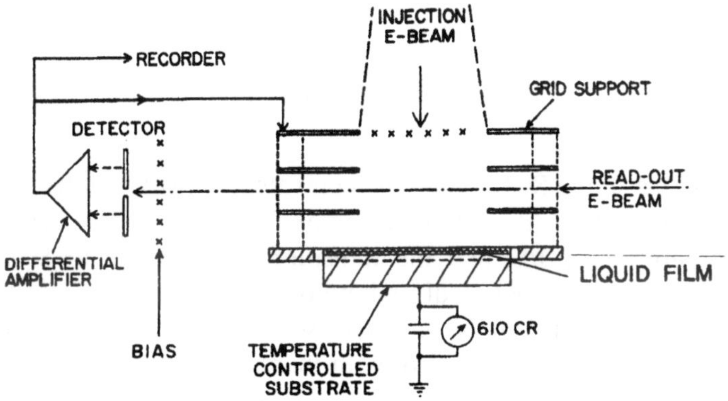

Fig. 1, Apparatus showing the location of the two electron beams with respect to the liquid film. The charging beam operates in a raster scan mode, in either constant current, or pulsed mode. Beam energy 2.2 kV.

In a conventional conduction measurement using metal electrodes one thinks of voltage and sample thickness as the controlled variables; here, in contrast the injected current is controlled, and surface potential and sample thickness are dependent variables. An analysis of data of the type shown in Fig. 2 shows that at low current densities the conduction is ohmic, due to impurity ions in the liquid. At higher currents there is a transition from ohmic behavior to space charge limited conduction.

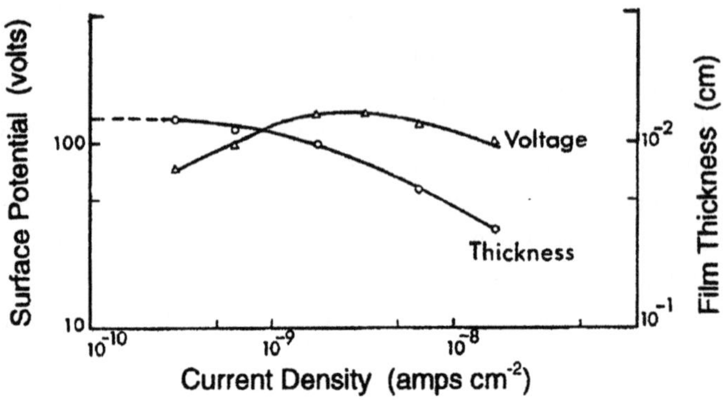

Fig. 2, Surface potential and sample thickness vs current density for a methyl phenyl siloxane fluid.

In order to emphasize this transition from ohmic to SCL conduction we rewrite Eq. (4) in the form equivalent resistivity,

$$E/J = \left[8L/9\mu\varepsilon J\right]^{1/2} \tag{11.5}$$

This equation shows that in the SCL region the equivalent resistivity falls with increasing J/L, so that log E/J versus log J/L yields a straight line of -½ slope. At low currents, in the ohmic region, E/J is independent of J/L, and the transition from ohmic to SCLC occurs at the injection level at which J/e is equal to qL, where q is the generation rate for impurity ions. Now, ion generation rate q is thermally activated, so both q and the transition current should increase with increasing temperature. Also, with increasing temperature the viscosity of the liquid decreases, and this leads to an increase in ion mobility and a decrease in SCL equivalent resistivity; both these effects are visible in Fig. 3. The mobility of the injected ions can be calculated from the space charge limited currents, and measurements of this type have been made on three members of this homologous series of methyl phenyl siloxanes (n = 2, 7 and 9) over the temperature range 0 to 50°C. In each case the results follow the same behavior, ohmic at low J/L and space charge limited at high J/L. From these results the charge carrier mobilities have been calculated, and in Fig. 4 the results are plotted versus viscosity.

If the charge carrier is an ion, then it should obey Walden's rule, which states that the mobility of an ion varies inversely as the viscosity of the liquid. In Fig. 4 we have plotted the data for carrier mobility vs liquid viscosity for three siloxane fluids and we find that the carrier mobility can be expressed in the form $\mu \propto 1/\eta^m$, where m ranges from 1.07 to 1.37. Although these exponents show departures from the theoretical value of unity, the magnitudes of the mobilities are consistent with the hypothesis that the charge carriers are negative molecular ions.

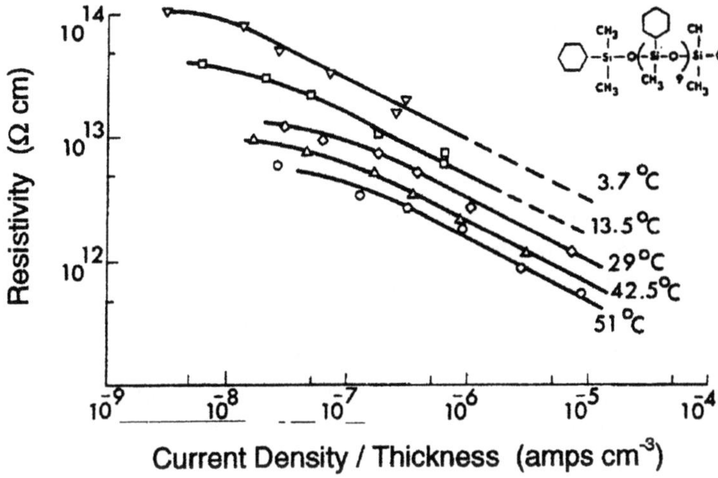

Fig. 3, Equivalent resistivity vs J/L for a methyl phenyl siloxane fluid (n = 9). At low current densities the liquid approaches ohmic behaviour (resistivity independent of current). At higher current densities the impurity carriers are saturated and the liquid is driven into the SCL region (resistivity varies as $(J/L)^{-1/2}$).

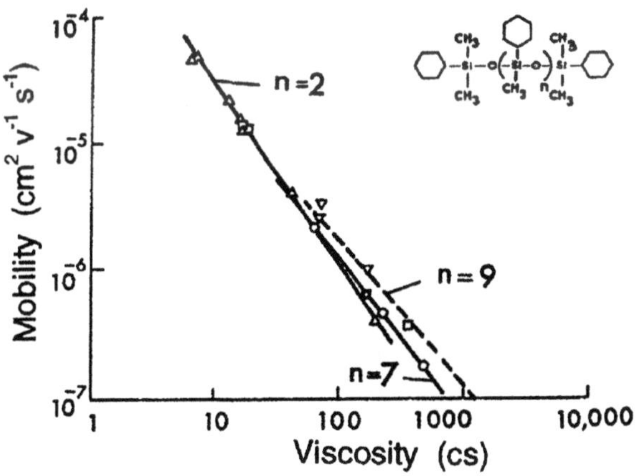

Fig. 4, Charge carrier mobility vs viscosity for three members of a methyl phenyl siloxane homologous series.

11.3 Transient Space Charge Limited Currents

(i) The Charging Transient: Under steady state conditions, the equations for electric field, E(x), and space charge distribution, q(x), in unipolar SCL conduction are given by:

$$E(x) = \sqrt{2Jx/\mu\varepsilon} \quad and \quad q(x) = \sqrt{J\varepsilon/2\mu x} \tag{11.6}$$

where J is the current density, μ the carrier mobility, and ε the permittivity of the dielectric. Now, consider the charging characteristics of an insulating film into which current is being injected - either by an electron beam or by corona discharge; there is no externally applied potential, so the charge carriers drift in their own space charge field. Following the initiation of current, the space charge propagates into the dielectric behind a front which is characterized by a position coordinate λ(t), and after time t there is an amount of charge, Jt, within the dielectric, occupying the volume between $x = 0$ and the space charge front at $x = \lambda$. In the region $\lambda \leq x \leq L$, therefore, the field is Jt/ε, so the velocity of the front is μJt/ε, and its position coordinate is given by

$$\lambda(t) = \mu J t^2 / 2\varepsilon \tag{11.7}$$

The field is increasing from $x = 0$ to $x = \lambda$ and the carrier concentration decreases in that same interval. It is not obvious at first sight, but the field distribution in the region $0 \leq x \leq \lambda$ is identical to the equivalent steady state SCL distribution: consider an element or slice of the space charge as it drifts across the gap; it is driven by the space charge behind the element, and is spread out by its own mutual repulsion. If the element leaves the cathode at time $t = t_o$, its subsequent position x is given by the equation

$$(t - t_o) = (2\varepsilon x/\mu J)^{1/2} \tag{11.8}$$

The charge Δq in the element Δx spreads by mutual repulsion, so the front of the element travels at a greater velocity than the rear by the increment Δu, where $\Delta u = \mu\Delta E = \mu\Delta q/\varepsilon$. The spread of the element is then given by

$$\Delta x = \Delta u(t - t_o) = (\mu\Delta q / \varepsilon)(t - t_o) \tag{11.9}$$

The space charge density is then

$$q(x,t) = \Delta q/\Delta x = \varepsilon/\mu(t - t_o) \tag{11.10}$$

Eliminating ($t-t_o$) gives the relationship between space charge density and position, $q(x)=\sqrt{J\varepsilon/2\mu x}$. This is identical to the steady state distribution, and it demonstrates that the charge distribution behind the front is independent of time and maintains the steady state value. This leads to the constant-current charging characteristic

$$V(t)=\frac{3}{2}\left(\frac{t}{\tau}\right)\left[1-\frac{1}{3}\left(\frac{t}{\tau}\right)^2\right]V_S \tag{11.11}$$

Here V_S is the steady state surface potential, and τ the transit time of the carrier front, defined by $V_S=\sqrt{8JL^3/9\mu\varepsilon}$, and $\tau=\sqrt{2\varepsilon L/\mu J}$. Typical charging transients, such as Figure 5, show good agreement with Eq. (11).

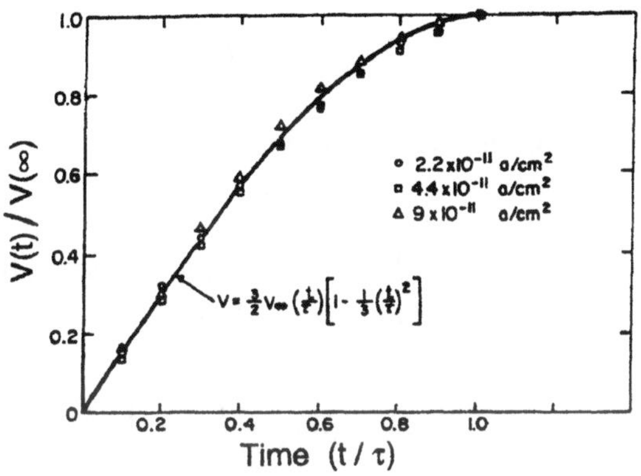

Fig. 5, Typical charging characteristics for constant current injection, in the absence of EHD effects. Comparison between theoretical curve (Eq. 11) and experimental results, which are normalized to the steady state voltage V(∞) and transit time τ.

(ii) Charge Decay from an Initial Distribution: To solve this problem, we require a relationship between the local charge density q(x,t) and the local field, which in turn is related to the integral of the charge density in the layer [15]. We define the charge between the surface, x = 0, and the plane x as Q (x,t). The field at (x,t) is Q (x,t)/ε, and the current is given by μq (x,t) · Q (x,t)/ε where q (x,t) is given by ∂Q(x,t)/∂x. Then from conservation of charge we have

$$\partial Q(x,t)/\partial t=-(\mu/\varepsilon)Q(x,t)\,\partial Q(x,t)/\partial x \tag{11.12}$$

The simplest, non-trivial solution to this equation is

$$Q(x,t) = \varepsilon x/\mu t; \text{ hence}, q(x,t) = \varepsilon/\mu t \tag{11.13}$$

This indicates that a uniform space charge is a stable distribution, so that once a uniform distribution is established, the decaying space charge retains that form. This result may be used to predict the voltage transient which occurs when a layer of thickness L is given an impulse of charge at t = 0. The initial charge distribution can be thought of as being uniformly distributed in a thin layer near the liquid surface, and by the above reasoning as the space change expands, the charge distribution remains uniform. Thus an initial potential, V_1=QL/ε will decay linearly to a value ½V_1 in a time τ_1 = εL/μQ. Once the expanding, uniform space charge fills the entire gap, the potential decay assumes a hyperbolic form, and this part of the potential decay curve is described by

$$V(t) = V_1\tau_1/2t \quad for \quad t \geq \tau_1 \tag{11.14}$$

(iii) Charge Decay from an SCL Distribution: Equation (12) is a form of Burger's Eq. (16), which can be solved explicitly. In its most general form this wave equation combines both nonlinear propagation and diffusive effects, but in the form we have to solve the diffusivity term can be neglected. Then, following Witham [16], the initial condition $Q(x,0) = [2\varepsilon x / \mu]^{1/2}$ leads to the required transient solution,

$$Q(x,t) = J\left[\left[(2\varepsilon x / \mu J) + t^2\right]^{1/2} - t\right] \tag{11.15}$$

Charge density, given by $q(x,t) = \partial Q / \partial x = (\varepsilon / \mu)\left[(2\varepsilon x / \mu J) + t^2\right]^{-1}$, is plotted in Fig. 6 as a function of x/L, with t/τ as a parameter and with the curves normalized to the magnitude of the space charge density at x = L, t = 0. The singularity at the cathode at t = 0 arises because we have neglected diffusion in our analysis. It is apparent from the curves that after a relatively short time (t > 2τ) the space charge density approaches a uniform value, and as a result the field becomes linear in x. (As already noted, other initial space charge distributions also collapse to a uniform distribution after a short time). The potential across the discharging layer is obtained from Eq. (15),

$$V(t) = V_s\left\{\left[1+\left(\frac{t}{\tau}\right)^2\right]^{3/2} - \frac{3}{2}\left(\frac{t}{\tau}\right) - \left(\frac{t}{\tau}\right)^3\right\} \tag{11.16}$$

where V_s and τ are defined as in Eq. (11). Experimental results show good agreement with this equation. A useful approximation to Eq. (16) was proposed by Schneider and Watson [15] based on an equivalent circuit; this equation has the form

$$V(t) = V_s\left[1 - \frac{2t}{\tau}\right]^{-1} \tag{11.17}$$

this approximate equation is 5% below the exact solution at $t/\tau \sim 0.3$, crosses Eq. (16) at $t/\tau = 0.9$, and is 8% above Eq. (16) at $t/\tau = 2$.

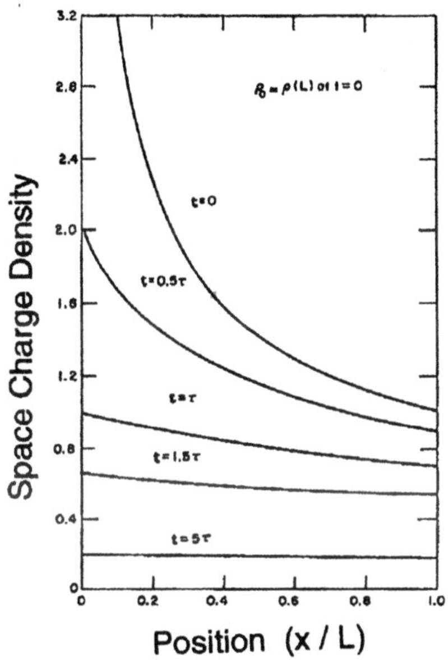

Fig. 6, Charge decay from a steady state SCL distribution. It is assumed that at t = 0 current injection ceases, and the space charge then decays. Note that the space charge is normalized to the initial value at the collector electrode. Also, the field at x/L = 0 is maintained at zero throughout.

References

1. Adamczewski, A., Ionization, Conductivity and Breakdown in Liquid Dielectrics, Taylor and Francis, 1969.

2. Dodelet, J. P. and Freeman, G. R., Electron Mobilities and Ranges in Liquid Ethers, Canad J Chem 53 (1975) 1263-74.
3. Schmidt, W. F., Nonpolar Liquids Ch. 9, Photoconductivity and Related Phenomena, Eds. J. Mort and D. M. Pai, Elsevier 1976.
4. Silver, Onn, Smejtek and Masuda, Hot-Electron Injection into Insulating Liquids from a Tunnel Junction, Phys Rev Lett, 19 (1967) 626-30.
5. Atten, P. and Gosse, J. P., Transients of One-Carrier Injection in Polar Liquids. J Chem Phys 51 (1960) 2804-11.
6. LeBlanc, O. H., Electron Drift Mobility in Liquid n-Hexane, J Chem Phys 30 (1959) 1443.
7. Hartmann, G. C. and Schmidlin, F. W., Transient Photostimulated Charge Transfer to an Insulating Liquid. J Appl Phys 46 (1975) 266-278.
8. Watson, P. K., Schneider, J. M., and Till, H. R., Electrohydrodynamic Stability of Space Charge Limited Currents in Dielectric Liquids, Phys Fluids 13 (1970) 1955-61.
9. Goodwin, D. W. and Macfadyen, K. A., Electrical Conduction and Breakdown in Liquid Dielectrics, Proc Phys Soc 66 (1953) 85-96.
10. See Lewis, T. J., the Electric Strength and High-Field Conductivity of Dielectric Liquids, Progress in Dielectrics, Volume 1, pp. 97-140, Eds. Birks, J. B. and Schulman, J. H., Wiley, 1959.
11. Watson, P. K. and Sharbaugh, A. H., High Field Conduction Currents in Liquid n-Hexane under Microsecond Pulse Conditions, J Electrochem Soc 107 (1960) 516-521.
12. Halpern, B. and Gomer, R., Field Emission in Liquids, J Chem Phys (1969) 51, 1031-1047 and 1045-1056.
13. Denat, A., Gosse, B., and Gosse, J. P., Electrical Conduction of an Ionic Surfactant in Hydrocarbons, J Electrostatics 12 (1982) 197-205.
14. Watson, P. K. and Clancy, T. M., Electron Injection Technique for Investigating Conduction Processes in Insulating Liquids. Rev Sci Instrum, 36 (1965) 217-222.
15. Schneider, J. M. and Watson, P. K., Transient Space Charge Limited Conduction in Liquid Dielectrics, 1970 Annual Report, Conf. on Electrical Ins. and Diel Phenom. (Nat. Acad. Sci. 1971) page 125-132.
16. Whitham, G. B., Linear and Nonlinear Waves, Wiley-Interscience, 1974.

Chapter 12

UNIPOLAR INJECTION: EHD ENHANCED SPACE CHARGE LIMITED CONDUCTION IN LIQUID DIELECTRICS

12.1 Introduction

Electrohydrodynamic processes can play a major role in ionic transport phenomena in liquids. Effects of this type were first studied by Avsec and Luntz [1] and later by Steutzer [2] who was interested in developing a dielectric pump. Ostroumov [3] seems to have been the first to analyze conduction processes in ionic liquids in the light of convective effects. Evidence of convective instabilities can be seen in the work of Watson, Schneider and Till [4] on electron-beam charged liquids; in these studies an electron beam was used to inject charge into the free surface of a thin film of liquid and the surface potential of the sample was measured as a function of the injected current. The results indicate that for sufficiently high current densities the liquid becomes EHD unstable and convection-enhanced conduction occurs. These experiments will be discussed below. Felici, Atten and their co-workers at CNRS, Grenoble [5-7] have investigated EHD phenomena over the past 30 years or so, and they have made extensive use of ion-exchange membranes as injecting electrodes into polar dielectrics such as nitrobenzene. By means of the Kerr electroptical effect they have been able to observe field distortion due to the charge injection process and to keep track of the resulting EHD effects.

Two types of electrical force can be distinguished that give rise to EHD effects in liquid dielectrics: the Coulomb force qE that the field E exerts on charge density q, and the dielectric or polarization force associated with the permitivity gradient $\Delta\varepsilon$. Two classes of problems can then be defined, depending on whether these forces are restricted to an interface, or applied to the bulk. As we are concerned with relatively low fields (10^4 to 10^5Vcm^{-1}) we can disregard the polarization force. We also neglect interfacial effects, and concentrate on Coulomb forces operating on bulk space charges. In this case when the liquid is set in motion changes occur in the distribution of space charge, and this modifies the field. Thus, the coupling between force and the resulting motion is non-linear, which complicates the analysis

[8], but there are ways of attacking such problems that sacrifice analytical rigor yet retain physical meaning, and it is this heuristic approach to the threshold of instabilities and to growth rate that concern us in this chapter.

12.2 Convective Instabilities

12.2.1 Thermal Convection

The first convective effects to be studied in detail were thermally driven processes such as the well known Bénard cells that develop when a layer of fluid is heated from below [9]. Most of the pioneering work on instability theory was done on thermal convection, and some of these ideas can be carried over to the electrical case, though the analogy is not exact. The similarity between the cell structure in electroconvection and thermal convection can be seen by comparing Figs. 1a and 1b; the former was obtained by corona charging the free surface of a 1 mm layer of DC-703, a methyl phenyl siloxane fluid. Figure 1b is a reproduction of a photograph from Bénard's paper of 1901, and is of a layer of spermaceti, heated from below. In Bénard's experiments the cells were made visible by suspended particles; in more recent work of this type the cells are photographed by interference optics [10]. What is not evident from Bénard's work is that there is a critical condition, an instability threshold, that specifies the onset of convective cellular motion.

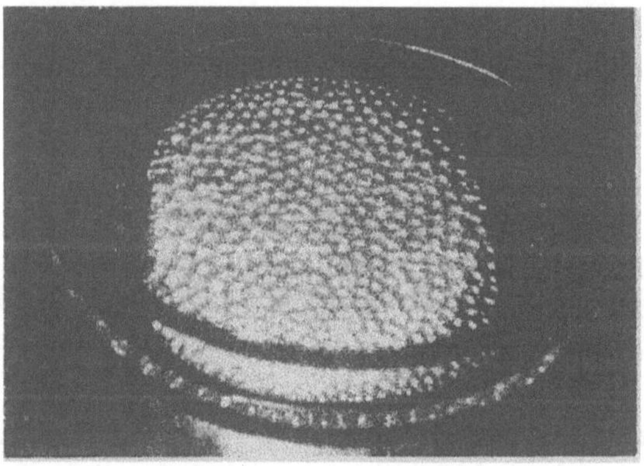

Fig. 1a, Convection cells in a 1 mm film of DC 703 charged by positive corona.

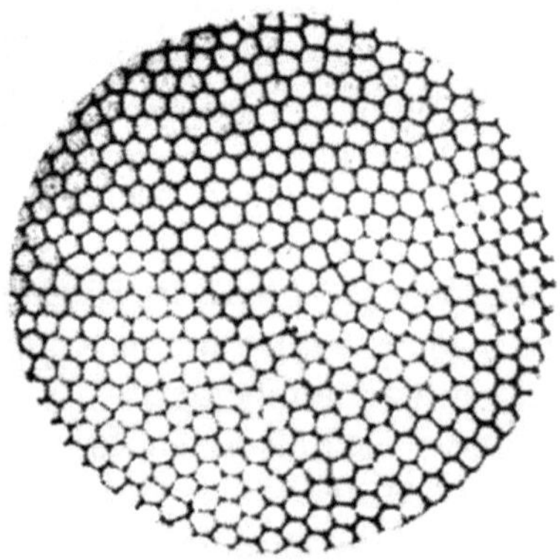

Fig. 1b, Bénard cells in spermaceti. The cells are made visible by particles in suspension. (From one of Bénard's original photographs.)

The analysis of instability phenomena, stimulated by Bénard's work, started with Rayleigh's paper of 1916, "Convection currents in a horizontal layer of fluid when the higher temperature is on the underside" [11]. The technique developed by Rayleigh is to set up the equation of viscous motion for the fluid, couple this to the equation for heat transport through the film, solve for equilibrium conditions, then examine the effect of a small sinusoidal departure from equilibrium and ask whether the disturbance grows or decays. The onset of growth then defines the onset of instability. Rayleigh's conclusion was that a fluid layer, thickness h, subject to an adverse thermal gradient $\beta = dT/dh$ is stable at low heat flux, but becomes unstable above a critical value given by

$$g\frac{\alpha\beta h^4}{\kappa\nu} \geq \frac{27\pi^4}{4} = 657.5 \qquad (12.1)$$

where ν, κ and α are the kinematic viscosity, thermal diffusivity and thermal expansion coefficient respectively. The dimensionless constant in Eq. (1) is now referred to as the Rayleigh number, R_a. The critical value that determines the onset of instability depends on the boundary conditions: for free upper and lower boundaries (the case analyzed by Rayleigh) $R_a = 657.5$; for one free and one fixed boundary $R_a = 1100$, and for both boundaries fixed R=1708. (There is an excellent treatment of the Rayleigh-Bénard problem by Chandrasekhar [12].)

A few years after Rayleigh's work, Low [13] introduced an interesting heuristic treatment of the criterion for thermal instability. Low was attempting to explain the regular geometric arrangements of rocks and pebbles found in the Arctic, such as the stone polygons of Spitsbergen, in terms of Bénard cells in the semi-liquid layer above the permafrost; he was basing his argument on the fact that water has a maximum density at 4°C, so the top, warmer layer will try to exchange with the colder fluid below -- the inverse of the usual unstable thermal gradient. Low pointed out that Eq. (1) can be rewritten in terms of density gradient, rather than thermal gradient, making use of the identities, $\beta h=\Delta T$, and $\alpha\beta h=\Delta\rho/\rho$, where ΔT and $\Delta\rho$ are the temperature difference and density difference across the layer. Substituting these values in Eq. (1), Low postulated that if the energy released on inverting the layer is greater than some critical value then the layer is unstable. (It is now generally agreed that Low's mechanism is not the one that is responsible for the stone polygons, but the idea is a stimulating one.)

12.2.3 EHD Instabilities

Following Low's argument, Watson and Schneider [8,14] showed that the Rayleigh equation (Eq. (1)) can be regrouped and used to derive the corresponding EHD stability criterion for space charge limited conduction in a fluid layer. This criterion states that if the electrostatic energy released on inverting a layer is greater than the viscous energy loss in the fluid in one relaxation time then the layer is unstable. In the thermal case the appropriate relaxation time is of order h^2/κ, whereas in the electrical case the relaxation time is $h^2/\mu V$. The rate of viscous energy loss in both cases scales with ηh per unit time. The electrostatic stored energy released on inverting an SCL space charge is $\Delta U_e = \left(\frac{3}{8}\right)\left(\varepsilon V^2/h\right)$. Combining these three terms we derive the appropriate stability criterion,

$$\frac{\varepsilon V_c}{\mu\eta} = R_e \sim 40 \tag{12.2}$$

Here V_c is the critical voltage for instability; the stability parameter $R_e \sim 40$ is for one fixed and one free boundary. It is interesting to compare this equation with the analytical result obtained by Schneider, $\varepsilon V_c/\mu\eta = 99$, for the instability criterion for one free and one fixed boundary [8]. A similar result was also obtained by Atten [6]. The general agreement between the approximate and the exact solutions emphasizes the parallel between the thermal and electrical instabilities, even though the conduction mechanisms are very different.

12.2.4 Convection-Enhanced Transport

The analogy between thermal and electrical convection can be extended by comparing the magnitudes of heat transfer through a fluid layer, and of space charge conductance in a layer of insulating liquid. The thermal case is illustrated in Fig. 2, taken from the work of Silveston [15]. In this figure the vertical axis is the heat flux per unit thermal conductance (the Nusselt number Nu) and the horizontal axis is the Rayleigh number, a dimensionless parameter proportional to thermal gradient. There is an abrupt break at the critical Rayleigh number, below which heat is entirely carried by thermal conduction, and above which the heat transfer is enhanced by the convective motion of the fluid. Clearly there is excellent agreement between the theoretical and experimental values of R_{crit}. Also, there is a marked increase in heat transfer coefficient Nu with increasing R_a above the critical Rayleigh number.

The corresponding electrical situation is best seen in the case of space charge limited conduction. In order to provide the required ohmic contact to the dielectric we have used an electron beam to inject charge into the free surface of the liquid layer; the injected electrons rapidly become thermalized, creating ion pairs as they do so, and then become trapped to form negative molecular ions. This thin layer of ions with its excess negative charge constitutes the ohmic contact to the fluid. There is no external field applied to the sample, the excess negative charge provides the internal field, and in the absence of bulk fluid motion the velocity of the charge carriers is determined by the space charge field (diffusion can be neglected under the conditions of these experiments). Then, provided the concentration of

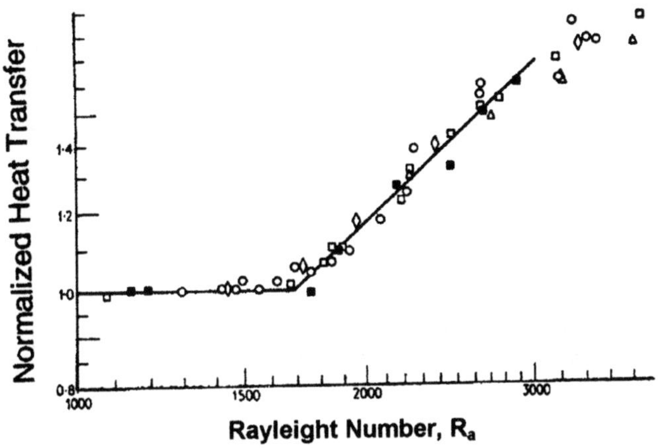

Fig. 2, Onset of thermally driven convective instabilities. Silveston's data [15] in the vicinity of the instability threshold, for a range of liquids including silicone oils, heptane and water. In all cases the instability sets in close to the theoretical Rayleigh number, 1708.

impurity ions is negligible compared to the injected carrier concentration, the system is driven into a region of pure space limited conduction, and the current is described by the familiar equation

$$J = (9/8)\mu\varepsilon V^2/L^3 \tag{12.3}$$

where J is the current density, V is the surface potential, ε is the permitivity and μ is the ion mobility. Rearranging this equation we can define the SCL conductance JL^3/V^2 which scales with με, and is therefore constant for a given liquid, provided mobility is constant.

Experimental results presented in an earlier lecture show the way surface potential and film thickness vary with current density. Similar sets of data, obtained with the silicone fluid DC550, are replotted in Fig. 3 in the form JL^3/V^2 versus surface potential, V. These results show that when the surface potential is low (i.e., when injected current is small) the conductance is constant, so the ion mobility is constant, and convection is therefore negligible. Above the critical voltage (about 135 V in this case) the conductance increases as voltage increases; this increase in conductance implies increasing carrier mobility and this is entirely due to bulk convection of the fluid. The similarity in form between Figs. 2 and 3, showing the onset of convection-enhanced conductance at a critical point, emphasizes the close parallel between these two classes of convection phenomena, in spite of the fact that the mechanisms of conduction are very different.

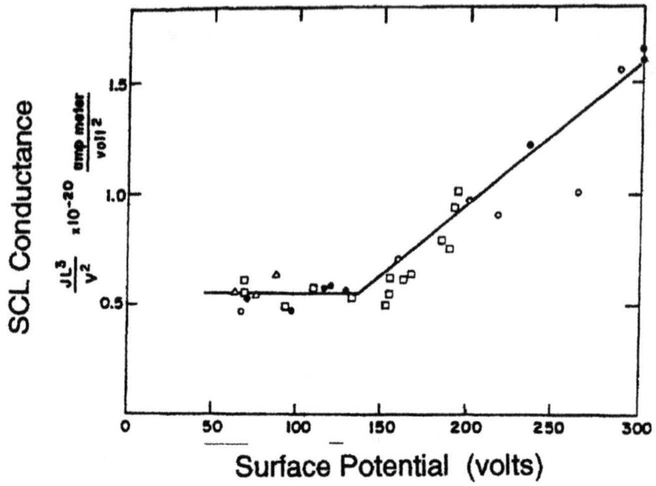

Fig. 3, Variation of SCL conductance with surface potential.
At low values of V_s (low injected currents) the conductance is constant (i.e., mobility is constant). Above the critical potential $V_c \sim 135V$ the conductance increases with V_s, indicating that charge transport has been enhanced by convection.

12.3 Time Dependence of EHD Instabilities

12.3.1 Charging Transient with Convection

A liquid film carrying a space-charge-limited current should be stable provided the surface potential does not exceed the critical value $V_c = 99\ \mu\eta/\varepsilon$. Therefore, if the surface voltage has not reached V_c when the advancing space charge front reaches the collector electrode the fluid layer will remain in its non-convecting state. At high currents, the potential across the space charge will exceed V_c before the carrier front reaches the counter electrode, and a convective instability can then build-up before steady state is attained. We have confirmed experimentally that at high currents the liquid layer can become unstable before the space-charge front crosses the gap [4]. In such cases a well-defined peak in the surface potential occurs before the transit time for space charge front; as the instability develops the conductance of the space charge layer increases and voltage passes through a maximum. Figure 4 shows typical charging transients: at $1\text{x}10^{-9}$ amp and below (i.e., surface potentials of 100 V or less) there is no evidence of convection; at higher currents the surface potential passes through a well-defined peak as the convective component builds up. We can explain the location of the voltage peak by the plausibility argument that the instability becomes dominant when the potential across the space charge, V_λ exceeds the critical potential V_c, and from this we obtain a relationship between current density and the convective time delay t_2.

$$J^2 t_2^3 / \varepsilon\eta \cong 300 \tag{12.4}$$

At low currents (i.e., at values of surface potential less than V_c) we should observe the build-up of a non-convecting SCLC, with transit time $\tau = (2\varepsilon L/\mu J)^{1/2}$, so the time from current onset to steady state voltage should be related to current density by $J\tau_1^{\,2} = C_1$. Here C_1 depends only on thickness and carrier mobility. A typical set of time delay measurements taken from transient such as Fig. 4 are shown in Fig. 5. The two types of behavior predicted by the model are clearly evident in the figure. At low currents the results follow the equation $J\tau^2 = C_1$, and at higher currents the convection model is followed (i.e., $J^2\tau^3 = C_2$). The break between the two types of behavior occurs at 4×10^{-6} A/cm^2, and we identify this as the point at which the transition between space-charge-limited conduction and convection-assisted conduction occurs. However, in spite of the general agreement between the experimental data and the threshold model, this method of approach does not deal with the rate at which the instability grows, and for many purposes that is as important as the instability threshold. We now turn to another approximate method that does deal with the growth rate, and provides insight into the form of the instability.

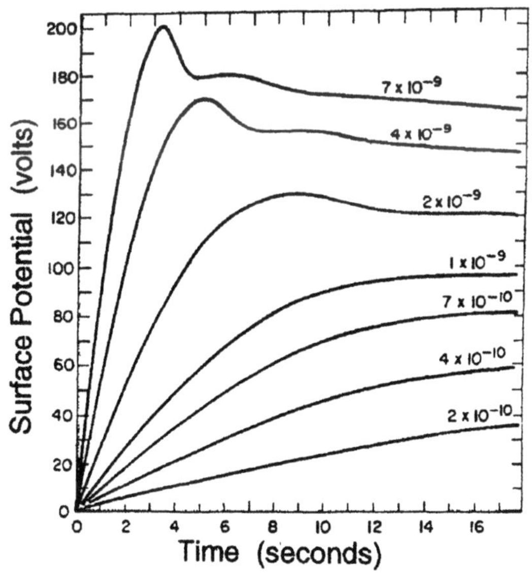

Fig. 4, Typical charging transients in electron beam experiments: DC-703, film thickness 0.45 mm, area 16 cm^2. At 1×10^{-9} amp and below there is no evidence of convection. At higher currents the voltage passes through a peak as the convective component builds up.

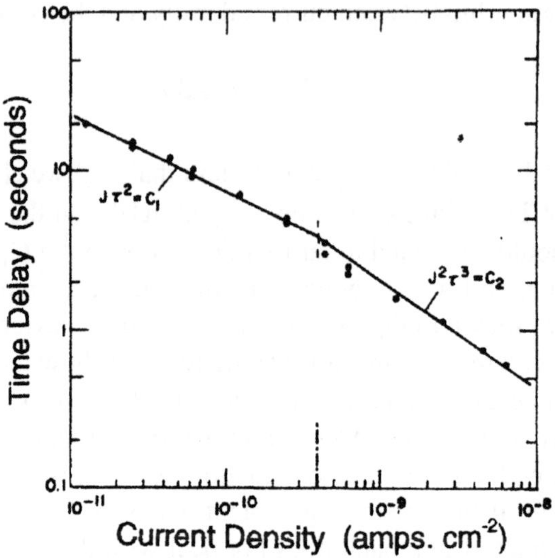

Fig. 5, Variation of time lag with current density for DC-703. Time lag in the period from initiation of current to the steady state (as in Fig. 4 at currents equal to and below 1×10^{-6} amp) or the voltage peak (as in Fig. 4 at 2×10^{-6} amp and above).

12.3.2 Growth of an Instability in Advancing Space Charge

To illustrate this type of growth rate calculation we consider the stability of an advancing space charge. We suppose that there is a perturbation in the carrier front of the form shown in Fig. 6, in which the position of the front λ is moved incrementally by the amount λ' in area A, consistent with the convection cell shown. Fluid flows into the volume $A(\lambda+\lambda')$ and carries space charge with it so that the perturbed volume gains space charge at the expense of the adjacent area. Now the average charge per unit volume behind the front is $Jt/\lambda = 2\varepsilon/\mu t$ so the net increase of charge in the volume subtended by A is approximately $\lambda' A\ 2\varepsilon/\mu t$. This leads to an increase in field in area A and a corresponding decrease in the adjacent area. Thus the perturbation in position gives rise to a differential increase in field, $E(\lambda)(\lambda'/\lambda)$, and a corresponding increase in force per unit area.

$$F' = \tfrac{1}{2}\varepsilon\left[E^2(1+\lambda'/\lambda)^2 - E^2\right] \cong \varepsilon E^2(\lambda'/\lambda) \qquad (12.5)$$

This extra force F′ will cause the perturbed area of the front to advance at an incrementally faster pace than the adjacent area; there will then be an extra component of viscous loss associated with the perturbation, and this loss will limit the rate at which the perturbation grows. The force F′ can be most simply related to dλ′/dt by means of an equation due to Frenkel [16], showing that the change in length of a viscous, cylindrical region is related to force by

$$F' = 3\alpha\eta A \tag{12.6}$$

where the coefficient $\alpha = (2/\lambda)(d\lambda'/dt)$, A is the cross sectional area, λ the length, and η the viscosity. If now we treat the volume Aλ in Fig. 6 as the region under stress we can solve for λ′(t). (Of course there will be other viscous losses in the fluid associated with the perturbation, but in the spirit of this approximate calculation we consider only leading terms.) Then from Eqs. (5) and (6) we have

$$\varepsilon E^2(\lambda'/\lambda) = (6\eta/\lambda)\,d\lambda'/dt \tag{12.7}$$

Now E = Jt/ε, and so $(J^2t^2/6\eta\varepsilon)dt = d\lambda'$. Integrating this equation gives the required equation for the growth of the perturbation

$$\lambda'(t) = \lambda'_0 \exp\left(J^2t^3/18\eta\varepsilon\right) \tag{12.8}$$

Two interesting conclusions follow from this equation: (i) For the instability to be observable, the amplitude λ′ must reach a critical amplitude λ'_c, and this occurs at time τ. Current and growth time are therefore related by $\left(J^2\tau^3/18\eta\varepsilon\right) = \text{const.}$ This is identical to Eq. (4), which we have shown to be in agreement with experimental results, as in Fig. 5.

(ii) At the instability threshold a perturbation must have grown to its critical size λ'_c when the carrier front reaches the collector electrode; substituting for ion transit time $L^2/\mu V$ in the equation $\left(J^2\tau^3/18\eta\varepsilon\right) = \text{const.}$, and eliminating J be means of Eq. (3) we obtain the criterion for the EHD instability threshold, $(\varepsilon V/\mu\eta) = \text{const.}$ This is identical to Eq. (2). The fact that we are able to derive these two well established results from this approximate calculation lends credibility to this method of approach.

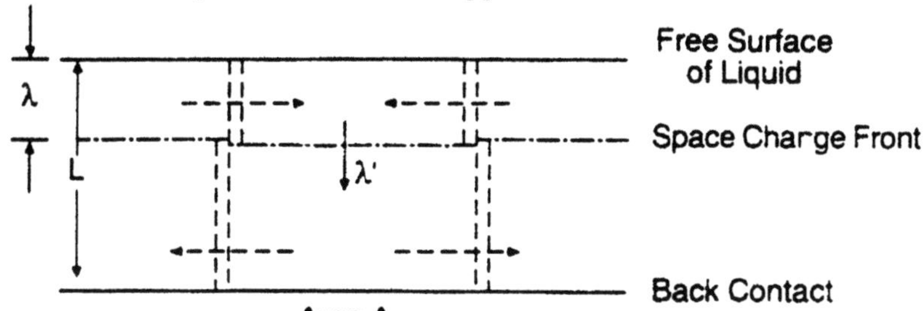

Fig. 6, Advancing space charge front in a liquid dielectric, showing a hypothetical perturbation, λ′ in the location of the front.

References

1. Avsec, D. and Luntz, M., Compt. Rend, 203, (1936), 1140, and 204, (1937) 470.
2. Steutzer, O. M., J App Phys, 30, (1959) 984, and 31, (1960) 136.
3. Ostroumov, G. A., Soviet Phys JETP, 14, (1962) 317.
4. Watson, P. K., Schneider, J. M., and Till, H. R., Phys Fluids, 13, (1970) 1955.
5. Felici, N., Rev. Gén Elect, 78, (1969) 717.
6. Atten, P., CR Acad Sci, 283B, (1976) 29.
7. Atten, P. and Lacroix, J. C., J. Méchanique, 18, (1979) 469.
8. Schneider, J. M. and Watson, P. K., Phys Fluids, 13, (1970) 1948.
9. Bénard, H., Ann de Chimie et de Physique, 23, (1901) 62.
10. Koschmieder, E. L. and Pallas, S, G., Int Heat Mass Transfer, 17, (1974) 991.
11. Rayleigh, Lord, Phil Mag, 32, (1916) 529.
12. Chandrasekhar, S., Hydrodynamic and Hydromagnetic Stability Chapter II, Oxford, 1961.
13. Low, A. R., Nature, 115, (1925) 299.
14. Watson, P. K. and Schneider, J. M., Ann Rep. 1969 Conf. on Elec. Ins. and Diel. Phenom., p. 97.
15. Silveston, P. L., see ref. 15, p. 69.
16. Frenkel, J., J. Physics (USSR), 9, (1945) 385.

Chapter 13

THE ROLE OF EHD PROCESSES IN THE ELECTRICAL BREAKDOWN OF LIQUID DIELECTRICS

13.1 Introduction

Our understanding of the electrical breakdown of liquid dielectrics depends largely on the use of pulsed schlieren and shadowgraph optical techniques to photograph the pre-breakdown events. Experimental work of this type began 35 years ago with the pioneering studies of Hakim and Higham [1] and of Farazmand [2], who were the first to publish schlieren photographs that optically resolved pre-breakdown events in liquids. Studies of this type were continued by Chadband and Wright [3] and many others, and a considerable amount of photographic data of this type is now available.

From such high speed schlieren and shadowgraph observations, electrical breakdown can be seen as a sequence of hydrodynamic processes involving the growth and distortion of a vapor cavity in the electric field, leading to streamers that cross the gap and cause the actual breakdown. The initiation of the pre-breakdown cavity is associated with a nanosecond current pulse (probably an electron avalanche in the high field region near the point cathode), and this initiating pulse is followed by a train of nanosecond pulses of somewhat regular spacing that accompanies the expansion of the cavity. A study of electrical breakdown in liquids should therefore focus on both electronic processes that initiate the cavity and the electrostatic and hydrodynamic effects that accompany the rapid expansion of the cavity, but it is the hydrodynamic and EHD processes that give rise to the characteristic form of the pre-breakdown events that are observed optically.

In this chapter we shall confine our attention to the negative point/plane geometry, voltages near breakdown, and short gaps. It is fortuitous that discharges in point/plane gaps are more amenable to analysis than discharges between parallel plates; not initially, of course, but once an ionized cavity is formed and space charge builds up, the point/plane configuration lends itself to approximations that are a great help in the analysis. By means of these approximations we can derive the scaling laws that describe the various steps in the breakdown process; our aim is then to derive these scaling laws, and to check them against the experimental data in order to test our understanding of the underlying physical processes.

13.2 The Negative Point, Pre-Breakdown Cavity

A typical sequence of flash shadowgraphs, shown in Fig. 1, illustrates the growth of a pre-breakdown cavity in a negative point gap, in 2.0 cSt, DC-200 silicone fluid [4]. From photographic observations of this type, and other evidence, it is clear that the formation of a vapor cavity is a critical step in the breakdown of liquids. Current measurements and simultaneous shadowgraph observations indicate that cavity formation is associated with a nanosecond pulse of current. The shadowgraphs also show that the cavity expands rapidly in a few microseconds, and that the expansion is field-dependent.

13.2.1 Cavity Expansion and Collapse

During the early stages of expansion the cavity approximates a constant kinetic energy system. Surface tension and viscosity effects are small and may be neglected. For an incompressible fluid $v(r)r^2$ = const. and the kinetic energy of the fluid surrounding a spherical cavity is given by $U = 2\pi\rho v^2R^3$ where R is the cavity radius, v = dR/dt and ρ is the fluid density. If the kinetic energy of the system is constant, U_0, then $(dR/dt)^2 = U_0/(2\pi\rho R^3)$, with the initial condition $R_0 = 0$,

$$\therefore R(t) = \left[\frac{5}{2}\left(\frac{U_0}{2\pi\rho}\right)^{0.5} t\right]^{0.4} + R_0 \cong \left[\frac{U_0}{\rho}\right]^{0.2} t^{0.4} \tag{13.1}$$

Two features emerge from this equation, (i) the rate of cavity growth decreases with time and (ii) cavity radius during the initial expansion is proportional to $U_0^{1/5}$, so the expansion is rather insensitive to initiating energy. As the cavity expands it does work against the external pressure P_a. Thus we can write the energy balance equation for a system with initial kinetic energy U_0.

$$U_0 = 2\pi\rho v^2 R^3 + (4/3)\pi R^3 P_a \tag{13.2}$$

The expansion of a vapor cavity ceases when the kinetic energy of the system passes through zero, at which point the internal pressure is far below ambient; the bubble is then fully expanded ($R = R_m$) and the external pressure P_a causes the bubble to contract. To describe the contraction we follow Rayleigh [5] and equate the work done by the external

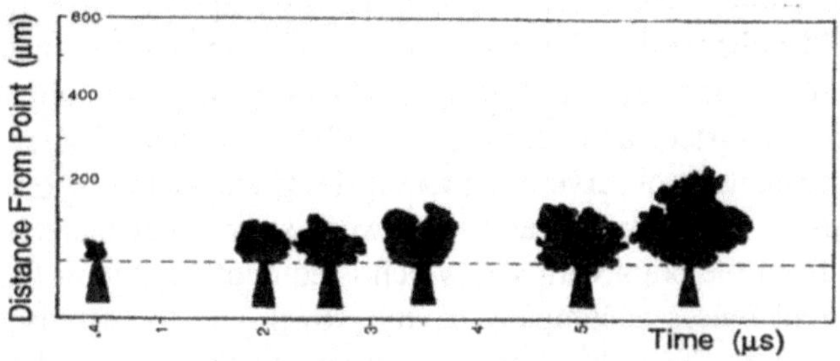

Fig. 1, Shadowgraph flash photograph of pre-breakdown cavities in 2.0 cSt DC-200, voltage 12 kV negative point-to-plane, showing cavity growth with time [4].

pressure, $(4\pi P_a/3)(R^3_m - R^3)$ to the kinetic energy of the fluid and hence derive the equation for velocity as a function of cavity radius. The integration of this equation gives the time for bubble collapse $\tau \cong 0.915\ R_m\ (\rho/P_a)^{0.5}$. By symmetry τ is also the time for bubble expansion, so the lifetime of a bubble is approximately $1.8\ R_m(\rho/P_a)^{0.5}$. This equation can also be expressed in terms of the energy of the cavity by making use of the identity $U_0 = (4/3)\pi R_m{}^3 P_a$. Hence, bubble lifetime is proportional to $U_0{}^{1/3}$. The generation of these cavities and their subsequent expansion and collapse has been studied by Kattan, Denat, and Lesaint [6] and by Qureshi, Chadband, and Watson [7]. These studies show that the energy to generate the cavity is provided by a current pulse from the point cathode. The cavity then behaves as a constant energy system, provided there is no further input of energy from the field; in the absence of any additional driving force the cavity closely follows the Rayleigh model.

13.2.2 Growth of a Cavity in an Electrostatic Field

If a cavity is to expand beyond the size set by the kinetic energy limit, some other factor must provide the driving force for expansion. Two possibilities come to mind: the vapor pressure of the liquid, and the force generated by the electrostatic field. After a cavity has been generated, only a relatively small amount of heat is required to evaporate more liquid, and hence to drive the expansion process. Therefore, if vapor pressure is the main driving force, then liquids of higher boiling point (i.e., lower vapor pressure) should show a much reduced rate of cavity growth. Experimentally, however, cavity expansion is not a strong function of liquid volatility. On the other hand, cavity expansion increases markedly with applied voltage. It therefore follows that vapor pressure is not an important factor driving the cavity expansion, and we adopt as a working hypothesis the idea that the electrostatic field provides the main driving force.

In order to analyze the electrostatic force, assumptions have to be made about the electrostatic field, the charge distribution in the vapor cavity and the space charge distribution in the liquid. Chadband and Wright [3] suggested that the region might be an ionized plasma, which would make the surface an equipotential with the cathode. Kelley and Hebner [8] used Kerr effect measurements to observe the internal fields associated with pre-breakdown events in point/plane gaps in nitrobenzene, and their work shows that there is a roughly spherical region, extending from the point cathode, which is equipotential with that electrode. Now, it is known that current pulses accompany cavity growth and these pulses are identified with discharges in the cavity; the field in the cavity and in the liquid adjacent to the cavity is therefore constantly changing, the cavity field rising between pulses and collapsing with each pulse.

The general problem relating applied field to cavity expansion is therefore a difficult one to analyse, but a limiting case can be defined that leads to a useful scaling law: We assume that the vapor in the cavity is sufficiently ionized that its surface is equipotential with the cathode throughout the expansion, and that the effect of the negative space charge which must be present in the fluid will have a negligible effect on the local field. These approximations enable one to reduce this complex problem to a simple form, and by this means we can set an upper bound to the electrostatic force acting on the cavity wall. To balance this force, kinetic energy is delivered to the liquid by the expanding cavity. To calculate the electrostatic pressure, we need to derive the equation for the electrostatic field at the cavity surface. The on-axis field at the surface of a spherical conductor, radius R, center distance a above a ground plane is given to a good approximation by the field in a hyperboloid/plane configuration,

$$E(R) \cong \frac{V}{R \ln 2(a/R)^{1/2}} \tag{13.3}$$

Now in the limited range of interest ($5<a/R<100$), the term $\ln 2(a/R)^{1/2}$ can be approximated to better than 5% accuracy by $(a/R)^{1/4}$ so the electrostatic pressure is given by

$$P_{es} = (1/2)\varepsilon\varepsilon_0 E^2(R) \cong (1/2)\varepsilon\varepsilon_0 \frac{V^2}{\left(a^{1/2}R^{3/2}\right)} \tag{13.4}$$

The work done by the electrostatic force in expanding the cavity is obtained by integrating P_{es} over the swept volume. Equating this work to the kinetic energy and integrating, with the initial condition $R = 0$ at $t = 0$, gives the required equation for cavity radius vs time.

$$R(t) \cong \left[\frac{2\varepsilon\varepsilon_0 V^2 t^2}{a^{1/2} \rho} \right]^{2/7} \tag{13.5}$$

According to this relationship cavity radius should scale with $(Vt)^{4/7}$. Extensive measurements have been made on a series of DC-200 silicone fluids [4] and some of these results are shown in Fig. 1. The values of cavity radius predicted by Eq. (5) have the correct magnitude for these fluids from 0.65 to 10 cSt viscosity at voltages near the threshold of breakdown in short gaps.

Considering the approximations involved in setting up this model, the agreement with experimental results is surprisingly close. In part, the analysis must have benefited from a few self-canceling errors, but it is also a result of the relative insensitivity of cavity growth to energy input, as discussed in section 2.1.

13.3 EHD Instability of the Vapor-Liquid Interface

In our discussion of the expansion of the pre-breakdown cavity, in order to define the field at the interface we have assumed that the cavity field has collapsed to zero (clearly this is an oversimplification, but the general agreement between theory and experiment indicates that the approximation is justified--at least to first order). Thus the expanding cavity wall is similar to an interface driven by a perpendicular field. Such an interface is unstable, and we expect to see evidence of an EHD instability in shadowgraph images of expanding cavities. Evidence can be seen of irregular interfaces in Fig. 1, and two unusually regular instabilities are shown in Fig. 2.

The case of a field-driven interface with laterally mobile charges has been considered by Melcher and Smith [9], Kath and Hoburg [10], and others. The Kath-Hoburg version of the theory is in the most convenient form for our purpose.

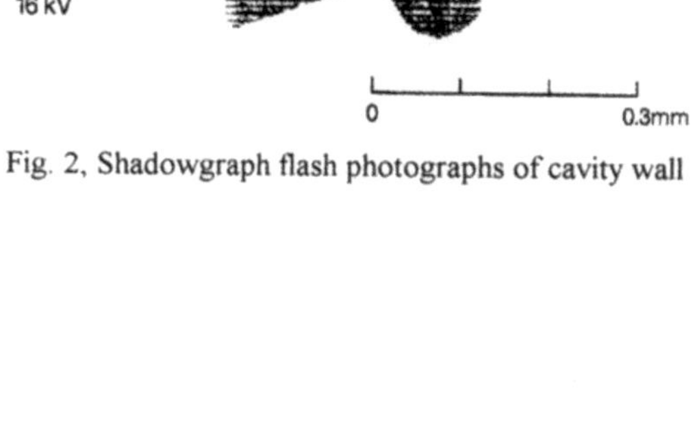

Fig. 2, Shadowgraph flash photographs of cavity wall

instabilities in 2 cSt and 1000 cSt DC-200.

13.3.1 The Kath-Hoburg Instability

Two time constants define this instability, $\tau_v = \rho/k^2\eta$, and $\tau_e = 2\eta/\varepsilon E^2$, where ρ is density and η is viscosity. By means of these time constants Kath and Hoburg distinguish two limiting cases, an inertia-dominated regime characterized by $\tau_v \gg \tau_e$ and a viscosity-dominated case, $\tau_v \ll \tau_e$. For small amplitudes the instability grows as $A(t) = A_0\exp(st-ikx)$ where s is the growth rate and k is the wavenumber. In the inertial limit the growth rate is given by

$$s_i = k\left[\varepsilon\varepsilon_0 E^2 / 2\rho\right]^{1/2} \tag{13.6}$$

In the viscous limit, growth rate is independent of wavenumber and is given by

$$s_\eta = \varepsilon\varepsilon_0 E^2 / 4\eta \tag{13.7}$$

An order of magnitude calculation shows that for most pre-breakdown cavities $\tau_v \gg \tau_e$, so the instabilities start in the inertial regime. Then, at first sight, the proportionality between s_i and k seems to imply a growth rate that increases without bound as k increases (i.e., shorter wavelengths propagate). However, as k increases viscous losses increase and there is therefore a limit beyond which k cannot increase without driving the instability into the viscous regime. This transition between inertial and viscous regimes has been analyzed by Castellanos and González [11]. They point out that the dispersion relation for this instability is unusual in that it does not have a maximum (i.e., there is no preferred wavenumber). They also show that the dispersion relation can be approximated by extrapolating from the two extreme cases, with the lines crossing at the wavenumber k_c given by

$$k_c = \left[\varepsilon\varepsilon_0 E^2 \rho / 8\eta^2\right]^{1/2} \tag{13.8}$$

This equation defines the boundary between the inertial and viscous regimes, and although it is not strictly a preferred wavenumber it does represent the condition of fastest growth. Substituting for k_c in Eq. (6) we obtain the magnitude of the growth rate,

$$s_c = \varepsilon\varepsilon_0 E^2 / 4\eta \tag{13.9}$$

This growth rate is equal to the viscous limit case, Eq. (7). Clearly this is an oversimplification but it provides a useful, approximate solution, with the following reservation: Castellanos and González point out that an interfacial tension--even a small, transient value--will have the effect of locking in the wavenumber to a preferred value. But in the case of pre-breakdown cavities there will be a high surface charge on the cavity wall which reduces the surface tension, and may even drive it negative, so, in the absence of information about the effective surface tension we take Eqs. (8) and (9) as the appropriate values for the cavity wall instability.

13.3.3 Experimental Values of EHD Wavenumber and Growth Rate

We estimate the experimental values of wavenumber from enlargements of pre-breakdown flash shadowgraphs, such as Figs. 1 and 2. Two particularly clear examples of developing instabilities are shown in Fig. 2, but in general the wave structures are much less uniform than these examples. Wavelengths scaled directly from prints of this type are used to generate the wavenumbers plotted in Fig. 3.

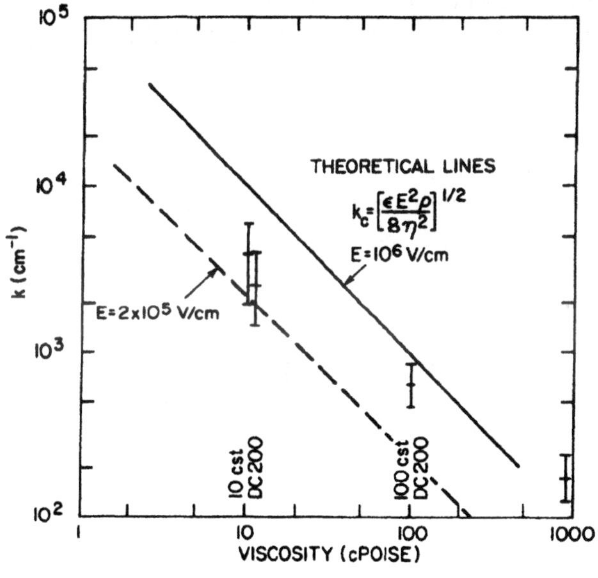

Fig. 3, Wavenumber of cavity wall instabilities in viscous fluids. Data obtained from prints such as Figs. 1 and 2.

Comparing the results shown in Fig. 3 with Eq. (8) we see that the experimental data for 10, 100 and 1,000 cSt silicone fluids are consistent with this relationship (though a better fit would be $k \propto \eta^{-2/3}$). The corresponding results for low viscosity fluids are however anomalous. The trend of results for low viscosity fluids indicates that the wavenumbers for these fluids are smaller than the predicted values (i.e., the preferred wavelengths are longer than predicted). Indeed for n-hexane and 0.65 cSt DC-200 there is more than an order of magnitude discrepancy between the experimental and theoretical values of wavenumber.

A likely cause of this anomaly is the presence of space charge in the liquid adjacent to the vapor/liquid interface, which will tend to screen out the high frequency components of the instability [12]. We know that charge carriers are readily injected through a vapor/liquid interface (though there will be shallow trapping of charge at the interface due to polarization effects) and that the injected charge will reduce the field at the interface.

As a result of this shielding effect, if a disturbance starts to propagate at a wavelength shorter than the space charge length the corresponding field component will be screened by the space charge; longer wavelengths will be less affected and the shortest wavelength that can propagate will be of the same order as the thickness of the space charge layer ($\lambda \cong \mu E \tau$) where μ is the negative-ion mobility and τ is the pulse length. Now ion mobility varies inversely with viscosity, so k_s should be proportional to η. This limiting trend for k_s is shown in Fig. 4, calculated for typical values of μ, E, and τ.

In high viscosity fluids the instability is not affected by this space charge screening effect because the screening distance is less than the preferred wavelength (i.e., $k_s > k_c$). However there is an intermediate viscosity at which k_s and k_c are equal, and at this viscosity the instability wavenumber should exhibit a maximum. Based on this conjecture the instability wavenumber should go through a maximum of order 10^4 cm^{-1}, and this should occur at about 2 cSt viscosity. Experimental results follow this general trend, as shown in Fig. 4, with the maximum located at 4 cSt, and a wavenumber of 5×10^3 cm^{-1}.

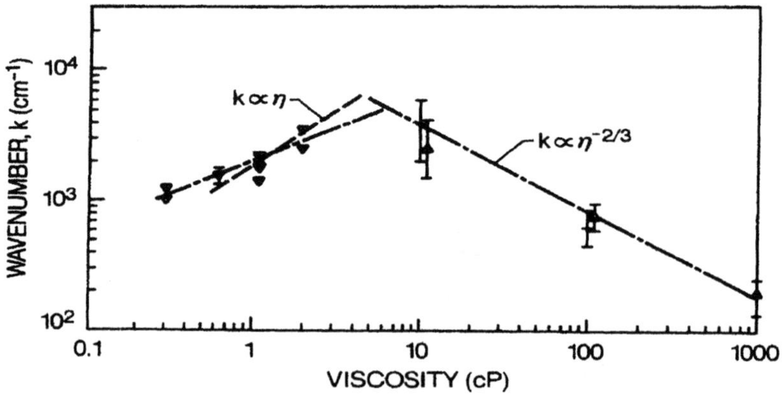

Fig. 4, Wavenumber of cavity wall instability. At low viscosities the wavenumber increases with viscosity due to space charge screening.

We now turn to the question of the growth rate of the instability. Kath and Hoburg's analysis is based on a two-dimensional geometry in which the applied field is constant. But as we have shown, the cavity interface field is significantly reduced by space charge in the liquid; moreover as the cavity expands, the geometric field decreases. Also, the assumption of laterally mobile charges is questionable as the interface charges are weakly trapped. We cannot hope to do better therefore than an order of magnitude agreement between experiment and theory.

Growth rates are difficult to measure experimentally. Ideally one needs a sequence of high speed photographs showing the growth of instabilities in time, but the only available experimental data [4] are restricted to single-shot photographs, one from each of a large number of experiments, so the results are subject to a large scatter. A set of experimental results showing the instability amplitude vs time for 2 cs DC-200 is shown in Fig. 5. The scatter in the results is considerable, as expected, but there is a clear trend in the results, roughly following the expected form, $A = A_0 \exp st$. From this we estimate the magnitude of the growth rate to be about $10^6 s^{-1}$.

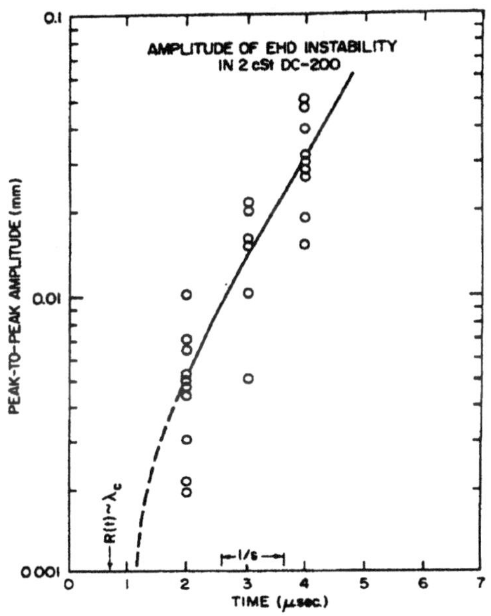

Fig. 5, Amplitude of cavity wall instability in 2 cSt DC-200. The general trend of the data indicates a growth rate of about $10^6 s^{-1}$.

In these experiments the applied voltage was 11.5 kV, so that at 3 μsec the cavity diameter (based on Eq. (5)) should be about 0.13 mm, hence the interface field is $10^8 vm^{-1}$, (ignoring space charge). The wavenumber, $k \cong 3\times10^5\ m^{-1}$, so Eq. (9) predicts a growth rate of

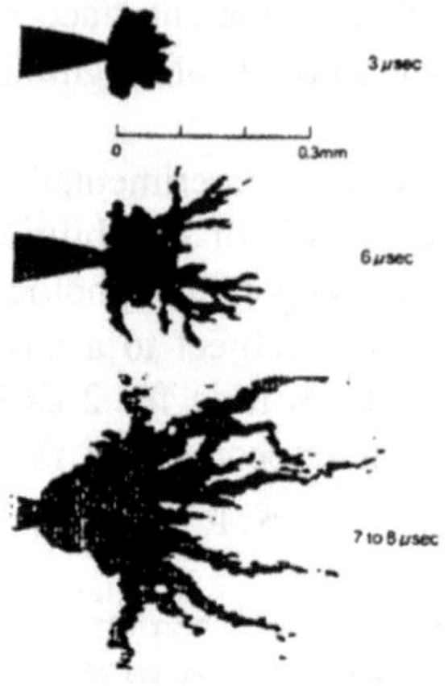

Fig. 6, Development of streamers from cavity wall instability in 2 cs DC-200, 0.6 mm gap, 11.5 kV. At 3 μs the instability is growing exponentially. By 6 μs the peaks of the instabilities have developed well defined streamers and by 7 or 8 μs the streamers are growing linearly in time.

$3\times10^{6}sec^{-1}$. Considering the uncertainty in the interface field and the scatter in the growth rate measurements, agreement between experiment and theory is reasonable.

13.3.4 Streamer Propagation

Finally we turn to the question, what happens to the instabilities when they extend beyond the range of exponential growth? One answer is given in Fig. 6 which shows a sequence of results obtained by Sadeghzadeh-Araghi on 2 cSt DC-200 [4]. This sequence indicates that as the EHD instability develops in amplitude, the wave structure evolves into finger-like protrusions from the ionized vapor cavity; the field will be concentrated at the tips of the fingers, which will then propagate as streamers that cross the gap and cause the actual breakdown.

A significant feature of streamer growth is the correspondence between the number of peaks in the EHD instability and the number of propagating streamers at later times. This correspondence between the waves of the instability and the propagating streamers explains an interesting regularity in streamer structure, for they are often characterized by an almost uniform, circumferential spacing which implies a coupling between them (electrostatic or hydrodynamic). The fact that they arise from an EHD instability provides a simple explanation for this regular structure.

We end with the question of how to model the streamer propagation process. The velocity of propagation of a streamer is generally found to be proportional to applied voltage; moreover, for a given voltage, velocity is relatively constant as the streamer crosses the gap (though with some increase as the streamer approaches the plane electrode). Also, streamer velocity is weakly viscosity dependent. Clearly we are far beyond the range of behavior that can be explained by simple, linear EHD theory, and the transition from instability to streamer is an area that requires careful study.

References

[1] Hakim S. S. and Higham, J. B., "A Phenomenon in n-Hexane Prior to its Electrical Breakdown," Nature 189, (1961) 966.

[2] Farazmand, B., "The Study of Electric Breakdown of Liquid Dielectrics Using Schlieren Optical Techniques," Brit J Appl Phys 12, (1961) 251.

[3] Chadband W. G. and Wright, G. T., "A Pre-Breakdown Phenomenon in Liquid Hexane," Brit J Appl Phys 16, (1965) 305.

[4] Watson, P. K., Chadband, W. G., and Sadeghzadeh-Araghi, M., "The Role of Electrostatic and Hydrodynamic Forces in Negative Point Breakdown of Liquid Dielectrics," IEEE Trans Elec Ins 26, (1991) 543.

[5] Rayleigh, Lord, "On the Pressure Developed in a Liquid During the Collapse of a Spherical Cavity," Phil Mag 34, (1917) 94.

[6] Kattan, R., Denat, A., and Lesaint, O., "Generation, Growth and Collapse of Vapor Bubbles in Hydrocarbon Liquids," J Appl Phys 66, (1989) 4062.

[7] Qureshi, M. I., Chadband, W. G, and Watson, P. K., "Pre-Breakdown Cavities in Viscous Silicone Fluids," 1992 DMMA (IEE Pub. 363), p. 89.

[8] Kelley, E. F. and Hebner, R. E., "The Electric Field Distribution Associated with Pre-Breakdown Phenomena in Nitrobenzene," J Appl Phys 52, (1981) 191.

[9] Melcher, J. R. and Smith, C. V., "Electrohydrodynamic Charge Relaxation and Interfacial Perpendicular-Field Instability," Phys Fluids 12, (1969) 778.

[10] Kath, G. S. and Hoburg, J. F., "Interfacial EHD Instability in Normal Electronic Fields," Phys Fluids 20, (1977) 912.

[11] Castellanos A. and González, A., Interfacial EHD Instability, Phys Fluids A4, (1992) 1307.

[12] Watson, P. K. and Chadband, W. G., "The Dynamics of Pre-Breakdown Cavities in Viscous Silicone Fluids," IEEE Trans Electrical Ins 23, (1988) 729.

Chapter 14

ELECTROSPRAY AND ATOMIZATION

14.1 Introduction

The electrostatic spraying of liquids is a well established technique [1] that offers several advantages over other methods of atomization. It appears in such diverse fields as spray painting, rocket propulsion, high intensity ion sources from liquid metals, etc. In many of these sources the liquid, dispersed from a high voltage capillary, forms a cone of semi-angle ~49.3°, as discussed by Taylor [2]. In this configuration the electrostatic forces are balanced by the surface tension of the liquid, and the spray arises from the tip of the cone where the field is a maximum and the equilibrium cannot be maintained.

This technique has been brought into scientific prominence in the past few years with the emergence circa 1988 of electrospray as a method for producing molecular ions in the gas phase from macromolecules in solution. These ions are required in order to extend the molecular weight range of mass spectrometers (indispensable tools for the chemist) so as to include large, polar organic molecules of interest in biology and medicine. A useful survey of this topic and its historical background is to be found in a review paper by Fenn et al. [3].

There were in effect two distinct problems that needed to be solved in this field, and it is a remarkable fact that the resolution of the first of these problems brought with it the answer to the second: (i) The technique of mass spectrometry requires as a preliminary step, the transfer of neutral molecules into the vapor phase, by evaporation, and their transformation into molecular ions in vacuo; the subsequent trajectories of these ions in a mass analyzer provides a measure of the normalized mass-to-charge ratio m/z. But macromolecules such as protein, for example, cannot be vaporized without extensive or catastrophic decomposition, so classical methods of ionization cannot be used. Moreover, even if one could transform macromolecules of say 10^5 daltons (units of molecular weight) into the corresponding singly charged ions, one would then run into the second problem:

(ii) Mass spectrometers are limited in the m/z range they are able to analyze. For example, a typical quadruple analyzer can resolve singly charged ions of up to 1500 daltons but not beyond that, and molecules of interest to the biochemist are one or two orders of magnitude larger than this (for example insulin ~5730 daltons, lysozyme ~14,300 daltons, bovine albumen dimer ~133,000 daltons). In their 1984 paper reporting on early experiments on electrospray mass spectrometry, Fenn and coworkers [4] called attention to the tendency of this spray technique to produce multiply charged ions, and pointed out that if one could promote multiple charging that would reduce m/z and thus extend the effective mass range of the mass analyzer. The fact that this was subsequently accomplished is a major reason for the success of the method.

The electrospray technique is particularly interesting in the context of EHD because two related phenomena are involved, viz, the spray process itself, and the instability of the resulting charged droplets. Before discussing the theory, however, it is useful to survey these physical processes in the context of mass spectrometry.

14.2 The Electrospray Technique

The original electrospray method was developed by Dole et al. [5], and published in 1968. It is based on the idea of creating ions of macromolecules by spraying droplets of a somewhat conducting, dilute solution of macromolecules from a high voltage capillary into an evaporation chamber containing an ambient gas such as nitrogen at atmospheric pressure, and then allowing the volatile solvent to evaporate from the charged droplets, leaving behind intact, gas-phase macroions. (Note that the charges on these macroions originate from extrinsic ions in the solution, not from the ionization of the macromolecules themselves.)

A typical electrospray apparatus is shown in Fig. 1. In order to form the electrospray a doped solution of the analyte flows through a hypodermic needle located on the axis of a cylindrical electrode. According to Dole et al. [5], the electric field at the tip of the needle should be about 4×10^4 Vcm^{-1}, the electrical conductivity of the solution should be in the range 10^{-8} to 10^{-9} Ω^{-1} cm^{-1}, and the concentration of a typical macromolecular species (say 10^4 daltons) should be of order 0.01 wt %, yielding an average of one macromolecule per 0.1 μm diameter droplet.

The reason for spraying into an ambient gas (such as nitrogen or carbon dioxide) rather than directly into vacuum is dictated by the strong evaporative cooling of the droplets. In the absence of a heat source the decrease in temperature of a droplet is

$$\Delta T = \left(\Delta H_v / c_p\right) \ln \beta \tag{14.1}$$

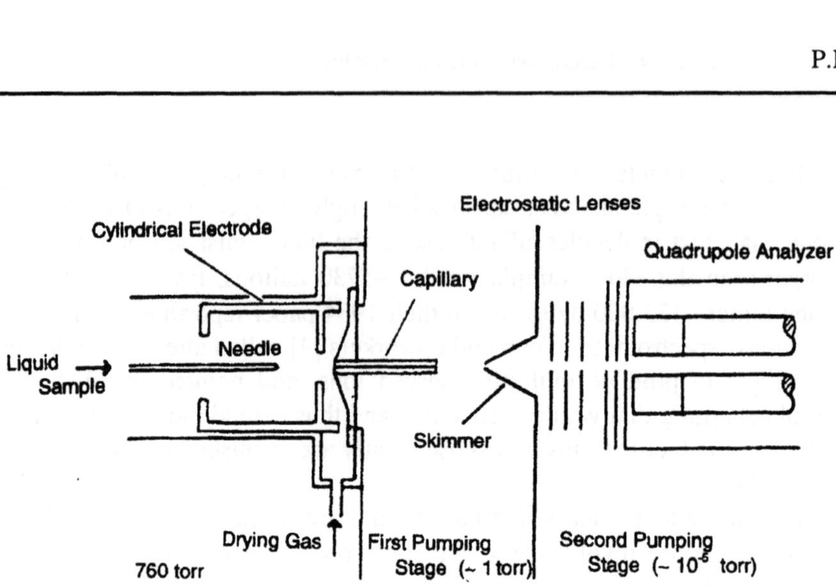

Fig. 1, Electrospray ion source, coupled to a quadrupole mass spectrometer via a differentially pumped vacuum manifold.

where β is the unevaporated droplet fraction, ΔH_v is the heat of vaporization and c_p is the heat capacity of the solvent. For benzene, if $\beta = 0.5$ then $\Delta T = -160°C$; therefore in order to maintain a high rate of evaporation heat must be transferred to the droplet, and that is the function of the ambient gas stream. As the droplet evaporates, the charge density on its surface increases until it approaches the Rayleigh limit [6],

$$q_S = 8\pi\left(\varepsilon_o \gamma R^3\right)^{1/2} \tag{14.2}$$

At this point Coulomb repulsion equals the surface tension, γ, the droplet becomes unstable and daughter droplets are ejected. This sequence, which has been studied by Gomez and Tang [7], is repeated until the radius of curvature becomes so small that the field due to the charge on the droplet is strong enough to desorb ions from the surface into the ambient gas; this occurs when the repulsion between charges exceeds the polarization force. (Ion desorbtion was first postulated by Iribarne and Thomson [8], and they have published drift spectrometer evidence of the desorbed ionic species.) When carried to completion this evaporation and desorbtion process yields so-called "quasi-molecular" ions suitable for mass analysis; each of these ions consists of one macromolecule plus an indefinite number of extrinsic ions. (Note that the actual number of charges per molecular ion is unknown at this point.) These gas-phase molecular ions are then ducted into the vacuum of the mass analyzer via a differentially pumped manifold as shown in Fig. 1. It is at this stage that the power of the method becomes evident, as the number of charges on the ions can be so arranged as to bring m/z into the range of the mass analyzer; the fact that z is at first unknown is immaterial, as an entire family of macro-ion species is generated, each member differing by unit charge from the next, and this shows up in the output signal from the mass spectrometer as a family of peaks such as the examples given in Fig. 2. Knowing that each peak differs from the adjacent peak by one unit charge it is then a routine matter to deconvolute these multi-line spectra.

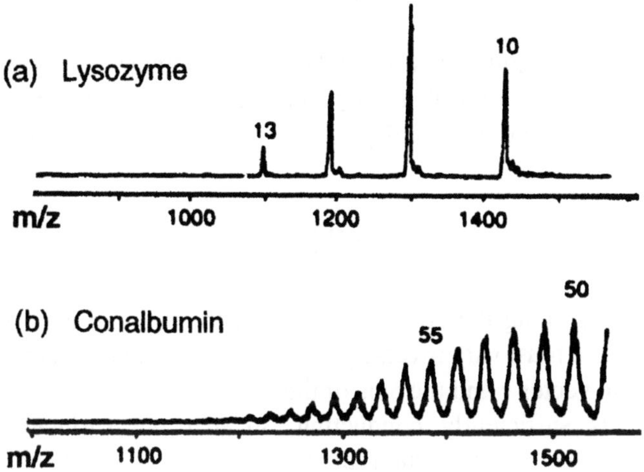

Fig. 2, Typical electrospray mass spectra from proteins in acidified water/methanol mixtures. (a) Lysozyme, 14,300 daltons, 10 to 13 unit charges (protons) per macroion. (b) Conalbumin, 76,000 daltons, 50 to 60 charges per ion.

14.3 EHD Aspects of Electrospray

As we have seen, the electrospray technique involves a succession of processes including generation of charged droplets at the tip of a high voltage capillary, rapid evaporation of the charged droplets to the Rayleigh limit, ejection of daughter droplets, and ion desorbtion in the droplet self-field.

Two of these steps involve EHD processes and it is these two related processes that concern us, namely the deformation of the liquid/gas interface at the capillary tip, and the instability of the evaporating, charged droplets.

14.3.1 Electrified Liquid Cones

In the presence of an electric field, conical protrusions form at the interface between a liquid conductor and an insulating ambient gas or liquid. These cones form when the surface charge density reaches the critical level at which the interface becomes unstable. In the case of a horizontal fluid interface in a vertical electric field, Frenkel [9] showed that the condition for stability is

$$\sigma_s^4 < \rho g \gamma \varepsilon_o / \pi \tag{14.3}$$

In the case of a meniscus, stabilized at the end of a high voltage capillary (a case first studied by Zeleny [10]), the best theoretical description is that put forward by Taylor [2]. In this analysis it is assumed that the electrical pressure, $p_E = (½)\, \varepsilon\, E^2$, is balanced by the surface tension pressure p_γ

$$p_\gamma = (\gamma / R) \cot \alpha \tag{14.4}$$

where R is the distance along the cone surface from the tip, and α is the semiangle of the cone. Clearly the E field must vary inversely with $R^{-½}$ and as some function of the polar angle θ between the position vector **r** and the axis of the cone.

If the fluid is sufficiently conducting, so that the surface of the cone is an equipotential, then the electric field which satisfies the stress condition has the potential

$$V = V_o + AR^{1/2} P_{1/2}(\cos\theta) \tag{14.5}$$

where the line $\theta = 0$, $\theta = \pi$ is the axis of the cone, $P_{½}(\cos \theta)$ is the Legendre function of order ½, and $A = (2\gamma/\varepsilon)^{½}$. Now on the surface of the cone $V = V_o$ and therefore $P_{½}(\cos \theta)$ must be zero; but this function has only one zero in the range $0 < \theta < \pi$ and that corresponds to an angle $\alpha = 49.3°$. Thus a conical interface between a conducting fluid and an insulating fluid can only exist in equilibrium in an electric field when the cone has this critical, semi-vertical angle.

14.3.2 The Apex of the Cone

Although Taylor's analysis provides an excellent description of the cone shape it does not deal with the tip of the cone when a microjet forms. But as de la Mora [11] has pointed out, the cone base diameter is several orders of magnitude larger than the jet diameter, and for typical electrospray condition the ratio of jet velocity to mean velocity in the cone can be as large as 10^7. The cone is therefore nearly hydrostatic (as assumed by Taylor) except for a small zoné in the vicinity of the apex, and so the analysis can be subdivided into two regions, the outer cone (as discussed in Section 3.1) and the region adjacent to and including the apex.

There are two, interesting, complimentary analyses of the cone tip, by Crowley [12] in 1977, and more recently by de la Mora, who took account of the space charge in the droplet stream emanating from the apex of the cone. Crowley took a different point of view; he was interested in the fact that, although most electrostatic sprays produce charged liquid droplets it is possible to produce ions from liquid metal cones, without accompanying charged droplets. These ion sources, which include various liquid metals and alloys such as sodium, NaK and Galn, form Taylor cones at high fields, and the ion beam is emitted from the tip of the cone where the cone is pulled out into a cusp. Currents of several hundred microamps have been reported, and Crowley pointed out that in the tip region the resulting high current density may cause a significant temperature rise due to Joule heating. To check this idea he solved the heat flow equation in the vicinity of the apex, and to avoid the singularity at R = 0 the cone was truncated at R = a (this distance cannot be less than an atomic radius). It was also assumed that heat flow into the jet is zero. The temperature inside the cone is then given by

$$\Delta T = \left(T - T_o\right) = \frac{I^2}{4\pi^2 \sigma k (1-\cos\alpha)^2}\left(\frac{1}{aR} - \frac{1}{2R^2}\right) \tag{14.6}$$

where σ is electrical conductivity, and k is thermal conductivity. For a Taylor cone, α=49.3°, and so the temperature rise at the tip is given by

$$\Delta T_{MAX} = 0.105\left(I^2/\sigma k a^2\right) \tag{14.7}$$

where the units of I, σ, k and a are amps, Ω^{-1} cm^{-1}, w $cm^{-1}s^{-1}$ and cm, respectively. From this analysis Crowley concluded that fluids used in electrostatic spraying may be divided into two groups: (i) Liquids such as water, glycerine, doped benzene, etc., which have low thermal conductivity and relatively low electrical conductivity. In these liquids the temperature rise at the cone tip is quite high (e.g., for salt water ΔT≥100°C at I = 1 μA and a = 1000 Å) and this temperature exceeds the boiling point of the liquid. Clearly such high temperatures cannot persist, and this suggests that droplet formation is due to this thermal effect rather than to an EHD instability, as is generally assumed.

(ii) Liquid metals such as NaK, Cs etc., which have high thermal and electrical conductivity. In such liquids the predicted temperature rise is six or seven orders of magnitude less than in the former case. (Even at a 1 Å tip a 1 μA current in a liquid metal only leads to a temperature increase of about 1°C.) If the thermal model is correct then high conductivity liquid metal cones should not produce liquid droplets, and this is confirmed in practice as such cones are utilized as pure ion beam sources. Liquids of intermediate conductivity however, such as mercury, are known to produce fine droplets, and this is consistent with the rank order implied by Eq. (7).

For a recent review of electrospray see [13] and references therein. The topic of liquid metal ion sources (LMIS) is an increasingly important one, and a useful survey of this field is to be found in a recent review by Forbes [14]. This paper includes a discussion of the shape of the cusp that forms at the tip of the LMIS cone, and the length of the cusp as a function of emission current.

References

1. Bailey. A. G., Electrostatic Spraying of Liquids. Wiley 1988.
2. Taylor. G. I., Disintegration of Water Drops in an Electric Field. Proc Roy Soc, Ser A 280 (1964) 383-97.
3. Fenn, J. B., et al., Electrospray Ionization for Mass Spectrometry. Science 246 (1989) 64-71.
4. Whitehouse, Dryer, Yamashita, and Fenn. Electrospray Interface for Liquid Chromatographs and Mass Spectrometers. Anal Chem 57 (1985) 675-9.
5. Dole, M., et al., Molecular Beams of Macroions. J Chem Phys 49 (1968) 2240-9.
6. Rayleigh, Lord, On the Equilibrium of Liquid Conducting Masses Charged with Electricity. Phil. Mag 14 (1882) 184-186.
7. Gomez, A. and Tang, K., Charge and Fission of Driplets in Electrostatic Spray, Phys Fluids 6 (1994) 404-14.
8. Iribarne, J. V. and Thomson. B. A., On the Evaporation of Small Ions from Charged Droplets, J Chem Phys 49 (1968) 2240-9.
9. Frenkel, Ya.I., (1935) See Landau and Lifshitz. Electrodynamics of Continuous Media p. 35, Pergamon 1960.
10. Zeleny, J., Instability of Electrified Liquid Surfaces. Phys Rev 10 (1917) 1-6.
11. de la Mora. F., Effect of Charge Emission from Electrified Liquid Cones. J Fluid Mech 243 (1992) 561-574.
12. Crowley, J. M., Role of Joule Heating in the Electrostatic Spraying of Liquids. J Appl Phys 48 (1977) 145-7.
13. Gañan-Calvo A. M., Dávila J. and Barrero A. Current and droplet size in the electrospraying of liquids. Scaling laws. J. of Aerosol Science 28 (1997) 249-275.
14. Forbes, R. G., Understanding how the Liquid-Metal Ion Source Works. To be publishd in "Vacuum" (1997).

PART IV

INSTABILITY AND CONVECTION INDUCED BY INJECTED SPACE CHARGE

P. Atten

CNRS - Electrostatic and Dielectric Mat. Laboratory, Grenoble, France

and

Joseph Fourier University, Grenoble, France

Abstract

The problem of unipolar injection from an electrode into an insulating fluid plays a fundamental role in electrohydrodynamics. The questions of instability and convection are investigated with particular emphasis on the analogies and differences with the Rayleigh-Bénard problem of a horizontal layer of fluid heated from below.

Chapter 15 is devoted to instability of a fluid layer between two plane parallel electrodes subjected to unipolar injection with non linear as well as linear analyses and instability study under transient conditions of injection. The convection of finite amplitude which arises above the critical conditions is considered in chapter 16. Its time dependent character for space charge limited injection is difficult to model. The hyperbolic nature of the retained charge conservation equation requires particular approaches and numerical techniques to obtain steady solution and to examine their stability.

Chapter 17 focuses on electroconvection and its analogy with thermoconvection (Rayleigh-Bénard problem) for very strong injection of charge. An order of magnitude analysis is recalled which accounts for the major properties (velocity amplitude and current) of the two regimes of convection for low and high values of the turbulent Reynolds number. Electroconvection during the transient regime of injection is also examined and results are presented on the influence of a forced flow and on the effect of electroconvection on heat transfer. The last chapter (18) concerns instability, electroconvection and superposition of unipolar injection with a forced flow and with a temperature gradient in the particular configuration of co-axial cylinders.

Chapter 15

UNIPOLAR INJECTION IN PLANAR GEOMETRY : LINEAR, NON-LINEAR AND TRANSIENT INSTABILITIES

For the problem of unipolar injection into an insulating fluid, the question of hydrodynamic instability arises in the three cases where the geometry of electrodes is of high symmetry : plane and parallel plates, co-axial cylinders and concentric spheres. It implies that the density of injected charge is uniform on the injecting electrode. In these conditions the motionless state $\mathbf{u} = 0$ is a solution because, when taking the curl of the Navier-Stokes equation we have the condition :

$$\nabla \times (q\mathbf{E}) = \mathbf{E} \times \nabla q = 0 \tag{15-1}$$

which is satisfied, the field **E** and the charge density gradient ∇q being colinear [1]. The instability problem is then to determine in what conditions this solution $\mathbf{u} = 0$ destabilises.

There is an analogy between the pure unipolar injection problem in plane parallel electrodes geometry and the Rayleigh-Bénard problem of a horizontal fluid layer heated from below. In the latter case the unstable density gradient brings about fluid motion when the temperature difference is large enough for buoyancy to overcome the damping action of viscous forces [2]. Above the critical value Ra_c of the Rayleigh number Ra, instability sets in and heat is partially transferred by convection [2]. In the unipolar injection problem, the charge distribution is also potentially unstable as it is shown below. The instability problem consists in examining the time evolution of perturbations of vanishingly small amplitude and to determine the critical conditions for which at least one perturbation begins to grow exponentially. Then convection sets in due to the action of Coulomb force perturbation which overcomes the viscous friction.

1. Linear instability mechanism

For the stationary state of pure unipolar injection, due to coulombic repulsion the charge density q decreases from the injector to the collector. The decrease of q in the direction of the force that the charge carriers experience results in the unstable character of the charge distribution. A qualitative argument very similar to the heuristic one developed by Lord Rayleigh in the case of a fluid layer heated from below has long been proposed [3,4], but it is not strictly correct because it does not take into account the finite drift velocity of the charge carriers with respect to the fluid. When accompanying the charge carriers during

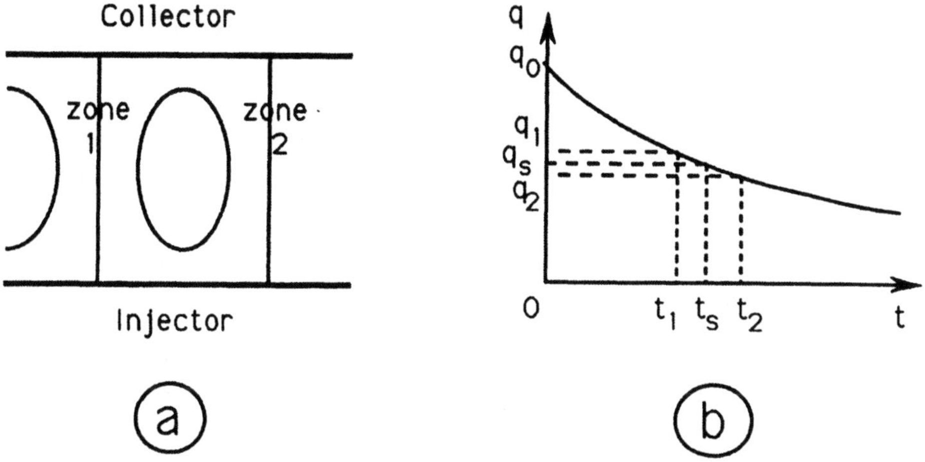

Fig. 15-1 – a) Cross-section of a roll-like flow perturbation and - b) values of the charge density as a function of the time of coulombic repulsion.

their movement, due to coulombic repulsion the charge density q decreases according to the law $q = q_0/(1+ t/\tau_0)$, where τ_0 is a time constant : $\tau_0 = \varepsilon/\mu q_0$, ε being the fluid permittivity, μ the charge carrier mobility and q_0 denoting the uniform charge density on the injector. In the absence of fluid motion, the time of flight t_s of the ions between the injector (at z = 0) and a point at height z is :

$$t_s = \int_0^z \frac{dz}{\mu\, E(z)} \tag{15-2}$$

Let us consider a convective motion in the form of two-dimensional rolls (Fig. 1-a). A positive velocity component w along the field line (zone 1) results in a decrease of the time required for the ions to reach the height z : $t_1 < t_s$. Coulombic repulsion therefore acts during a smaller time and q_1 is higher than the value q_s which exists in the motionless state (Fig. 1-b). In zone 2, conversely, the liquid velocity diminishes the resulting charge carrier velocity and we have $t_2 > t_s$ and $q_2 < q_s$. Therefore the convective motion gives rise to a net torque which tends to accelerate it. Because of this positive coupling between velocity and charge perturbations, the stationary charge distribution is potentially unstable [1]. The destabilising force is proportional to q_1-q_2 and consequently to the mean charge gradient [1,3].

This unstable character of the space charge distribution resulting from a unipolar injection into an insulating liquid does not imply that any velocity perturbations will amplify. Indeed a sustained motion implies that an energy source can provide for the dissipation due to viscous forces and it is clear that for low enough applied voltages the electric forces cannot overcome the viscous friction. In analogy with the Rayleigh-Bénard problem [2], it was possible to show, in a particular case, that instability sets in when the extra current δI due to convection is such that the extra energy $V\delta I$ becomes at least equal to the energy dissipated by friction [5].

The fundamental parameter which arises in the instability problem is the non dimensional number T which formally represents the ratio of driving Coulomb force and damping viscous force :

$$T \sim \frac{qE}{\eta \nabla^2 w} \sim \frac{\varepsilon VE / d^2}{\eta w / d^2} \sim \frac{\varepsilon VE}{\eta \mu E} \quad \Rightarrow \quad T = \frac{\varepsilon V}{\mu \eta} \qquad (15\text{-}3)$$

Note that the stability parameter T does not explicitly depend on any length. Moreover, in liquids, T is simply proportional to the applied voltage V according to Walden's rule relative to ions ($\mu\eta \cong$ Cst). In the three geometries of high symmetry, the instability criteria have been obtained by using the standard mathematical technique of linear instability analysis [2,5,6]. In the absence of precise laws concerning the injection rate, it was further assumed that injection is autonomous [5], i.e. q_o does not depend on variables like the fluid velocity or the electric field (this assumption leads to lower bounds for the instability criteria). The critical values T_c depend on the parameter C which measures the intensity of charge injection. In the weak injection limit, $C \ll 1$, which corresponds to a very small field distortion, the charge gradient is proportional to C^2 and the criterion takes the asymptotic expression $T_c \cong Cst/C^2$ [5].

Let us recall briefly that in the case of two plane parallel rigid electrodes, the weak injection limit ($C \ll 1$) is characterised by the criterion $T_cC^2 = 220.7$. In the other asymptotic case of space charge limited (SCL) injection ($C \to \infty$), $T_c \to 161$. The critical values have been determined for any injection strength [5]. The critical non dimensional wave-number k_c takes values close to 5 and only slightly depends on C. No information on the preferred convective planform (two-dimensional rolls or hexagonal cells) can be gained from the linear analysis. The instability conditions have also been obtained when one or both electrodes are free surfaces [5,6]. An experimental study has been carried out in planar geometry [7] only in the case of SCL injection into viscous liquids. Focusing here on the linear instability voltage as given by Fig. 5, we see that V_c is independent of the distance d as predicted. The criterion, however, takes a value $(T_c)_{exp} \approx 100$ lower than the predicted one $(T_c)_{th} = 161$ [7]. Part of this departure between experiments and theory arises from diffusion being neglected whereas it plays a role in a layer very close to the injector where $|\nabla q|$ takes very high values ($C \to \infty$) [8].

2. Hydraulic model of non linear instability

The analogy with the Rayleigh-Bénard problem is interesting from a general physical viewpoint but is only partial as appeared in the foregoing section where the argument valid for the thermal problem could not be transposed directly to the unipolar injection problem. The analogy fails when considering the non-linear behaviour for T close to T_c. Here again the physical reason is the difference in the conduction mechanism of the scalar quantity in the motionless state which is drift with a finite velocity μE for charge carriers and diffusion for heat. In the thermal problem, for a value of the Rayleigh number $Ra > Ra_c$, two-dimensional rolls appear in the case of high Prandtl number liquids and the amplitude of steady convective motion varies in a continuous way with Ra. In the unipolar injection problem, conversely, there is a discontinuity at T_c for the steady amplitude of the convective velocity and also for the steady current. We recall first a simplified treatment which highlights this phenomenon.

The great features of the non linear instability mechanism have been derived by Félici who used a hydraulic model in the case of weak injection ($C \ll 1$) between two parallel plates [9]. He modelled a convective cell by two columns of equal radius R in which the liquid velocity is assumed to be constant in magnitude and to have a direction respectively identical (column 1) and opposite (column 2) to that of the force exerted by the electric field on the charge carriers. With these approximations it is easy to deduce the charge density at any height z. The charge conservation equation leads to the relation :

$$\frac{dq}{dt} = -\frac{\mu}{\varepsilon}q^2 \qquad (15\text{-}4)$$

which expresses that, when moving with the charge carriers, the charge density q decreases due to Coulomb repulsion. From (15-4) we deduce :

$$\frac{1}{q_1} = \frac{1}{q_o} + \frac{\mu t_1}{\varepsilon} \qquad \frac{1}{q_2} = \frac{1}{q_o} + \frac{\mu t_2}{\varepsilon} \qquad (15\text{-}5)$$

with $t_1 = z/(\mu E + w)$ and $t_2 = z/(\mu E - w)$. The charge density difference at z is :

$$q_1 - q_2 = 2\,q_1 q_2 \frac{\mu z w}{\varepsilon\left(\mu^2 E^2 - w^2\right)} \approx 2 q_o^2 \frac{\mu z w}{\varepsilon\left(\mu^2 E^2 - w^2\right)} \qquad (15\text{-}6)$$

in the weak injection case ($C \ll 1$) provided w is not close to μE. As $q_1 - q_2$ increases more than linearly as a function of w, for a viscous liquid a stable steady solution cannot exist with $w < \mu E$ [1,9]. This can be seen by calculating first the driving pressure difference Δp_d between the two channels which is obtained by multiplying (15-6) by E and integrating from $z = 0$ to $z = d$:

$$\Delta p_d = \frac{q_o^2 d^2 w}{\varepsilon \mu E\left(1 - A^2\right)} \qquad (15\text{-}7)$$

$A = w/\mu E$ denotes the non-dimensional velocity amplitude. The viscous pressure drop Δp_v is proportional to the liquid viscosity η, to the velocity w and to the length d of the channels. It writes :

$$\Delta p_v = \alpha \eta w \frac{d}{R^2} \qquad (15\text{-}8)$$

where R is the half width of the channels and α is a non-dimensional constant on the order of 10. The linear instability criterion is obtained by balancing both pressure differences for vanishingly small velocity ($w \to 0$, $A \to 0$) :

$$T_c C^2 = \alpha \left(\frac{d}{R}\right)^2 \tag{15-9}$$

For a steady state convection with $A < 1$, we have :

$$A^2 = 1 - \frac{T}{T_c} \tag{15-10}$$

A steady solution with $A < 1$ can only exist for $T < T_c$ but as the $A^2(T)$ curve has a negative slope, it is easy to see that such a solution is unstable [7,9]. A stable solution will therefore be characterised by a liquid velocity greater than the drift velocity μE. Indeed as soon as the liquid velocity has an amplitude w which overcomes μE, in the "return" column (#2) the injected charge carriers cannot move towards the collecting electrode and the column is free of charge. In column #1, due to convection the charge occupies only a part of width λ which can be determined from conservation of the current injected in the convective cell of width 4R. This current per unit length is $J = 4Rj_o = 4R\mu q_o E$ (for a justification of the current conservation in that case, see §16-1). At any height z ($0 < z < 1$) we have $J = q_o \lambda (\mu E + w)$ and therefore $\lambda = 4R/(1+A)$. The driving pressure difference then takes the simple expression :

$$\Delta p_d = 2 \frac{q_o E d}{1+A} \tag{15-11}$$

which, when balanced with the viscous pressure drop finally leads to :

$$A(1+A) = 2 \frac{T}{T_c C} \tag{15-12}$$

In this hydraulic model the non-linear criterion T_{nl} is defined as the smallest T value for which a stable steady convection exists. It corresponds to $w = \mu E$ ($A = 1$) and takes the value :

$$T_{nl} = T_c C \tag{15-13}$$

From a mathematical viewpoint, the non linear instability problem is characterised by a subcritical bifurcation : in the interval $[T_{nl}, T_c]$ below T_c there exist two finite amplitude solutions, one stable and the other one unstable (Fig. 3). This is related with a hysteresis loop and jumps of the steady values of both the velocity amplitude and the current at the linear T_c and non-linear T_{nl} criteria [7].

The linear instability analysis gives the critical conditions but cannot predict the planform of the convective cells which will appear. A non-linear analysis is required to gain information on the flow organisation. The hydraulic model can be extended to the case of hexagonal convective cells and this leads to a prediction concerning the planform. A hexagonal convective cell can be modelled by a central channel #1 and a outer annular

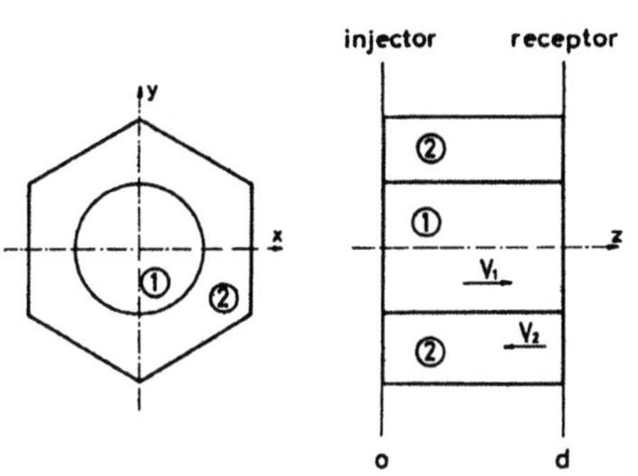

Fig. 15-2 – Hydraulic model for an hexagonal convective cell. Arrows define positive velocities V_1 and V_2 (*from Atten & Lacroix* [7]).

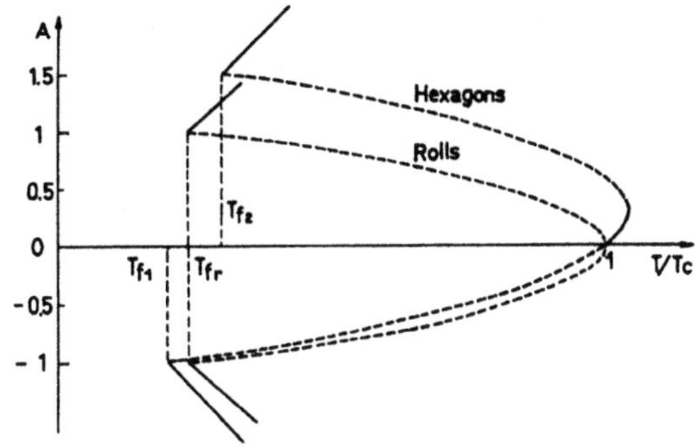

Fig. 15-3 – Velocity parameter A versus T/T_c. Simplified model : C = 0.2 (*from Atten & Lacroix* [7]).

region #2 of respective cross section areas S_1 and S_2 and uniform velocities w_1 and w_2 (Fig. 2). Conservation of mass gives :

$$\frac{w_2}{w_1} = \frac{S_1}{S_2} = \beta \tag{15-14}$$

From Christopherson's solution [2], $\beta \approx 0.66$. Considering again a very weak injection ($C \ll 1$), the charge densities in the channels again take the expressions (15-5) where now $t_1 = z/(\mu E + w_1)$ and $t_2 = z/(\mu E - w_2) = z/(\mu E - \beta w_1)$. The driving pressure difference arising from the imbalance of Coulomb force in the 2 channels then writes :

$$\Delta p_d = \frac{q_o^2 d^2 (1+\beta)\, w_1}{2\varepsilon\mu E\,(1+A)(1-\beta A)} \tag{15-15}$$

with $A = w_1/\mu E$ ($A < 1$). The viscous pressure drop associated with the flow can be written :

$$\Delta p_v = \alpha' \eta w_1 \frac{d}{R^2} \tag{15-16}$$

(R is related to the cell transverse dimension). Balancing the two pressure terms gives :

$$T_c C^2 = \frac{2\alpha'}{1+\beta} \left(\frac{d}{R}\right)^2 \tag{15-17}$$

for the linear instability criterion and the relation :

$$\beta A^2 - (1-\beta)A = 1 - \frac{T}{T_c} \tag{15-18}$$

for finite values of A $(-1 < A < \beta)$. Remark that (15-10) is recovered for $\beta = 1$ which corresponds to the case of two-dimensional rolls. Fig. 3 shows that there is then a transcritical bifurcation with both supercritical and subcritical branches in the vicinity of T_c. Relation (15-18) is valid for values of w_1 and w_2 lower than μE and not close to μE. When the liquid velocity in a channel compensates for the drift velocity $\mu \mathbf{E}$, this channel becomes free of charge and the driving pressure difference is $(w_i = \mu E)$:

$$\Delta p_d = q_o E d \tag{15-19}$$

In the case of the central channel being free of charge $(A = -1)$, the non linear critical value is [7] :

$$(T_{nl})_1 = \frac{1+\beta}{2} T_c C = \frac{1+\beta}{2} (T_{nl})_{rolls} \tag{15-20}$$

In the other case of no charge in the annular channel $(w_2 = \mu E \;\Rightarrow\; w_1 = \mu E/\beta)$, the expression is :

$$(T_{nl})_2 = \frac{1+\beta}{2\beta} T_c C = \frac{1}{\beta} (T_{nl})_1 \tag{15-21}$$

Relation (15-20) shows that there exist finite amplitude motions of hexagonal planform in conditions where all roll motions are damped (Fig. 3). Therefore we expect the hexagonal convective cells with flow in the central zone towards the injector to be the most unstable perturbations. This conclusion on the preferred convection mode was confirmed by an analysis of the time evolution of perturbations of small amplitude : these hexagonal convective cells have the greatest growth rate [7].

3. Approximate non linear instability analysis

In order to perform an analysis, we have first to derive the equations governing the evolution of disturbances. The classical way is to decompose the different variables into mean and fluctuating parts :

$$\begin{array}{llll} \text{electric potential} & \phi & \rightarrow & \overline{\Phi} + \phi \\ \text{electric field} & \mathbf{E} & \rightarrow & \overline{E}\,\mathbf{i}_z + \mathbf{e} \\ \text{charge density} & q & \rightarrow & \overline{q} + q \end{array} \tag{15-22}$$

Here the bar denotes the average on a horizontal plane (parallel to the electrodes) and $\mathbf{i}_z$ is the vertical unit vector. The dimensionless equations are then :

$$\frac{\partial^2 \overline{\Phi}}{\partial z^2} = -\frac{\partial \overline{E}}{\partial z} = -\overline{q} \tag{15-23}$$

$$\nabla^2 \phi = -\nabla . \mathbf{e} = -q \tag{15-24}$$

$$\frac{\partial \bar{q}}{\partial t} + \frac{\partial}{\partial z}\left[\bar{q}\bar{E} + \overline{q\left(w - \frac{\partial \phi}{\partial z}\right)}\right] = 0 \tag{15-25}$$

$$\frac{\partial q}{\partial t} + \bar{E}\frac{\partial q}{\partial z} + 2\bar{q}q + \frac{\partial \bar{q}}{\partial z}\left(w - \frac{\partial \phi}{\partial z}\right) = \nabla.[q(\mathbf{u}+\mathbf{e})] - \frac{\partial}{\partial z}\left(\overline{q\left(w - \frac{\partial \phi}{\partial z}\right)}\right) \tag{15-26}$$

$$\nabla.\mathbf{u} = 0 \tag{15-27}$$

$$\frac{T}{M^2}\left[\frac{\partial \mathbf{u}}{\partial z} + \mathbf{u}.\nabla\mathbf{u} - \overline{\mathbf{u}.\nabla\mathbf{u}} + \nabla p\right] = \nabla^2\mathbf{u} + T\left[\bar{E}q\mathbf{i}_z + \bar{q}\mathbf{e} + q\mathbf{e} - \overline{q\mathbf{e}}\right] \tag{15-28}$$

where $M = \sqrt{\varepsilon/\rho}\,/\mu$ is the mobility parameter equal to the ratio of hydrodynamic and charge carrier mobilities (see §17-1). The boundary conditions for rigid walls are :

$$\bar{\Phi} = 1, \quad \bar{q} = C, \quad w = \frac{\partial w}{\partial z} = \phi = q = 0 \qquad \text{at } z = 0 \tag{15-29}$$

$$\bar{\Phi} = 0, \quad w = \frac{\partial w}{\partial z} = \phi = 0 \qquad \text{at } z = 1 \tag{15-30}$$

Following a standard technique [10], the horizontal variations of fluctuations of the variables are expanded in orthonormal functions f_k satisfying :

$$\nabla_1^2 f_k(x,y) = -a_k^2 f_k(x,y), \qquad \overline{f_j f_k} = \delta_{jk} \tag{15-31}$$

where $\nabla_1^2 = \nabla^2 - \partial^2/\partial z^2$ and a_k is the horizontal wave number. The expansions are :

$$\begin{pmatrix} \phi \\ q \\ w \end{pmatrix}(x,y,z,t) = \sum_k f_k(x,y)\begin{pmatrix} F_k \\ G_k \\ W_k \end{pmatrix}(z,t) \tag{15-32}$$

By expressing the horizontal velocity components u and v, substituting the different expansions in the equations (15-23) to (15-28), taking the appropriate averages and eliminating the pressure term, a set of equations is obtained relating $\bar{E}$ and $\bar{q}$ with the F_k, G_k and W_k [7]. At steady state the system reduces to ordinary differential equations which can be transformed into integral equations. In the asymptotic case of liquids characterised by very high values of the mobility parameter M, the inertial terms are negligible and the system reduces to [7] :

$$\bar{E}(z) = \left[2\bar{J}z + \frac{\bar{J}^2}{C^2} - 2\int_0^z \sum_m G_m\left(W_m - \frac{dF_m}{dz}\right)dz\right]^{1/2} \tag{15-33}$$

$$\bar{q}(z) = \frac{1}{\bar{E}}\left[\bar{J} - \sum_m G_m\left(W_m - \frac{dF_m}{dz}\right)\right] \tag{15-34}$$

$$F_k(z) = \int_0^1 H_{1k}(z,u)\, G_k(u)\, du \tag{15-35}$$

$$G_k(z) = \frac{1}{\bar{E}^2(z)} \int_0^z \left\{ -\bar{E} \frac{d\bar{q}}{dz}\left(W_k - \frac{dF_k}{dz} \right) - \bar{E}\, S_k \right\} du$$

$$\text{with} \quad S_k = \sum_{p,m} B_{kpm} \left[G_p \left(\frac{c_{kpm}}{a_m^2} \left(\frac{dW_m}{dz} - a_m^2 F_m \right) + 2G_m \right) + 2\frac{dG_p}{dz}\left(W_m - \frac{dF_m}{dz} \right) \right] \tag{15-36}$$

$$W_k(z) = Ta_k^2 \int_0^1 H_{2k}(z,u) \left[\bar{E} G_k + \frac{d\bar{q}}{dz} F_k + \frac{1}{a_k^2} \sum_{p,m} B_{kpm}\, c_{kpm} \left(F_m \frac{dG_p}{dz} - G_m \frac{dF_p}{dz} \right) \right] du \tag{15-37}$$

with the following notations :

$$B_{kpm} = \frac{1}{2}\, \overline{f_k f_p f_m}\, , \qquad\qquad c_{kpm} = a_p^2 + a_m^2 - a_k^2 \tag{15-38}$$

$H_{1k}(z,u)$ and $H_{2k}(z,u)$ are the Green functions of the operators $\mathcal{D}_k$ and $\mathcal{D}_k^2$ associated with the boundary conditions respectively $F_k = 0$ and $W_k = \partial W_k/\partial z = 0$ at $z = 0, 1$ ($\mathcal{D}_k = \partial^2/\partial z^2 - a_k^2$). The value of the mean current density $\bar{J}$ is adjusted taking into account the condition on the field. The system of eqs. (15-33) to (15-37) is solved by an iterative procedure with interpolation between successive approximations [7]. Above a threshold for the amplitude of the fundamental mode of vertical velocity component w, the iterative

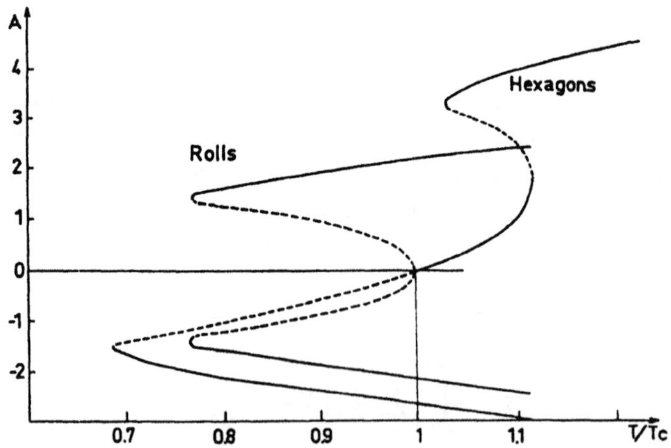

Fig. 15-4 – Nondimensional velocity A versus T/T_c. One mode theory in the case $C = 10$ and wave number $a = 5.1$ (*from Atten & Lacroix* [7]).

process diverged due to the amplitude of charge perturbation increasing without limit; this would correspond to the charge density taking the opposite sign in some regions, which is not possible from the physical viewpoint. Convergence of the iterative process was then obtained by imposing the extra condition $\overline{q}+q(x,y,z) \geq 0$ [7]. Note that these difficulties in solving eqs. (15-33) to (15-37) arise from the inadequacy of using Fourier expansions in treating the first order equation (15-26) (this equation is derived from the charge conservation equation and is hyperbolic in nature).

The results obtained for small values of the injection level C fully confirm those given by the hydraulic model. For moderate and high C values, the qualitative picture of phenomena remains the same as shown by Fig. 4 relative to strong injection (C = 10). We can then conclude that the finite amplitude convection which appears organises in the form of hexagonal convective cells with flow towards the injector around the axis of the cells. For very strong injection ($C \rightarrow \infty$), the non linear criterion is $T_{nl} \approx 110$ with $A \cong -1.6$ and a jump of about 7 % in the mean current density [7].

4. Experimental results on non linear instability

The crucial point in experiments on convection induced by unipolar charge injection is to control the injection of charge carriers. In gases the usual way to inject ions is to use the corona effect which appears in the vicinity of zones of high curvature of the electrodes when the applied voltage overcomes a threshold value. In liquids, the problem is much more complex because the creation of ions at the electrodes generally is a phenomenon of

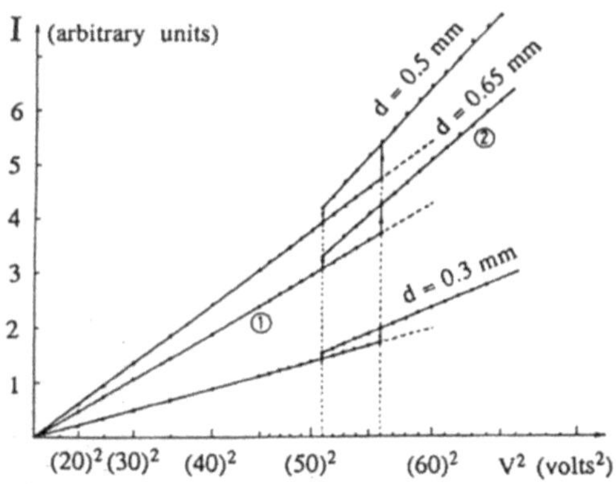

Fig. 15-5 – Steady state current versus the square of applied voltage U for 3 values of the thickness (space charge limited injection in "pyralene 1460") (*from Atten & Lacroix* [7]).

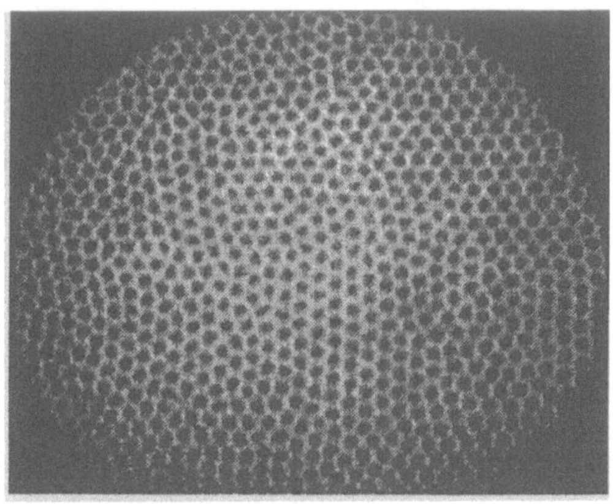

Fig. 15-6 – Schlieren photograph (using transparent electrodes) of the convective pattern. SCL injection in "pyralene 1460" (U = 60 V, d = 0.3 mm) (*from Atten & Lacroix* [7]).

electrochemical nature [11,12]. With plane electrodes, one of the most reliable methods to inject ions into polar liquids is to use permselective membranes covering the metal. These membranes which yield as many ions as the system can receive give approximately SCL injections ($C \to \infty$) in highly purified liquids of dielectric constant $\varepsilon_r \geq 4$ [13,7].

Experiments were carried out in the case of space-charge-limited (SCL) injection into viscous mildly polar liquids. The steady current measured by increasing and then decreasing slowly the applied voltage by steps clearly revealed the existence of a hysteresis loop (Fig. 5). Independently of the distance d between electrodes, there occur jumps of the steady current at the well characterised linear and non-linear instability voltages. For any point of the upper branches in Fig. 5, strioscopic observation provided evidence of a convective motion of hexagonal planform (Fig. 6). By considering the electric Nusselt number Ne defined as the ratio of effective $\bar{J}$ and motionless J_0 mean current densities :

$$Ne = \frac{\bar{J}}{J_0}, \qquad J_0 = \frac{9}{8}\mu\varepsilon\frac{V^2}{d^3} \tag{15-39}$$

a unique curve Ne(V) is obtained (Fig. 7). At the non-linear critical voltage the electric Nusselt number takes a value $Ne \cong 1.06$ very close to the predicted one : $Ne = 1.07$ [7]. Although a direct velocity measurement was not performed in these experiments, a good estimate could be obtained from the observation of small particles cycling along stable closed trajectories of approximate elliptic shape. From the mean rotation time (averaged on several particles) the variation of the velocity amplitude was found to be quadratic in V (Fig. 8). The calculation of the rotation time for hexagonal convective cells led to [7] :

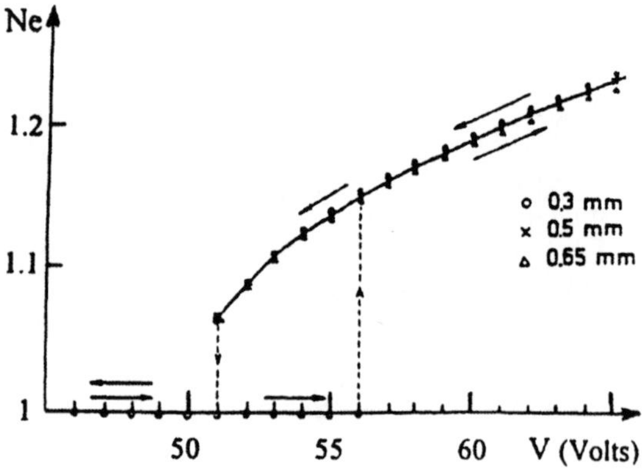

Fig. 15-7 – Electric Nusselt number Ne versus the applied voltage V for SCL injection in "pyralene 1460" (*from Atten & Lacroix* [7]).

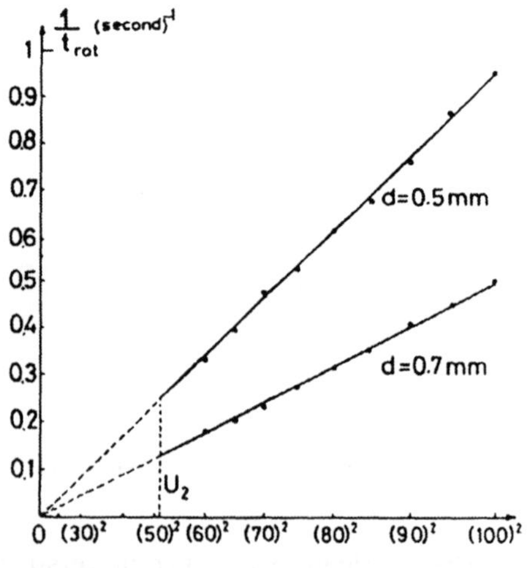

Fig. 15-8 – Mean frequency of particles cycling versus the square of the applied voltage V (*from Atten & Lacroix* [7]).

$$|A| = \frac{|w_{max}|\, d}{\mu V} = 1.5\ (\pm\ 0.2)\ \frac{T}{T_{nl}} \tag{15-40}$$

This is in good agreement with the theoretical prediction $A \cong -1.6$ at $T = T_{nl}$ (see Fig. 4).

The experiments performed on various liquids and mixtures give for the non-linear criterion the value $(T_{nl})_{exp} \cong 90$ lower than the calculated one $(T_{nl})_{th} \cong 110$ [7]. There is no explanation for this difference. Finally notice that, to our knowledge, there is no experimental study on the non-linear instability in the case of weak and moderate injections.

5. The instability problem for transient unipolar injection

The transient current technique was first used in the unipolar injection problem to determine the mobility of charge carriers. By applying a step voltage, the current variation exhibits a transient regime of increase up to a maximum which corresponds to the arrival of the first injected charges on the collector [14,15]. The transit time allows to recover the ion mobility provided that the liquid is at rest. For large enough applied voltages V (or strong enough injections), however, a motion sets in before the first injected charge carriers have reached the opposite electrode. This developing convection accelerates the transit of charges and leads to apparent mobility values μ_a greater than the true one. Generally μ_a increases with V and further takes a saturation value in the case of SCL injection when the motion is fully turbulent (see the detailed review by Tobazeon [16]). An apparent instability voltage V_a can be defined as the voltage above which the apparent mobility begins to increase [17]. The question which then arises is to know what is the relation between V_a and V_c .

The first detailed experimental study of the hydrodynamic instability induced by unipolar injection was performed by Watson et al. [18] who determined the instability criterion through transient space charge limited current experiments. These authors implicitly assumed that $V_a = V_c$. Such a relation has no theoretical justification and to some extent can be fortuitous because the instability problems are conceptually different. In stationary conditions we have instability once a perturbation has a positive growth rate, whatever its magnitude may be, and by waiting a long enough time the convection will become detectable. In transient conditions the instability is not strictly characterised; it is revealed by the effects of “manifest convection” and therefore depends on both the sensitivity of the detection technique of the motion and the magnitude of initial perturbations.

Experimentally, in most liquids the transient current is the best variable to detect manifest convection : for times t lower than the "stability time" t_l , the current I follows the law for motionless transient regime (Fig. 9); for $t > t_l$, I increases more steeply because of the contribution of convection to charge transport. An expression for t_l can be obtained by assuming that manifest convection is detected for a well defined value T_{al} of the stability

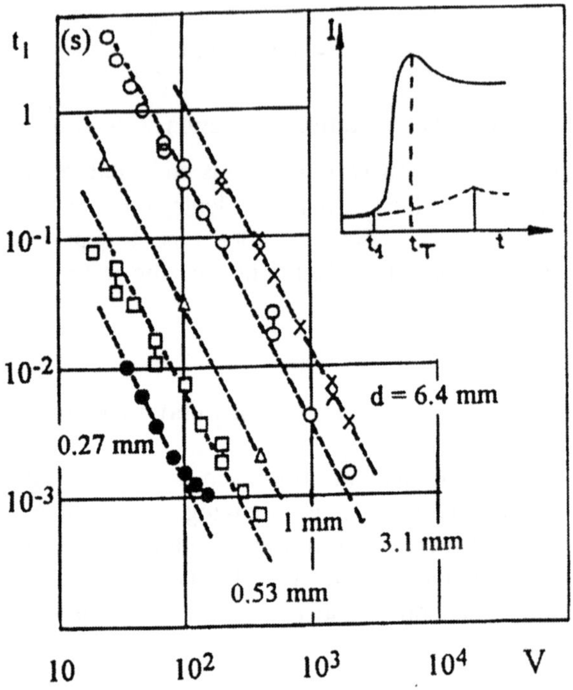

Fig. 15-9 – Transient SCL unipolar injection in nitrobenzene. Delay time t_l versus step voltage V. Insert : measured and motionless (dashed line) currents *(from Filippini et al.* [19]).

parameter characterising the space charge layer of thickness $\overline{z} \approx \mu E t_l$, i.e. : $T_{al} = \varepsilon V_a'/\mu\eta$, with $V_a' \approx E\overline{z}$:

$$t_l \approx T_{al} \frac{\eta}{\varepsilon E^2} \tag{15-41}$$

Here the ratio $\eta/\varepsilon E^2$ is the electroviscous time. Experiments [19,20] fully confirm the variation law (15-41) for SCL injection (Fig. 9).

An attempt to justify the observation of the E^{-2} variation of the stability time t_l was developed by considering the mathematical initial value problem corresponding to the particular case of constant current space charge limited injection [21]. After linearizing the mathematical problem and simplifying the resolution by the technique of "frozen time" approximation, solutions were obtained by further neglecting the deformation of the front of first injected charge carriers due to the developing convective flow [21]. However the analysis failed to account for the quantitative observations. Neglecting the front deformation is too a drastic approximation because, as will be seen in the next section, it appears that the deformation provides the major contribution to the driving pressure difference responsible for the convective motion.

More recently Cross et al. proposed a new treatment of the instability problem in transient conditions of unipolar injection and gave another expression (in $E^{-5/3}$) for the variation of the stability time [22]. These authors based the analysis on the analogy with the Rayleigh-Taylor instability, i.e. the instability of a horizontal layer of fluid lying above another fluid of smaller density [2] and they applied some results of the interfacial EHD instability of two adjacent layers of different conductivities [23]. This approach is interesting because it gives a picture where clearly the deformation of the interface plays the major role. The treatment, however, appears questionable because the equivalent conductivity μq has not all characteristic properties of a true conductivity and also because, unlike the layer in Rayleigh-Taylor instability, the charged layer has a thickness which increases with time : $\bar{z} = \mu Et$. This problem of instability in transient conditions is examined through a very crude analysis taking into account the deformation of the interface.

The starting point of this approach [24] is to take into account the geometric imperfections of the planar injecting electrode which always exhibits a certain roughness independently of its nature (metal or permselective membrane). In a detailed study of development of convective motion during the transient regime of unipolar injection, Rigo showed that this roughness effect fully dominates over the velocity perturbations induced by thermodynamic fluctuations [25]. At time t = 0 a step voltage is applied which induces an injection of charge carriers. The roughness of the electrode implies corresponding perturbations in the local electric field and this results in a slightly increased roughness of the front of first injected carriers (which is the interface between charge and charge free regions in the liquid) even in the absence of motion. The problem now is to analyse the evolution of these initial perturbations.

The space charge layer has a mean thickness $\bar{z}$ which, at least initially, increases linearly with time : $\bar{z} = \mu Et$. The interface is defined by $z = \mu Et+\xi$, ξ being a function of the horizontal coordinates and of time. Any velocity perturbation induces a charge perturbation δq in the charged layer and contributes to the perturbation ξ of the front of charges. These perturbations in turn give rise to force perturbations which modify the

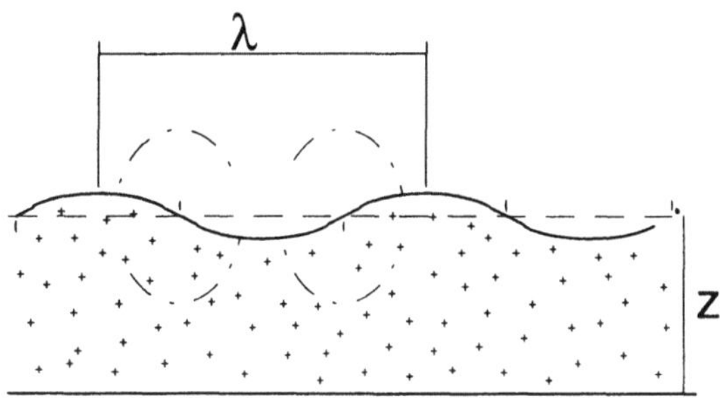

Fig. 15-10 – View of a sinusoidally deformed interface between charged and charge free regions of the liquid at a given time t and of the streamlines of the associated convective motion.

velocity field. In the weak injection case ($C << 1 \Rightarrow q_o << \varepsilon V/d^2$), only the "pressure" term associated with the charge excess and depletion due to the interface deformation is important for determining the growth of ξ. For strong injection both interfacial and bulk contributions should be taken into account but we content ourselves here to retain only the interfacial term.

Let us consider a sinusoidal deformation, of wave-length λ (see Fig. 10), of the interface between charged and charge free regions (as usual in the linear instability analyses the perturbations are decomposed into their Fourier components). The driving pressure difference is then :

$$\Delta p_d = 2\, q(\bar{z})\, E(\bar{z})\, \xi\, . \tag{15-42}$$

When the mean thickness $\bar{z}$ is much smaller than the wave-length λ, the viscous damping due to the presence of the rigid injector is strong and the liquid velocity is very weak. When conversely $\bar{z} >> \lambda$ the convective cells centred on the interface are no more influenced by the rigid injecting boundary. The transition between these two cases occurs when the width of the convective cell is close to $\bar{z}$ (Fig. 10). We therefore make the following rough approximation concerning the effect of viscous damping :

i) for $\bar{z} < \lambda/2$ the damping is strong and the effect of convection is neglected so that $\xi \cong Cst$;

ii) for $\bar{z} > \lambda/2$ the velocity field on both sides of the plane $z = \bar{z}$ remains similar and in liquids of high enough viscosity the viscous pressure drop takes the expression (analogous to (15-8) and (15-16)) :

$$\Delta p_d = \alpha \eta \frac{w}{\lambda} \tag{15-43}$$

where α is a non-dimensional constant and w the vertical velocity component which modifies the interface deformation : $w = d\xi/dt$. By balancing the two pressure differences (as was done in §15-2) we obtain the following equation for the evolution of ξ [24] :

$$\frac{d\xi}{\xi} = \frac{2}{\alpha}\frac{q(\bar{z})E\lambda}{\eta}\, dt \tag{15-44}$$

valid for times $t > t_o$, t_o being such that :

$$\mu E t_o = \frac{\lambda}{2} \tag{15-45}$$

In the weak injection case ($C << 1$), the Coulomb repulsion can be neglected to a first approximation and $q(\bar{z}) \approx q_o$. Integrating (15-44) by taking (15-45) into account leads to :

$$\xi = \xi_o \exp\left[\frac{2}{\alpha}\frac{q_o \mu E \lambda}{\mu \eta}\left(t - t_o\right)\right] = \xi_o \exp\left[\frac{4}{\alpha}\frac{q_o \mu^2 E^2}{\mu \eta}\, t_o\left(t - t_o\right)\right] \tag{15-46}$$

This expression allows us to see what is the wave-length of the perturbation which has been the most amplified at time t. As t_o is proportional to λ, we seek for the t_o value which makes ξ maximum at time t. From (15-46) this obviously happens for $t_o = t/2$ and we then obtain [24] :

$$\max \xi = \xi_o \exp\left[\frac{1}{\alpha}\frac{q_o\mu^2E^2t^2}{\mu\eta}\right] = \xi_o \exp\left[\frac{1}{\alpha}\frac{q_o\bar{z}^2}{\mu\eta}\right] = \xi_o \exp\left[\frac{1}{\alpha}(TC)_{layer}\right] \qquad (15\text{-}47)$$

Assuming that the initial perturbation ξ_o is such that it gives a white spectrum for its amplitude as a function of the wave-number $2\pi/\lambda$ and that the convection is detectable when ξ reaches a given amplitude ξ_{app}, it appears that the criterion corresponding to a magnification factor is characterised by the product TC relative to the charged layer at the instability time [24] :

$$(TC)_a = \alpha \ \ln\frac{\xi_{app}}{\xi_o} \qquad (15\text{-}48)$$

This criterion is similar to the non linear instability criterion for weak injection (§15-2). An estimate of the numerical value based on $\alpha \cong 10$ and an amplification factor ranging from 100 to 1000 gives $(TC)_a \cong 50$ to 70.

In the case of SCL injection (C >> 1) it is also possible to solve eq. (15-44). The charge density just behind the front is $q(\bar{z}) = q(t) = \varepsilon/\mu t$ which expresses that the decrease in q is due to Coulomb repulsion. The equation then writes :

$$\frac{d\xi}{\xi} = \frac{2}{\alpha}\frac{\varepsilon E\lambda}{\mu\eta}\frac{dt}{t} \qquad (15\text{-}49)$$

which gives after integration :

$$\xi = \xi_o \exp\left[\frac{2}{\alpha}\frac{\varepsilon E\lambda}{\mu\eta}\ln\frac{t}{t_o}\right] = \xi_o \exp\left[\frac{4}{\alpha}\frac{\varepsilon E^2}{\eta}\ t_o \ln\frac{t}{t_o}\right] \qquad (15\text{-}50)$$

Again the maximum amplitude magnification leads to the wave-number selection through the time t_o such that $t_o = t/e$. This gives :

$$\max \xi = \xi_o \exp\left[\frac{4}{e\alpha}\frac{\varepsilon E^2}{\eta}t\right] = \xi_o \exp\left[\frac{4}{e\alpha}\frac{\varepsilon E\bar{z}}{\mu\eta}\right] = \xi_o \exp\left[\frac{4}{e\alpha}(T)_{layer}\right] \qquad (15\text{-}51)$$

The criterion in terms of an amplification factor is a certain value of T [24] :

$$T_a = \frac{e\alpha}{4}\ \ln\frac{\xi_{app}}{\xi_o} \qquad (15\text{-}52)$$

This justifies that the criterion for transient conditions of SCL unipolar injection is characterised by the same parameter than for steady conditions. Retaining a magnification factor between 100 and 1000 gives T_a ranging from 31 to 47 [24].

Fig. 9 shows typical results for the variations of the stability time t_1 as a function of the applied voltage in the case of space-charge-limited injection in deionized nitrobenzene. Clearly the variation law is in E^{-2}. This law was also obtained for other liquids [16,19,20]. The critical values T_{a1} range from 20 to 50 which is in fair agreement with the above order of magnitude estimation.

In the case of SCL injection under AC fields, a similar derivation leads to a critical field [20] :

$$E_a \approx 2.2 \left[T_{a2} \frac{\eta}{\varepsilon} f \right]^{1/2} \quad E_a \approx 2.2 \, [T_{a2} \, (\eta/\varepsilon) \, f]^{1/2} \tag{15-53}$$

for frequency $f > f_c \sim 1/t_T$, $t_T = d^2/\mu V$ being the transit time of the charge carriers. Experiments confirm the $f^{1/2}$ dependence (Fig. 11) and give T_{a2} on the order of T_{a1} [20].

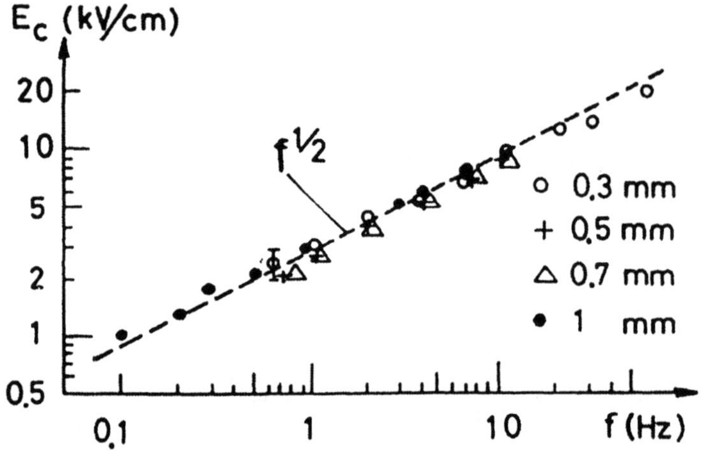

Fig. 15-11 – Critical field for manifest convection versus the frequency f of the applied AC field for SCL injection into a viscous mildly polar liquid ("pyralene 1460"). (*from Atten et al.* [20]).

6. Effect of a forced flow

A flow with a mean velocity parallel to the electrodes generally interacts with the convection generated by unipolar injection. Let us consider a Poiseuille flow between two parallel plates with an injection from one of the plates. The instability problem which then arises has been studied through linear analysis. Similarly to the problem of horizontal Poiseuille flow superimposed on a vertical temperature gradient (heating from below) [26], the flow has a stabilising effect for the velocity perturbations in the form of two-dimensional rolls with their axis normal to the flow direction [27,28], provided the Reynolds number is not too high : $Re \leq Re_1 \cong 10\,000$. But the forced flow has no influence on the destabilising of convective rolls whose axis is parallel to the flow direction until Re reaches the critical value $Re_c \cong 11\,500$ of instability induced by the Reynolds stresses (linear theory) [27,28].

In practice, there is always a finite extension for the fluid system and it is necessary to consider the effects of lateral walls or entering section. One interesting problem is that of a Poiseuille flow between two parallel plates entering a region where a SCL injection occurs beyond a certain abscissa x_0. For a low enough applied voltage V ($V < V_c$), the ions injected at $x = x_0$ have two orthogonal velocity components, U(z) due to the forced flow and μE due to the action of the electric field. Their trajectories determine the separatrix surface between charged and charge free zones and the ions reach the opposite electrode at the abscissa $x_0 + L_E$, L_E being the electric entrance length ($L_E \approx Ud/\mu E$).

For $V > V_c$, the instability problem bears some similarity with the one under transient conditions : the abscissa difference $x - x_0$ plays a role similar to that of time t. Manifest convection is detected for $x \geq x_0+L_1$, L_1 corresponding to the stability time t_1 : $L_1 \cong U\,t_1 \cong T_{a1}\,U\eta/\varepsilon E^2$ [29]. The flow pattern for $x \geq x_0+L_1$ depends on the ratio of ionic and mean forced flow velocities : for U lower than a few times μE, the pattern is three-dimensional (modulated transverse rolls or hexagonal cells) whereas for higher U values, there appear first longitudinal rolls which further destabilise [29,30].

REFERENCES

[1] N. Felici : Dir. Current, **2** (1972), 147-165.

[2] S. Chandrasekhar : "*Hydrodynamic and hydromagnetic stability*", Oxford University Press, 1961, chapter 2.

[3] P. Atten : J. Mécanique, **14** (1975), 471-521.

[4] A. Castellanos : IEEE Trans. Electr. Ins., **IE26** (1991), 1201-1215.

[5] P. Atten and R. Moreau : J. Mécanique, **11** (1972), 471-521.

[6] J.M. Schneider and P.K. Watson : Phys. Fluids, **19** (1970), 1948-1954.

[7] P. Atten and J.C. Lacroix : J. Mécanique, **18** (1979), 469-510.

[8] P. Atten : C.R. Acad. Sci. Paris, **283B** (1976), 29-32.

[9] N. Felici : Rev. Gén. Electr., **78** (1969), 717-734.

[10] D.O. Gough, E.A. Spiegel and J. Toomre : J. Fluid Mech., **68** (1975), 695-719.

[11] N. Felici : Dir. Current, **2** (1971), 90-99.

[12] N. Félici, J.P. Gosse and B. Gosse : Rev. Gén. Electr., **85** (1976), 860-869.

[13] J.C. Lacroix, P. Atten and E.J. Hopfinger : J. Fluid Mech., **69** (1975), 539-563.

[14] P. Atten and J.P. Gosse : J. Chem. Physics, **51** (1969), 2804-2811.

[15] A. Denat, B. Gosse and J.P. Gosse : J. Electrostatics; **7** (1979), 205-225.

[16] R. Tobazéon : J. Electrostatics, **15** (1984), 359-384.
[17] J.C. Lacroix and R. Tobazeon : in Proceedings 4th Intern. Conf. Cond. Breakdown in Dielectric Liquids (Ed. T. Gallagher), Dublin, 1972, 93-96.
[18] P.K. Watson, J.M. Schneider and H.R. Till : Phys. Fluids, **13** (1970), 1955-1964.
[19] J.C. Filippini, J.C. Lacroix and R. Tobazéon : C. R. Acad. Sci. (Paris), **B271** (1970), 936-939.
[20] P. Atten, B. Malraison and S. Ali Kani : J. Electrostatics, **12** (1982), 477-488.
[21] P. Atten : Phys. Fluids, **17** (1974), 1822-1827.
[22] J.D. Cross and H.T. Wang : IEEE Trans. Electr. Ins., **EI26** (1991), 641-646.
[23] G.S. Kath and J.F. Hoburg : Phys. Fluids, **20** (1977), 912-916.
[24] P. Atten : IEEE Trans. Diel. & Electr. Insul., **DEI3** (1996), 1-17.
[25] M. Rigo : "*Etude du régime transitoire d'un système électrohydrodynamique*", Ph. D. Thesis, University of Geneva, 1995.
[26] K.S. Gage and W.H. Reid : J. Fluid Mech., **33** (1968), 21-32.
[27] A. Castellanos and N. Agräit : IEEE Trans. Ind. Appl., **28** (1992), 513-519.
[28] J.L. Lara, A. Castellanos and F. Pontiga : Phys. Fluids, **9** (1997), 399-405.
[29] P. Atten and T. Honda : J. Electrostatics, **11** (1982), 225-245.
[30] P. Atten and B. Malraison : in "*Symmetries and broken symmetries in condensed matter physics*", (Ed. N. Boccara), IDSET, Paris 1981, 503-515.

Chapter 16

UNIPOLAR INJECTION : FINITE AMPLITUDE CONVECTION

The study of non linear instability showed that the convection which arises above the instability criterion is characterised by an amplitude greater than the ionic velocity μE. This result implies that the liquid has a high enough viscosity because it was derived by taking into account only the viscous effects. In this part, we consider again the case of the viscous effects being dominant, i.e. we assume that the convective motion is characterised by a low enough value of the Reynolds number :

$$Re \cong \frac{w'd}{\nu} \leq Re_{max} \sim 1 \qquad (16\text{-}1)$$

At the non linear instability criterion T_{nl}, the maximum of the convective velocity is such that $w' \cong \mu E$ and we have :

$$Re \cong \frac{\mu Ed}{\nu} = \frac{\mu V_{nl}}{\nu} = \frac{\varepsilon V_{nl}}{\mu\eta}\frac{\rho\mu^2}{\varepsilon} = \frac{T_{nl}}{M^2} \leq Re_{max} \qquad (16\text{-}2)$$

Condition (16-2) is fulfilled for the mobility parameter M taking high enough values. Note that this condition on M number is more and more stringent as the injection level C is decreased because $T_{nl} \sim 220/C$ in the weak injection case $(C \ll 1)$.

1. A solution for steady convection under weak injection

An approximate analytical solution has been obtained in the case of weak injection $(C \ll 1)$. Before giving this solution, it is worthwhile to mention that, in this case of weak injection, the liquid motion has a very weak influence on the enhancement of total current for an autonomous injection, i.e. for an injection with a charge density q_o which is not affected by the convection of the liquid and its subsequent perturbations (this is indeed the case with metallic electrodes subjected to rather moderate electric fields [1]). The weak injection case $(C \ll 1)$ is characterised by a weak effect of the coulombic repulsion between charge carriers. Therefore, the decrease in charge density q along the trajectories of charge carriers can be neglected and q can be prescribed as constant : $q = q_o$ [2,3]. With a uniform density q_o of injected charge at the injector, in the absence of liquid motion the field E_o on the injector is $(E_o = E(z{=}0))$:

$$E_o = \frac{V}{d}\left(1 - \frac{C}{2}\right) \tag{16-3}$$

At steady state, the mean current density $\bar{J}$ takes the same value at any height; this value calculated at the injector is :

$$\bar{J} = \mu q_o E_o \tag{16-4}$$

A liquid motion which convects the charge tends to diminish the mean space charge in the core and, therefore, to slightly increase the field E_o on the injector. With the assumption that q_o is independent of the liquid movement, we deduce the following relations for E_o and for the electric Nusselt number Ne [4] :

$$\frac{V}{d}\left(1-\frac{C}{2}\right) \le E_o < \frac{V}{d}, \qquad 1 \le Ne < 1 + \frac{C}{2} \tag{16-5}$$

The increase in E_o and in $\bar{J}$ induced by the convection is on the order of C. For $C \ll 1$ we can then make the approximation that the current is independent of the state of motion.

This approximation has already been used to derive relation (15-12) which gives the expression of the velocity amplitude in the hydraulic model of convection rolls. We consider again this geometry of two-dimensional (2-D) convective rolls in the layer bounded by rigid electrodes. At steady state, there are 3 points where the velocity of ions $\mathbf{v}_i = \mathbf{u}+\mu\mathbf{E}$ is zero (Fig. 1). Two points, P_1 and P_2, are stagnation points for the ions and P_3 is the centre of ion trajectories which would be cycles [5,6]. The singular trajectory joining the singular points P_1 and P_2 is the separatrix between the trajectories from injector to collector and the closed trajectories inside the zone that it delineates (Fig. 1). There is no ion entering this zone where the trajectories are cycles; at steady state after a very long time, this zone is free of charge as a result of the action of Coulomb repulsion.

An approximate analytical solution can be obtained by assuming that the velocity field is self-similar, which gives for the stream function ψ :

$$\psi(x,z) = A\, \psi_o(x,z) \tag{16-6}$$

where A is the velocity amplitude and ψ_o is such that the norm of the velocity field $\mathbf{u}_o$ is 1. In this type of problems, choosing for ψ_o a good approximation generally leads to accurate estimates of the velocity amplitude. We take here :

$$\psi_o(x,z) = \frac{L}{2\pi}\,(1-\cos 2\pi x)\,\sin \pi\frac{x}{L} \tag{16-7}$$

where L is the width of the 2-D rolls (from the value of the critical wave number $k_c \approx 4.57$ we deduce $L \approx 0.66$). In the charged zone Ω_1, we have $q = q_o$ whereas $q = 0$ inside the zone bounded by the separatrix. Expressing that the steady current takes the same value at any height z leads to the determination of the abscissa $x_{sep}(z)$ of the separatrix (Fig. 1) for $z_o < z < 1-z_o$ $\;(z_o = (1/2\pi)\,\mathrm{Arcos}[1-2/A])$:

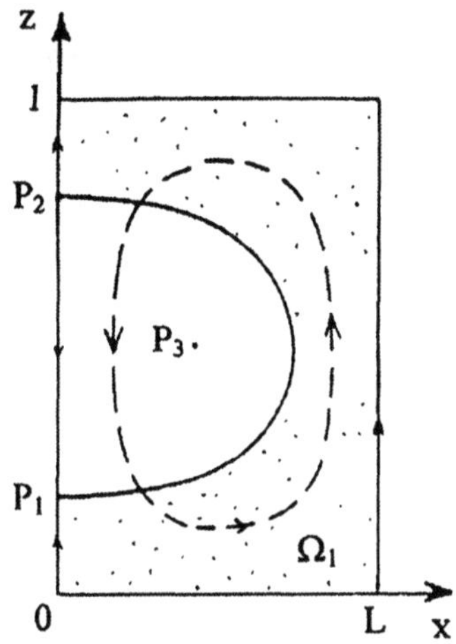

Fig. 16-1 – Cross section of a steady state convective 2-D roll. The separatrix (full line) joins the two stagnation points for ions P_1 and P_2. Ω_1 denotes the charged zone. The dotted line represents a streamline.

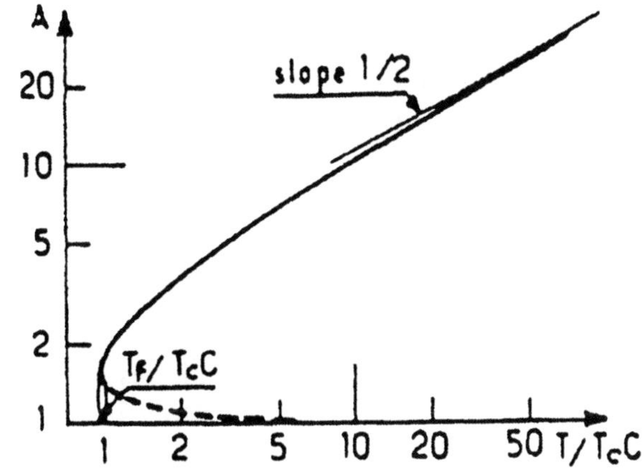

Fig. 16-2 – Amplitude of steady state approximate analytic solution as a function of the parameter T/T_cC in the case of weak injection ($C \ll 1$). The non linear criterion is $T_f = 0.933\ T_cC$ and corresponds to $A \approx 1.60$ (*from Castellanos & Atten* [6]).

$$x_{sep} = \frac{AL}{2\pi}\left(1-\cos 2\pi z\right)\sin \pi \frac{x_{sep}}{L} \qquad (16\text{-}8)$$

Relation (16-8) defines the separatrix for a given A value. It remains to relate A with the instability parameter T. This is done by multiplying the Navier-Stokes equation by $\mathbf{u} = A\,\mathbf{u}_o$ and integrating over the roll cross section Ω :

$$A\int_{\Omega} \mathbf{u}_o \nabla^2 \mathbf{u}_o \, dS + T\int_{\Omega_1} q\, \mathbf{u}_o \mathbf{E}\, dS = 0 \qquad (16\text{-}9)$$

The expression (16-7) for ψ_o leads to the following relation between A and T [5,6] :

$$A^2 = \frac{16}{L}\frac{T}{T_cC}\int_{z_o}^{1-z_o} x_{sep}(z)\, dz \qquad (16\text{-}10)$$

Fig. 2 shows the variations of T/T_cC as a function of A. The minimum obtained for $A = 1.595$ corresponds to the non linear criterion T_{nl} :

$$\frac{T_{nl}}{T_c C} = 0.933 \tag{16-11}$$

This ratio is close to the value 1 given by the hydraulic model. Relation (16-10) and the results as shown by Fig. 2 give the asymptotic behaviour of velocity amplitude for high T values already predicted by the hydraulic model. The asymptotic behaviour can be approximated by :

$$A \cong 3\left(\frac{T}{T_{nl}}\right)^{1/2} \approx 0.2\,(TC)^{1/2} \tag{16-12}$$

The approximate analytical solution is of theoretical rather than of practical interest. Independently of the fact that it concerns 2-D rolls whereas convection organises in the

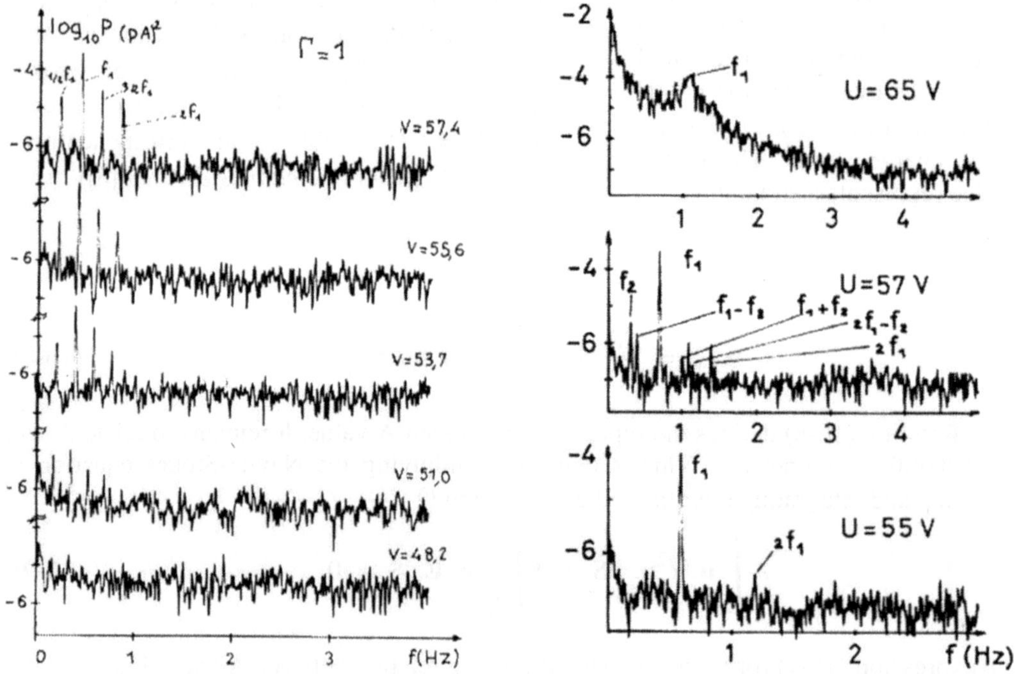

(a) (b)

Fig. 16-3 – Frequency spectra of the fluctuations of total current through a cylindrical domain of small aspect ratio Γ for different applied voltages. Space-charge-limited injection in Pyralene 1460 with liquid thickness $d \cong 0.25$ mm. a) $\Gamma = 1$, $V_c \cong 50$ V. b) $\Gamma = 1.4$, $V_c \cong 53$ V (*from Malraison & Atten* [8,9]).

form of hexagonal cells, there are two basic questions raised by this approach : i) How this asymptotic solution is reached (for very long times) ? ii) Is such a solution stable despite the very strong charge density gradients close to the separatrix ? Before examining these points, it is interesting to have a look at experimental results.

2. Unsteady character of electroconvection

Experiments were performed on finite amplitude electroconvection in the case of space-charge-limited (SCL) injection for which we know that the motion organises in the form of polygonal cells. As for various fluid systems this study was motivated by identifying the successive instabilities of the convective motion and eventually determining the route to chaos. A close examination of the fluctuations of the total current flowing through the liquid layer revealed that the finite amplitude convection is not strictly steady

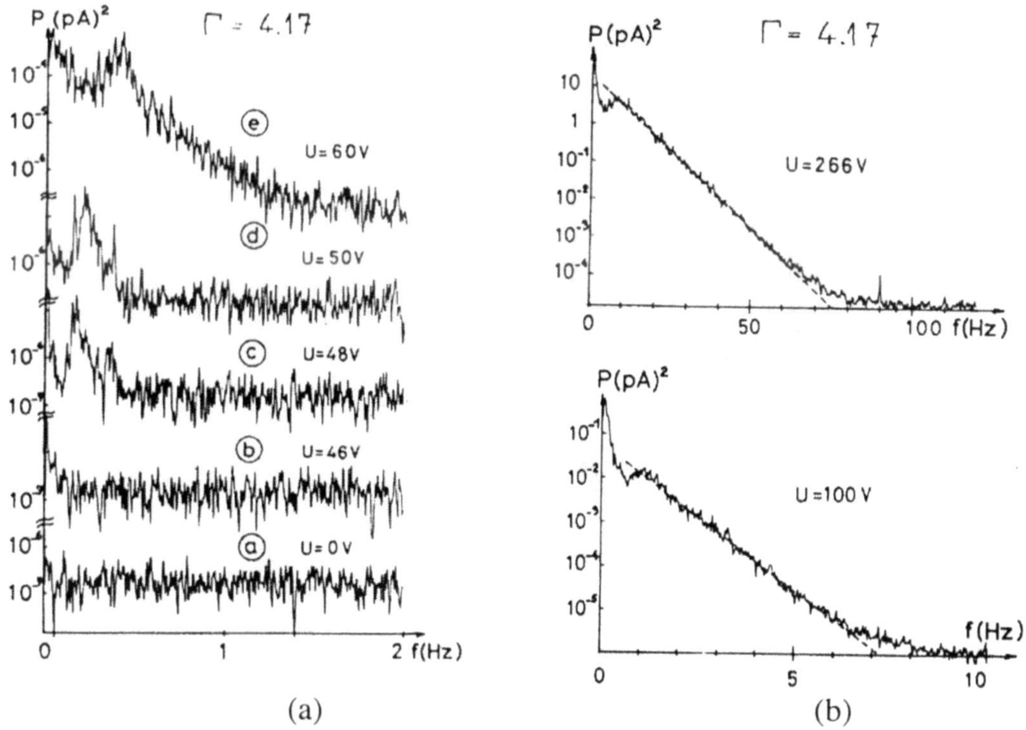

Fig. 16-4 – Frequency spectra of the fluctuations of total current through a cylindrical domain of aspect ratio $\Gamma \cong 4.2$ (space-charge-limited injection in Pyralene 1460 with liquid thickness $d \cong 0.25$ mm). a) around the critical voltage $V_c \cong 47$ V. b) at higher applied voltages (*from Atten et al.* [6]).

but always exhibits a time dependent character independently of the aspect ratio Γ of the test cell [7,8]. As the applied voltage V is very slowly raised above the critical value V_C, the liquid passes directly from rest to a time dependent motion.

For an axisymmetric domain of radius approximately equal to the electrode distance d (Γ ≅ 1), there is only one convective cell. In a voltage interval above the critical one, the total current has been found to exhibit a periodic fluctuation which can be induced only by a velocity field with a time periodic vacillation [8,9]. Fig. 3a shows typical power spectra of current fluctuations with discrete peaks at a well defined frequency. By further increasing the voltage, a second incommensurate frequency appears and for higher V values the power spectra become continuous (see Fig. 3b relative to a slightly higher aspect ratio Γ ≅ 1.4). The continuous power spectra indicate that the liquid motion is chaotic [7,8,10]. In the case Γ ~ 1, there is a transition to turbulence through quasi-periodicity.

For greater values of Γ, the power spectra of current fluctuations are always continuous (see Fig. 4a), which indicates that the time dependent motion is chaotic [7-10]. For any Γ value the continuous spectra always exhibit a broad peak with a well defined frequency f_1 (Fig. 4b). This frequency of current fluctuations to a first approximation does not depend on the value of Γ and, in particular, takes fully similar values in the case Γ ≅ 1 which corresponds to a unique convective cell. It varies like V^2/d^2 (Fig. 5) in a certain

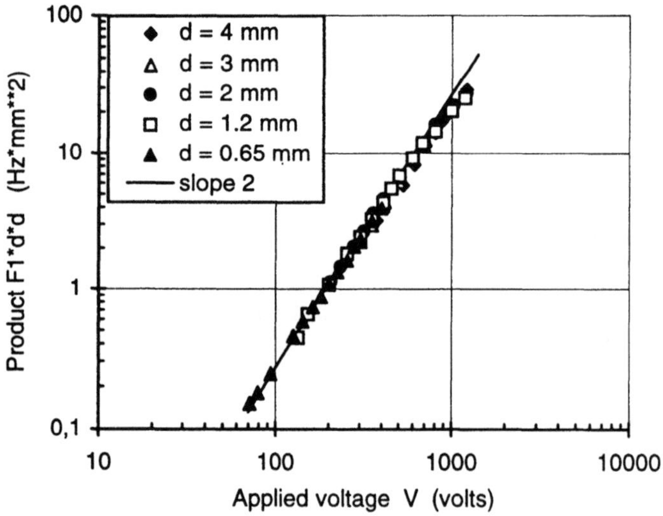

Fig. 16-5 – Variations versus the applied voltage V of the product $f_1 \times d^2$ of the frequency f_1 of the broad peak in the power spectrum of current fluctuations and the square of interelectrode spacing d. SCL injection in Pyralene 1460 for different values of the liquid thickness in convective regime where the viscous effects are dominant.

interval $V_c \leq V \leq V_{tr}$ which corresponds to a convection regime where the viscous effects are dominant (see chapter 17). The frequency f_1 is closely related to a modulation of the convective velocity around a mean value (vacillation). Indeed the period $T_1 = 1/f_1$ is close to the mean rotation time in the convective cells which is proportional to d/w' where w' is the typical liquid velocity. By taking expression (17-9) for w' (see §17-3-a) we have $T_1 \sim d/w' \sim (\eta/\varepsilon)\, d^2/V^2$ which is fully consistent with the results as shown in Fig. 5 [7-10].

Another interesting observation is that, in the spectra, the power P of current fluctuations has an exponential decrease as a function of frequency for $f > f_1$: $P \propto \exp(-f/f_c)$ (see Fig. 4-b). The characteristic frequency f_c is proportional to the peak frequency f_1 [7,8,10].

3. Theoretical approaches to unsteady convection

From the evidence of the time dependent character of the convection induced by SCL injection, we can conjecture that the steady solution for convective motion obtained in the case of weak injection is unstable and cannot be observed. Although the existence of very steep gradients of charge density on the separatrix makes it very likely that the steady solution is unstable, a direct proof through a stability analysis has not been performed due to the mathematical difficulty of the analysis. Only a much simpler problem could be studied. In this problem, the convective flow is assumed to have a small periodic vacillation, i.e. its non dimensional velocity amplitude has the following variation :

$$A(t) = A_o + \varepsilon \sin \omega t \qquad (16\text{-}13)$$

where ε is a small parameter. In the asymptotic case of very weak injection ($C \ll 1$), a mathematical study of the variations of the charge distribution and of the associated motive torque (which should sustain the convective motion) leads to the conclusion that the response is chaotic independently of the values of ε and ω [11,2]. This suggests that the convective motion cannot be steady and exhibits a chaotic time behaviour.

Another way of investigating the transient regime of convection and the time dependent or steady character of the solution for very long times is to use numerical modelling. In the asymptotic case of very weak injection ($C \ll 1$), a first approach based on a finite difference scheme led to very bad results [5,6]. The basic reason is that the finite difference scheme introduces an important diffusion-like term in the first order equation expressing charge conservation and thus gives an unrealistic solution [5,6].

A second approach leads to much more satisfactory results. It is based on a particle type method where one follows the trajectories of numerous injected particles simulating the individual charge carriers. By taking again the assumption of a self similar velocity field, the particles method gives a good description and provides a way of solving, with a reasonable accuracy, the first order equation which defines a problem of "hyperbolic" nature from the mathematical viewpoint [5,6]. Fig. 6 shows typical distributions of particles at different times illustrating the progressive build-up of the nearly charge free zone bounded by the separatrix.

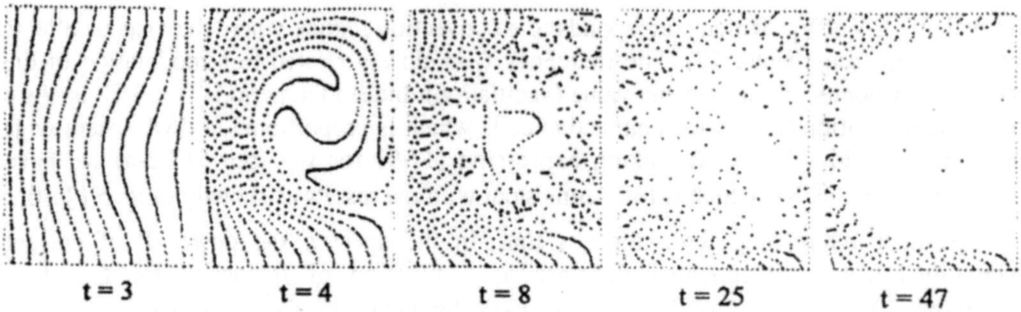

Fig. 16-6 – Distribution of particles in a convective 2D cell at different non dimensional times t. Weak injection (C = 0.1) in supercritical conditions T = 25 000 ($T_c \approx$ 19 460). There are 11 injecting sites on the injector (bottom electrode) (*from Castellanos et al.* [5]).

The time evolution of the velocity amplitude A exhibits a long transient regime [5] with large amplitude fluctuations around a slowly increasing average (Fig. 7-a). At long times, A tends toward a nearly steady value very close to the one predicted by the analytical steady solution (see Fig. 2). The fluctuations are irregular but appear to have a rather well defined frequency [5,6]. The power spectra of the velocity amplitude in a time interval where the averaged A value has reached its saturation value within a few percent have the

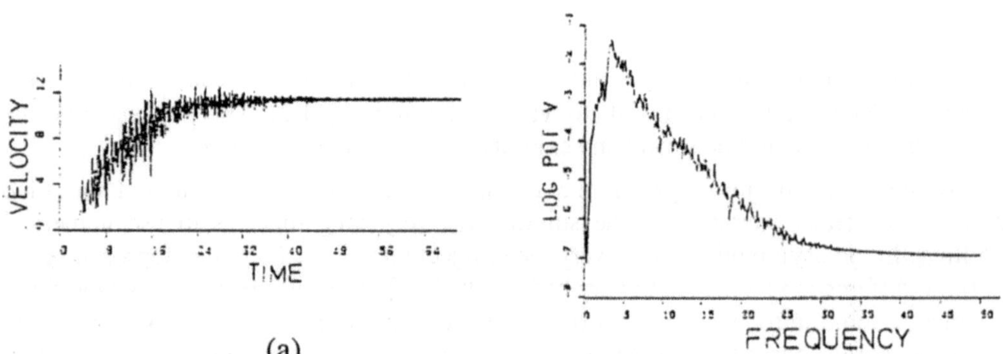

(a) (b)

Fig. 16-7 – Transient and time dependent behaviour of convection in a 2-D convective roll given by the particles method. Weak injection (C = 0.1) in supercritical conditions T = 25 000. a) Velocity amplitude versus time. b) Power spectrum of velocity amplitude filtered with a numerical high-pass Kaiser filter (*from Castellanos et al.* [5]).

typical broad-band shape shown on Fig. 7-b. Except for the low frequency noise in a small interval $0 < f < f_{max} \cong f_1/2$, these spectra are very similar to those obtained from current fluctuations in SCL injection (see Fig. 4-b). The peak frequency is on the order of A/2, a value close to the rotation frequency in 2-D rolls [12].

The only qualitative difference with the experimental observations (§16-2) is that the solution given by this particles method tends to a steady state at very long times. This very likely arises from the approximation done when expressing the boundary condition $q = q_o$ on the injector in terms of injection of particles. A recent work improving the particles method to make it "robust" shows that the chaotic behaviour of the system does not vanish [13]. Moreover, this work extends the simulation and the results to the case of moderate and strong injections and gives a picture closely analogous to the observations [13].

REFERENCES

[1] A. Denat, B. Gosse and J.P. Gosse : J. Electrostatics; **7** (1979), 205-225.

[2] A. Castellanos : IEEE Trans. Electr. Ins., **IE26** (1991), 1201-1215.

[3] N. Felici : Rev. Gén. Electr., **78** (1969), 717-734.

[4] P. Atten, F.M.J. McCluskey and A.T. Perez : IEEE Trans. Electr. Ins., **EI23** (1988), 659-667.

[5] A. Castellanos, P. Atten and A.T. Perez : J. Physico-Chem. Hydrodyn., **9** (1987), 443-452.

[6] A. Castellanos and P. Atten : IEEE Trans. Ind. Appl., **IA23** (1987), 825-830.

[7] P. Atten, J.C. Lacroix and B. Malraison : Phys. Lett., **79A** (1980), 255-258.

[8] B. Malraison and P. Atten : in "*Symmetries and broken symmetries in condensed matter physics*" (Ed. N. Boccara), IDSET, Paris 1981, 439-449.

[9] B. Malraison and P. Atten : C.R. Acad. Sci. Paris, **292** ser. II (1981), 267-270.

[10] B. Malraison and P. Atten : Phys. Rev. Lett., **49** (1982), 723-726.

[11] A.T. Perez and A. Castellanos : Phys. Rev. A, **44** (1991), 6659-6664.

[12] P. Atten and J.C. Lacroix : J. Mécanique, **18** (1979), 469-510.

[13] R. Chicon, A. Castellanos and E. Martin : *Numerical modelling of Coulomb-driven convection in insulating liquids*, J. Fluid Mech., **344** (1997), 43-66.

Chapter 17

UNIPOLAR INJECTION INDUCED ELECTROCONVECTION

1. General considerations

Up to now we considered the convection of rather limited amplitude and we implicitly assumed that the convective pattern is well defined so that we deal with nearly identical convective cells of long life time compared with the mean rotation time in a cell. The studies recalled in chapter 2 focused on the time dependent character of injection induced convection and the transition to chaotic behaviour for convection in one convective cell only. This picture no longer applies when the applied voltage is high enough : the pattern destabilises ("phase turbulence" [1,2]) and the life time of convective cells becomes so short that they can no longer be considered as individual and well defined cells (see for instance the pattern in [3]). The time behaviour of the measured current then reflects the complex evolution of the spatial distribution of the velocity field as well as the vacillation of velocity amplitude in individual convective cells (the "phase turbulence" results in low and very low frequency noise in the power spectra as it appears in Fig. 16-4).

In this part we examine the main properties of electroconvection induced by pure unipolar injection, in particular in the case of strong injection. A first point to mention is that this convection generally contributes to the transfer of charge and that the problem bears an analogy with thermal convection where the advection of hot and cold blobs in the fluid increases the heat transfer. Here the convection advects "blobs" which are highly charged in forward columns (where the velocity is in the direction of the force exerted by the electric field) but which are less charged in backward columns. The mean current density $\overline{J}$ due to both drift and convection of charge carriers takes the following expression in planar geometry (plane parallel electrodes) [3] :

$$\overline{J} = \mu \, \overline{q} \, \overline{E} + \overline{qw} \tag{17-1}$$

Here $\overline{q}$ and $\overline{E}$ denote the mean charge density and the electric field in a plane at height z and $\overline{qw}$ represents the correlation between velocity and charge density fluctuations and is the convective part of the current. There is a close analogy with the expression of heat flux $\overline{H}$ in the Rayleigh-Bénard problem :

$$\overline{H} = -\, k_T \frac{d\overline{T}}{dz} + \rho c_p \, \overline{\theta w} \tag{17-2}$$

where k_T is the thermal conductivity, $\overline{T}$ the temperature averaged in a horizontal plane, ρ the mass density, c_p the specific heat and $\overline{\theta w}$ the correlation between temperature (θ) and vertical velocity (w) fluctuations. When the motion is of high enough intensity, the core of the convecting fluid layer is at a constant mean temperature so that, from (17-2), the transfer of heat is due to convection alone. We have a very similar situation when the typical velocity w' is much greater than the mean drift velocity $\mu V/d$. Then $\overline{qw} \sim q'w' >> \mu\,\overline{q}\,\overline{E}$ (q' and w' denote the characteristic values of the perturbations q and w) and the convective current fully dominates over the current associated with the drift of charge carriers. In this case we deduce from (17-1) : $\overline{J} \approx \overline{qw}$. From this analogy with the thermal problem, we can expect to derive relations for the current density by using similar analyses, in particular by defining different zones in the layer, characterised by different properties in charge transfer.

A second fact to point out is the vigorous character of the motion induced by charge injection which reflects the intensity of the Coulomb force. In the case of liquids having a high enough viscosity (M >> 1), the finite amplitude convection is characterised by a maximum velocity always greater than the mean drift velocity $\mu V/d$ of charge carriers. More generally, an order of magnitude of the typical fluid velocity w' can be obtained by balancing the densities of kinetic energy of the fluid $(1/2)\rho w'^2$ and electrostatic energy $(1/2)\varepsilon E^2$ [4,5,6]. This gives :

$$w' \sim \left(\frac{\varepsilon}{\rho}\right)^{1/2} \frac{V}{d} \qquad \text{or} \qquad \frac{w'd}{\mu V} \sim M = \frac{1}{\mu}\left(\frac{\varepsilon}{\rho}\right)^{1/2} \tag{17-3}$$

The hydrodynamic mobility $(\varepsilon/\rho)^{1/2}$ is on the order of a few 10^{-7} m^2/Vs for dielectric liquids and the typical liquid velocities are in the range 10^{-2} - 1 m/s. The non dimensional number M is a parameter which plays an important role in characterising EHD flows. For M high enough, the fluctuating fluid velocity is greater than the drift velocity of charge carriers. Then the charge carriers can be locally entrained against the action of the electric field and the charge distribution depends drastically on the motion it induces. This is the case of most liquids and also of gases (at pressures $\geq 10^5$ Pa) with charged particles of small size [3,7]. For M < 1, conversely, the motion only slightly affects the trajectories of charge carriers which, in practice, only very slightly depart from the field lines. This occurs in the case of air (and gases) with ions or big particles as charge carriers [7] (in the example of the electric wind, the gas flow has no significant influence neither on the ion trajectories nor on the passage of current).

In the following, we examine only the case where the mobility parameter M takes high enough values. Another non dimensional parameter which plays a role is the number C which gives a measure of the intensity of injection. Electroconvection of course depends on the injection strength. Moreover two regimes of convection must be distinguished depending on which of the inertial or the viscous effects are dominating. The regime of dominant viscous effects corresponds to low values of the Reynolds number Re characterising the big eddies of length scale on the order of d, the interelectrode distance.

But although Re is low, the flow is not laminar and can be considered, to some extent, as "turbulent" if we take into account its mixing and transport properties. For high values of the Reynolds number, the inertial effects dominate and there are eddies of scales ranging from large (a fraction of d) to small or very small values [3,8,9]. This fully turbulent regime is characterised by different properties of transport of charge. The transition between both regimes corresponds to a particular value Re_{tr} of the Reynolds number.

2. Weak injection electroconvection

Although there has been no reliable experimental study of electroconvection under weak injection of charge, this case has nevertheless some practical interest because, for injections by metallic electrodes into a dielectric liquid, the amount of charge created at the metal/liquid interface is limited by the electrode process, which gives a finite value for q_o. In the asymptotic case of weak injection characterised by $C << 1$, the mathematical problem can be simplified and qualitative properties can be easily deduced. In conditions where the viscous effects are dominant, the hydraulic model approach and the determination of an approximate analytical solution (§16-1) are examples of such tractable problems.

We have shown (§16-1) that for $C << 1$ the increase in electric Nusselt number Ne is smaller than C/2 and can be neglected to a first approximation. The problem then reduces to determine the order of magnitude of the liquid velocity and we have to distinguish between the viscous and inertially dominated regimes of convection. In the case where the viscous effects are dominant ($Re < Re_{tr} \sim 1$) the expression (16-12) was obtained for a conjectured steady solution. It is interesting to recover the functional dependence given by (16-12) through order of magnitude arguments which can apply for the very chaotic state of convection. Because viscous dissipation dominates at large scale, it is natural to expect the typical fluid velocity to vary in proportion to the force. In the Navier-Stokes equation, to a rough approximation the viscous friction term balances the term of motive Coulomb force :

$$\eta \nabla^2 w \sim qE \qquad \Rightarrow \qquad \eta \frac{w'}{d^2} \sim q' \frac{V}{d} \qquad (17\text{-}4)$$

(q' and w' denote the characteristic values of the perturbations q and w). Multiplying (17-4) by w' and taking into account the remark that the product of fluctuations q'w' is related to the current density ($q'w' \sim \overline{qw} \approx \bar{J}$), we have [10] :

$$\eta \frac{w'^2}{d^2} \sim q'w' \frac{V}{d} \sim \overline{qw} \frac{V}{d} \cong \bar{J} \frac{V}{d} \qquad (17\text{-}5)$$

By expressing the current density on the injector, $\bar{J} \approx Cst \approx \mu q_o V / d$, we finally obtain :

$$w' \sim \left(\frac{\bar{J} V d}{\eta} \right)^{1/2} = \mu \frac{V}{d} \left(\frac{q_o d^2}{\mu \eta} \right)^{1/2} = \mu \frac{V}{d} (TC)^{1/2} \qquad (17\text{-}6)$$

which is equivalent to (16-12). Notice that for a constant injection level C, the typical velocity w' would vary as $V^{3/2}$. For a constant injected charge density (which is the case expected with metallic electrodes under limited field strength), w' would increase linearly with the applied voltage V.

In the fully turbulent regime (Re >> 1) where eddy viscosity is everywhere dominant, an order of magnitude of the liquid velocity can be obtained by stating that the kinetic energy of a "blob" of charged liquid leaving the immediate vicinity of the injector comes from the work of the Coulomb force. At mid-height this gives : $(1/2)\rho w'^2 \sim q'Ed/2$ (the same relation is obtained by balancing the inertial and Coulomb terms in Navier-Stokes equation). By assuming that the charge perturbation q' to a first approximation is proportional to q_o and by using the expression of the parameter C, one finally obtains [10] :

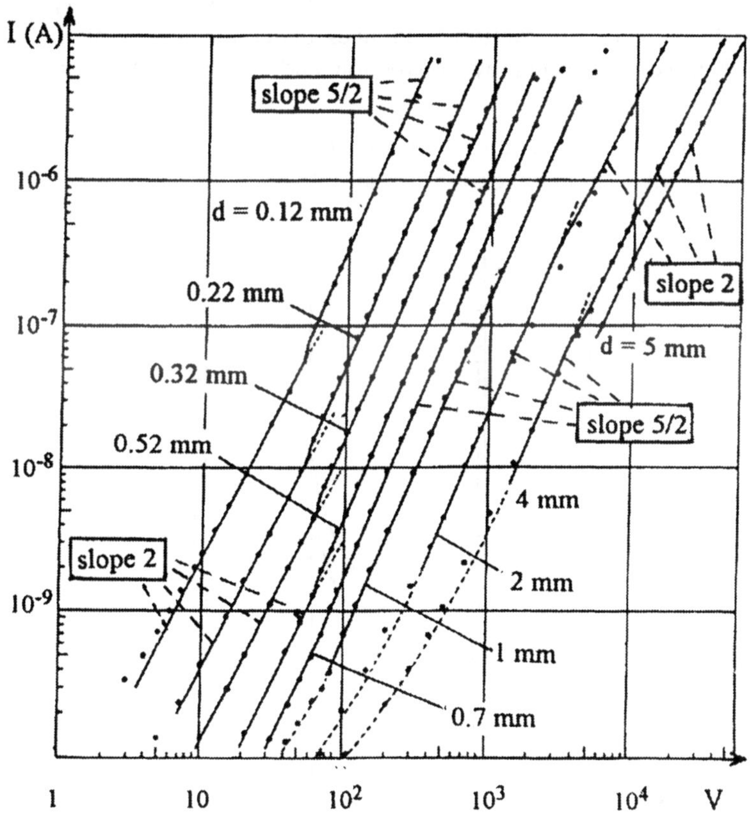

Fig. 17-1 – Space-charge-limited injection in a polychlorobiphenyl (pyralene 1500) of high viscosity ($\eta \cong 50$ mPa.s). Variations of the steady current (electrodes area $S \approx 8$ cm^2) as a function of the applied voltage V for various electrode spacings (*from Lacroix* [8]).

$$w' \sim \left(\frac{Jd}{\rho\mu}\right)^{1/2} \approx \left(\frac{q_o V}{\rho}\right)^{1/2} = C^{1/2}\left(\frac{\varepsilon}{\rho}\right)^{1/2}\frac{V}{d} \quad \text{or} \quad \frac{w'd}{\mu V} \sim C^{1/2} M \qquad (17\text{-}7)$$

For a constant injection level C, w' would increase linearly with the applied voltage V. For a constant q_o value, (17-7) predicts a square root law : $w' \propto V^{1/2}$. Note that the $C^{1/2}$ dependence of w' implies a rather vigorous convection in the case of weak injection.

3. SCL injection induced electroconvection

The second asymptotic case of electroconvection which concerns the space-charge-limited injection was studied in detail [3,5]. It has characteristics which make it, to some extent, comparable with thermal convection. In particular, the volume charge density plays a role very similar to temperature and can be considered as a scalar quantity both active and passive, active because of the motion it induces under the action of the electric field, passive because it is transported by convection and its distribution can be drastically modified.

In SCL conditions, the injection of charge carriers is limited only by the ability of the system to transport them and convection has an influence on the passage of current. The

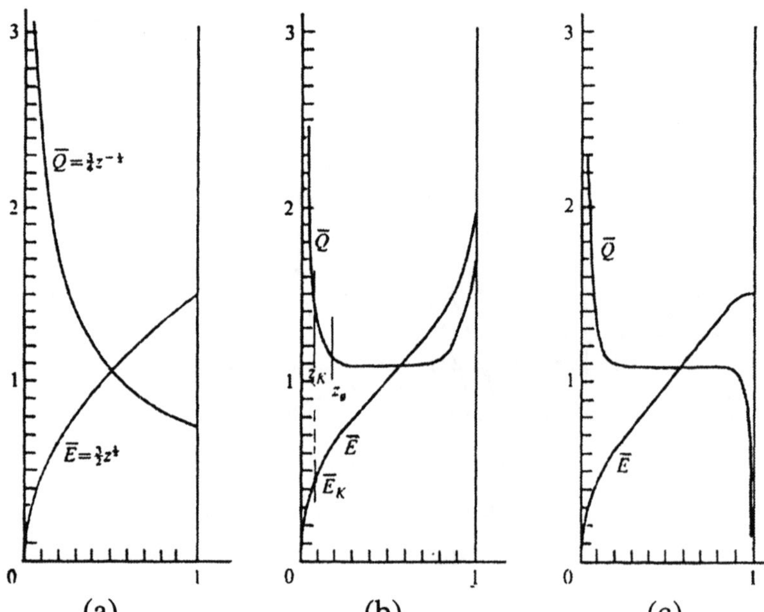

Fig. 17-2 – Distributions of mean electric field and charge density for different regimes of charge transfer in case of SCL injection). a) motionless state; b) turbulent state of motion without molecular diffusion; c) turbulent state with molecular diffusion (*from Lacroix et al.* [3]).

efficiency of the induced motion in increasing the current is illustrated by the characteristics I(V) of Fig. 1. Due to the space charge limitation of the current, I has a quadratic dependence on V in subcritical conditions $V < V_{nl}$ where there is no liquid motion ($V_{nl} \cong 60$ V in the case of Fig. 1). For $V_{nl} < V < V_{tr}$ (V_{tr} : transition voltage ≅ 3 kV here), there is a variation $I \propto V^{5/2}$. At high voltage ($V_{nl} > V_{tr}$) the current again exhibits a quadratic dependence ($I \propto V^2$), but with a level clearly greater than the one relative to the motionless state when only the drift of charge carriers takes place. These two well defined variation laws correspond to the two different electroconvective regimes.

The analysis of these regimes cannot simply extrapolate the approaches used in similar problems. For example, the assumption of a predominant gradient-type charge transport was justified for the problem of transient SCL injection. In this case a turbulent layer appearing first in the vicinity of the injector progressively invades the motionless liquid and there is a marked gradient of mean charge density $\overline{q}$ [11]. But at steady state, for both convection regimes the concept of gradient-type transport cannot account for the charge transport. Fig. 2 clearly shows that the gradient of $\overline{q}$ is nearly zero in the core and, close to the collector, there is a transfer of charge against the gradient of mean charge density. This fact together with the asymmetrical nature of charge and field distributions make the analysis complicated and specific [3].

3.- *a.* Low Reynolds number SCL electroconvection

The regime of dominant viscous effects corresponds to low values of the Reynolds number Re. When viscous dissipation dominates at large scale, as already stated it is natural to expect the typical fluid velocity to vary in proportion to the force., Balancing the Coulomb force and viscous terms in the Navier-Stokes equation gives the following relation (formally identical to (17-4)) :

$$\eta \frac{w'}{d^2} \sim q' \frac{V}{d} \qquad (17\text{-}8)$$

In this regime there is a big contrast between the charged forward flowing zones and the weakly charged backward flowing regions (the limit case of steady solution – see §16-1 – gives a picture of this contrast). The typical charge perturbation q' scales as the charge density (averaged in the bulk) : $q' \sim \overline{q} \sim \varepsilon V/d^2$ and we obtain from (17-8) :

$$w' \sim \frac{\varepsilon V^2}{\eta d} = T \frac{\mu V}{d} \qquad (17\text{-}9)$$

Taking into account the characteristics of non linear instability finally leads to [3] :

$$\frac{w'd}{\mu V} \cong \frac{T}{T_c} \qquad (17\text{-}10)$$

The results presented in Fig. 3 fully confirm the quadratic variation of the typical velocity amplitude with the applied voltage in this convective regime and give a relation $w'd/\mu V \cong 1.1\ (T/T_c)$ close to (17-10) [8].

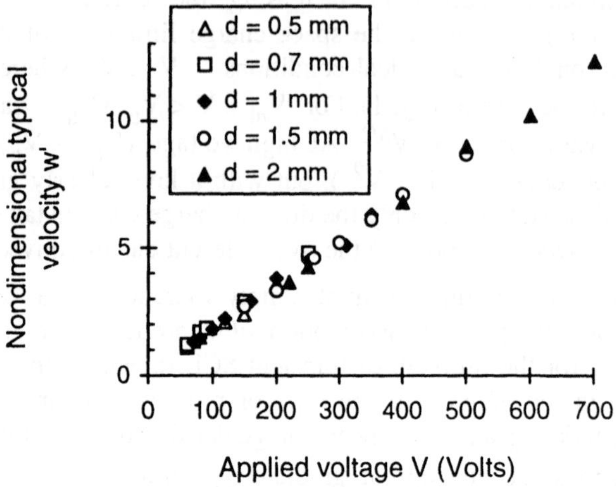

Fig. 17-3 – SCL injection in a polychlorobiphenyl (pyralene 1500). Variations of the ratio w'd/μV of typical liquid velocity w' and mean ionic velocity μV/d as a function of the applied voltage V for various electrode spacings (*from Lacroix* [8]).

The derivation of the variation law for the mean current density is not so straightforward and was obtained in a way analogous to Kraichnan's treatment for thermal convection [12]. We use here non dimensional variables. The z-component of the fluid velocity is assumed to be self similar (in the mean) :

$$w(z) = A\, w_o(z) \qquad (\|w_o(z)\| = 1) \tag{17-11}$$

We consider the layer on the injecting electrode in which the drift of charge carriers gives the dominant contribution to the electric current (Fig. 2). The layer thickness λ_e is defined by the condition that the drift ionic velocity and the vertical liquid velocity take the same value :

$$w(\lambda_e) = \overline{E}(\lambda_e) = \overline{E}_\lambda \tag{17-12}$$

(note that at $z = \lambda_e$ the conduction and convective components of the current density are nearly equal : $\overline{q}\,\overline{E} \cong \overline{qw}$). We further assume that w(z) has a quadratic variation with z in this layer adjacent to a rigid electrode : $w(z) = A\, w_o(z) = \alpha\, A\, z^2$ for $z \leq \lambda_e$ (α is a numerical constant). We then obtain from (17-12) :

$$\lambda_e = \left(\frac{\overline{E}_\lambda}{\alpha A}\right)^{1/2} \tag{17-13}$$

Determining the approximate distributions of field and charge density in the boundary layer leads to the following expression for the mean current density [3] :

$$\overline{J} \cong \frac{1}{2}\,\frac{\overline{E}_\lambda^2}{\lambda_e} \cong \frac{\alpha^{1/2}}{2}\,\overline{E}_\lambda^{3/2}\,A^{1/2} \tag{17-14}$$

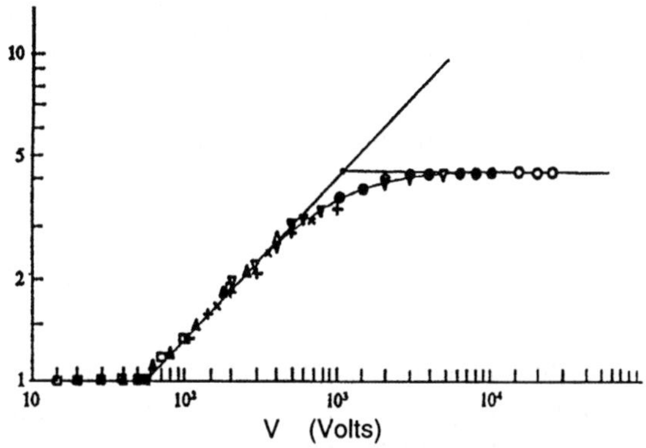

Fig. 17-4 – Electric Nusselt number Ne as a function of applied voltage for SCL injection in pyralene 1460 (η = 10 mPa.s). Values of the liquid thickness range from 0.16 to 3.4 mm *(from Lacroix et al.* [3]).

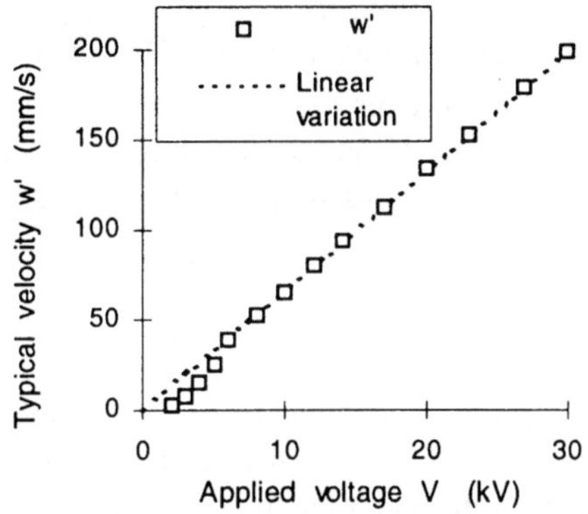

Fig. 17-5 – Variations of the r.m.s. vertical velocity component at mid-height versus applied voltage V for SCL injection in pyralene 1460 (η = 10 mPa.s). Gap d = 8 mm.

From the analysis of the different zones in the liquid layer and from the subsequent estimates, it is possible to show that $\overline{E}_\lambda \cong$ Cst in a reasonable range of variation of the velocity amplitude A [3]. From (17-9), A is proportional to T and this finally gives for the electric Nusselt number Ne (Ne is the ratio of the effective current density J and the current density J_0 which would exist without liquid motion) :

$$Ne = \frac{J}{J_o} \approx \left(\frac{T}{T_c}\right)^{1/2} \qquad (17\text{-}15)$$

(the two results of $T_{nl} \approx 0.9\, T_c$ and $Ne(T{=}T_{nl}) \approx 1.06$ have been used to obtain (17-15)). This relation fully accounts for the observed $V^{5/2}$ dependence of the current in this regime of dominating viscous effects (Fig. 1). It is very well verified by experiments in the voltage interval where the viscous effects indeed are dominant (on Fig. 4 this interval is between 60 V and about 500 V). Let us remark that although the current is transported mainly by convection, there is no proportionality between the current and the typical liquid velocity w' (in non dimensional form we have $Ne \cong A^{1/2}$).

3.- *b.* Fully turbulent SCL electroconvection

In the regime where the inertial effects are dominant (Re >> 1), an order of magnitude expression (17-3) has already been given. We can recover this variation law by using again the derivation of §2. This states that the kinetic energy of a "blob" of charged liquid leaving the immediate vicinity of the injector comes from the work of the Coulomb force. At mid-height it leads to : $(1/2)\rho w'^2 \sim q'Ed/2$ [3]. Here it is no more possible to state that q' is on the order of $\overline{q}$. Because of the strong mixing associated with the turbulent state of motion, the charge density has fluctuations of much smaller amplitude than in the viscous dominated regime. However the ratio $q'/\overline{q}$ is assumed not to depend on the applied voltage V (this implies that the large scale structure of the turbulent motion does not depend on V and induces charge density perturbations proportional to the mean charge density). This assumption then gives :

$$w' \cong \alpha_w \left(\frac{\varepsilon}{\rho}\right)^{1/2} \frac{V}{d} \qquad \text{or} \qquad \frac{w'd}{\mu V} \cong \alpha_w M \tag{17-16}$$

α_w being a constant to be determined from experiments (or from a much more detailed analysis). Fig. 5 shows results of measurement of the typical velocity in a viscous liquid. Clearly in the fully turbulent regime w' varies linearly with the applied voltage ($V > V_{tr}$) and $\alpha_w \cong 0.3$.

In order to evaluate the current, it is necessary to obtain an estimate for the ratio of typical value of charge perturbation q' and averaged charge density $\overline{q}$. Such an estimate is derived below, using non dimensional variables and equations. The charge conservation equation (15-25) averaged in a horizontal plane can be written in the steady state of fully turbulent convection :

$$\frac{d\overline{J}}{dz} = 0 \qquad \Rightarrow \qquad \frac{d}{dz}\left[\overline{q}\overline{E} + \overline{qw}\right] = 0 \tag{17-17}$$

In the core of the liquid layer in convection, to a first approximation the mean charge density is constant (see Fig. 2b). Therefore we deduce from (17-17) :

$$\frac{d}{dz}\left(\overline{q}\overline{E}\right) = \overline{q}^2 \sim \left|\frac{d}{dz}(q'\,w')\right| \sim q'\left|\frac{dw'}{dz}\right| + w'\left|\frac{dq'}{dz}\right| \tag{17-18}$$

Balancing the two main terms in the Navier-Stokes equation (15-28) :

$$\frac{T}{M^2}\,\mathbf{u}.\nabla\mathbf{u} \sim T\overline{E}q\mathbf{i}_z \tag{17-19}$$

gives for the typical velocity :

$$w'\left|\frac{dw'}{dz}\right| \sim M^2\, q'\,\overline{E} \tag{17-20}$$

From (17-16) we deduce [3] :

$$\left|\frac{dw'}{dz}\right| \sim M\,q' \tag{17-21}$$

The order of magnitude of $|dq'/dz|$ can be obtained from the charge conservation equation (15-26) written for the fluctuating parts of the variables. In the core $d\overline{q}/dz \approx 0$ and retaining the main terms only of (15-26) leads to :

$$\overline{q}\,q' \sim w'\left|\frac{dq'}{dz}\right| \tag{17-22}$$

Using (17-21) and (17-22), relation (17-18) then writes :

$$\bar{q}^2 \sim M q'^2 + \bar{q} q' \qquad (17\text{-}23)$$

In the case of high values of the mobility parameter M (M >> 1), we have :

$$\frac{q'}{\bar{q}} \sim M^{-1/2} \qquad (17\text{-}24)$$

which indicates that q' is clearly lower than $\bar{q}$. The current density then writes [3] :

$$\bar{J} \approx \overline{qw} \sim q' w' \sim M^{1/2} \mu \bar{q} \frac{V}{d} \qquad (17\text{-}25)$$

For the electric Nusselt number we deduce :

$$Ne = \frac{\bar{J}}{J_o} \sim M^{1/2} \qquad (17\text{-}26)$$

This estimate of Ne is an expression which does not depend on the stability parameter T. Relation (17-26) justifies the observation of a current varying as V^2 for high V values (see Fig. 6 which gives results obtained with various liquids). The variation of the saturation value Ne_s with M (Fig. 7) was found to be consistent with (17-26). The experimental results can be summarised by the relation [3] :

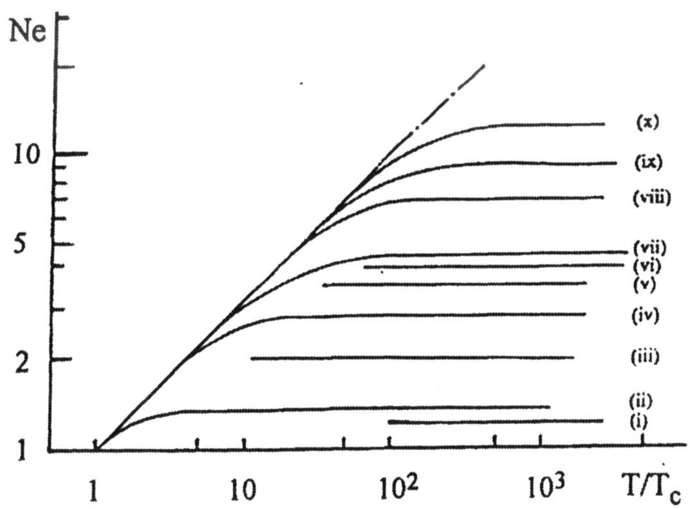

Fig. 17-6 – SCL injection in various liquids. Electric Nusselt number Ne as a function of the stability parameter $T/(T_c)_{exp}$ for various couples of liquids and injected ion species. (i) methanol/H^+; (ii) chlorobenzene/Cl^-; (iii) ethanol/H^+; (iv) nitrobenzene/Cl^-; (v) ethanol/Cl^-; (vi) propylene carbonate/Cl^-; (vii) pyralene 1460/Cl^-; (viii) pyralene 1500/Cl^-, 35°C; (ix) pyralene 1500/Cl^-, 20°C; (x) pyralene 1499/Cl^-. (*from Lacroix et al.* [3]).

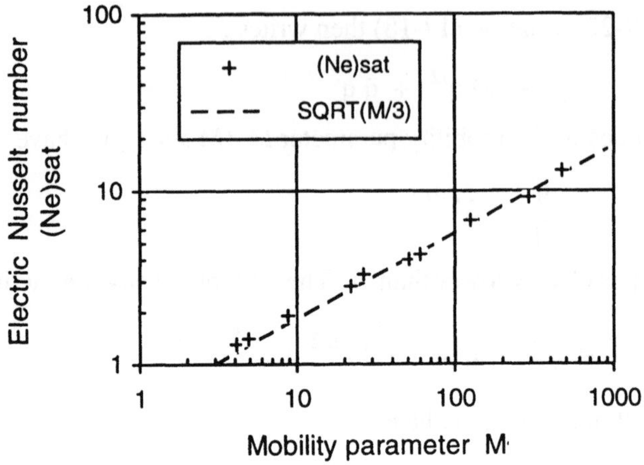

Fig. 17-7 – SCL injection in various liquids. Saturation value Ne_{sat} (for $V >> V_{tr}$) of the electric Nusselt number as a function of the mobility parameter M. (*from Lacroix et al.* [3]).

$$Ne \approx \left(\frac{M}{3}\right)^{1/2} \tag{17-27}$$

which is valid even for very moderate M values ($M \geq 4$) and not only for $M >> 1$. Let us remark that, again, there is no proportionality between the current and the characteristic velocity : when varying the M number, the electric Nusselt number varies as the square root of the non dimensional velocity.

3.- *c.* Transition in the states of SCL electroconvection

The parameter characterising the transition between the two convective states where respectively the viscous effects or the inertial effects are dominant can be obtained from the requirement that both expressions of the electric Nusselt number take the same value. This gives :

$$\frac{T_{tr}}{M} \cong \frac{T_c}{3} \cong 30 \tag{17-28}$$

The number T_{tr}/M is related with the transition Reynolds number Re_{tr} characterising the big eddies of scale d. By retaining expression (17-9) for the velocity amplitude in the viscous dominated regime and using (17-28), one obtains :

$$Re_{tr} = \frac{w'_{tr}\, d}{\nu} \cong \frac{T_{tr}}{T_c}\frac{\mu V_{tr}}{\nu} = \frac{1}{T_c}\left(\frac{T_{tr}}{M}\right)^2 \cong \frac{T_c}{9} \cong 10 \tag{17-29}$$

This value of the transition Reynolds number is identical to that characterising the Rayleigh-Benard convection [13] (note that this underlines the strong analogy between

SCL injection problem and thermal convection). An interesting consequence of this transition value is that it provides an estimate for α_w. By stating that at T_{tr} the two expressions (17-10) and (17-16) of w' give the same value, we have [3] :

$$\frac{w'd}{\mu V} \cong \frac{T_{tr}}{T_c} \cong \alpha_w M \qquad \Rightarrow \qquad \alpha_w \cong \frac{1}{3} \tag{17-30}$$

An EHD mobility can be defined from this estimate :

$$\mu_{ehd} = \frac{1}{3}\left(\frac{\varepsilon}{\rho}\right)^{1/2} \tag{17-31}$$

and, introducing an apparent mobility μ_{app}, the steady state current in the fully turbulent regime writes :

$$\bar{J} \approx \frac{9}{8}\,\mu_{app}\,\varepsilon\,\frac{V^2}{d^3} \qquad \text{with} \qquad \mu_{app} = \left(\mu\,\mu_{ehd}\right)^{1/2} \tag{17-32}$$

It appears that the more viscous the liquid, the greater will be the magnification of SCL current due to electroconvection.

4. Transient regime of electroconvection under SCL injection

The problem associated with injection under a step voltage has been examined in §15-5 mainly from the point of view of instability in these transient conditions. Let us recall that this technique of step voltage was first used to determine the mobility of charge carriers [14]. It has been shown in §15-5 that measuring the true value of mobility μ is possible only if the applied step voltage V is lower than the critical voltage V_a of manifest convection. For $V > V_a$ the induced convection contributes to the transfer of charge carriers. The apparent time of flight t_T defined as the time at which the current exhibits a maximum then leads to an apparent mobility $\mu_a > \mu$.

In §15-5 we focused on the stability time t_1 and on the criterion for manifest convection in the transient unipolar injection problem. Here we consider the development of convection and the associated current. At $t = t_1$ convection begins to be detectable and is restricted to a layer adjacent to the injector and of thickness $\bar{z} \approx \mu\,(V/d)\,t_1$. For $t > t_1$ the motion in the layer makes this layer to expand at a rate closely related to its typical velocity because convection transports the charge :

$$\frac{d\bar{z}}{dt} \cong w' \tag{17-33}$$

In the case of SCL injection, the typical velocity w' in the charged layer should be similar to that (given by expression (17-10)) characterising the steady convection in a layer across which the potential difference is the same ($\approx V\bar{z}/d$). This gives :

$$\frac{d\bar{z}}{dt} \cong \frac{T}{T_a} \mu \frac{V}{d} \frac{\bar{z}}{d} \tag{17-34}$$

This relation is valid only when the viscous effects are dominant ($V\bar{z}/d < V_{tr}$) and gives an exponential expansion of the zone invaded by the space charge and the motion. The rapid growth of the expanding layer (and of the typical convective velocity) has a limited duration t_{acc} because the inertial effects also grow rapidly. The end of this acceleration

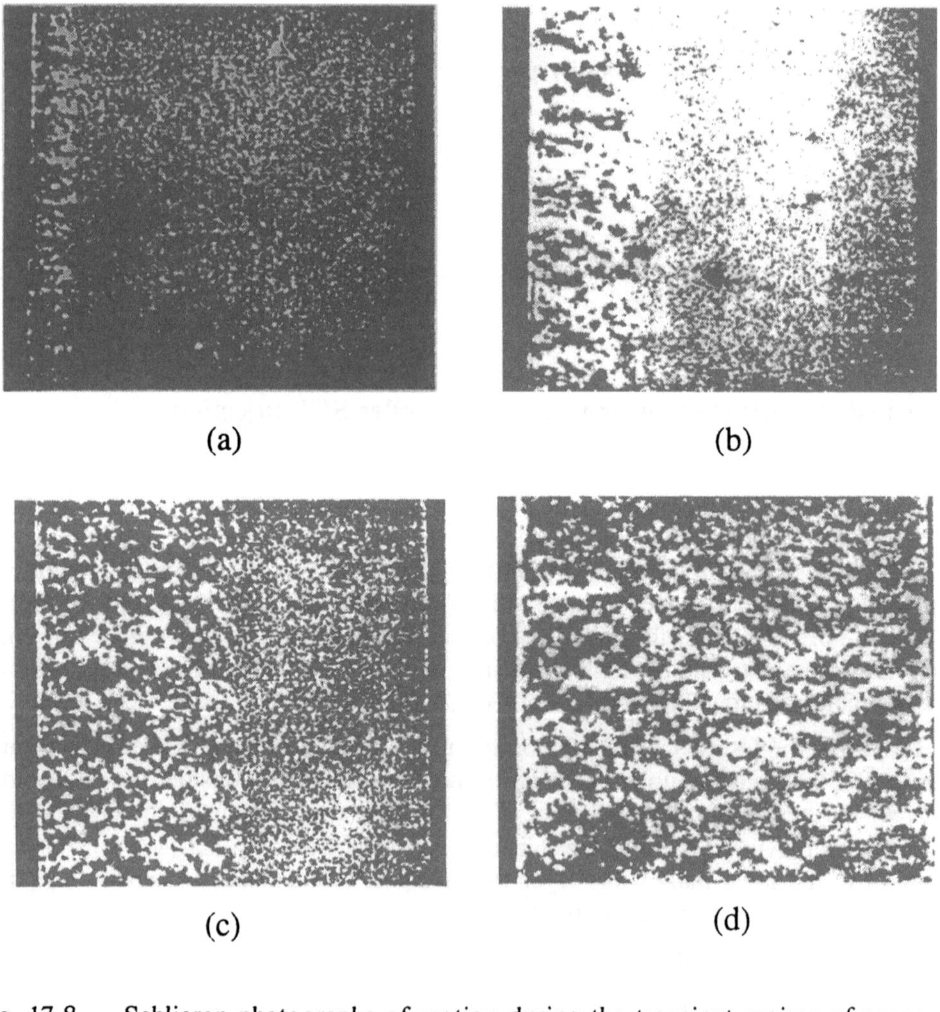

Fig. 17-8 – Schlieren photographs of motion during the transient regime of space-charge-limited injection (from the left electrode) in nitrobenzene (d = 5.6 mm; V = 18 kV). Pictures taken at times : a) t = 0.95 ms; b) t = 1.9 ms; c) t = 2.9 ms; d) steady state (*from Hopfinger & Gosse* [11]).

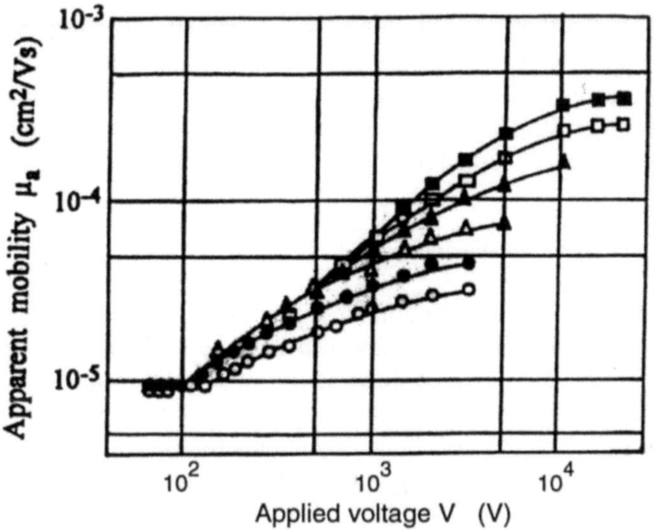

Fig. 17-9 – Transient SCL injection in a polychlorobiphenyl (pyralene 1500). Variations of the apparent mobility μ_a deduced from transit times t_T as a function of the step voltage value for different electrode spacings (*from Lacroix & Tobazéon* [16]).

stage should occur when $V\bar{z}/d \cong V_{tr}$. After this time, the applied voltage across the charged layer is such that the inertial effects dominate and the typical velocity is then expected to take a constant value on the order of $(\varepsilon/\rho)^{1/2}$ V/d. The layer then expands with this constant velocity [14-16].

In the asymptotic case of an applied step voltage much higher than the transition voltage ($V >> V_{tr}$), the front between the turbulent layer and the motionless liquid is rather sharp (Fig. 8) and moves at a constant velocity [11]. By neglecting the stability and the acceleration times, we obtain for the apparent mobility μ_a :

$$\mu_a \cong \frac{w'd}{V} \cong \alpha_a \left(\frac{\varepsilon}{\rho}\right)^{1/2} \tag{17-35}$$

The experiments give α_a close to α_w so that, to a first approximation [8] :

$$\mu_a \cong \mu_{ehd} \cong \frac{1}{3}\left(\frac{\varepsilon}{\rho}\right)^{1/2} \tag{17-36}$$

This implies that the typical turbulent velocity w' during the transient regime is not significantly higher than that characterising the steady state. Such a property is not surprising if we refer to considerations of electric energy being converted into kinetic energy [17,6].

When examining the variations of the apparent mobility μ_a as a function of the step voltage amplitude V, we see that $\mu_a = \mu$ for $V \leq V_a$; then μ_a increases with V and for high enough V values tends to a saturation value $\mu_a \cong \mu_{ehd}$. This is illustrated by Fig. 9 (the differences in the curves can be ascribed to the intensity of injection which is strong but not strictly space-charge-limited [16]). The variations of μ_a in the intermediate regime ($V_a < V < V_{tr}$) are reasonably accounted for by the following formula [14,18] :

$$\mu_a \cong \mu \frac{V / V_a}{1 + \ln(V / V_a)} \tag{17-37}$$

In transient conditions of SCL injection the current exhibits the typical shape shown in Fig. 10-b with an initial slow increase followed by a rapid growth up to a maximum I_m at time t_T and, then, by a marked decrease down to the steady state value which is reached after a rather long time (which can be up to 10 t_T). The magnitude I_m of the transient current is clearly higher than the one of the steady state current. In the convection regime where the viscous effects are dominant, I_m varies like V^3 for $V<V_{tr}$ [19,8,20] and this gives :

$$Ne_m \sim T \tag{17-38}$$

instead of $Ne \sim T^{1/2}$ in the steady state. For high applied step voltages, the inertial effects are dominating and as in the steady state $I_m \propto V^2$, but the magnitude is clearly higher than in the steady state and, to a first approximation [14,19] :

$$Ne_m \sim M \tag{17-39}$$

In both cases, during the first transient stage the current approximately varies in proportion to the typical velocity w'. We can give the following picture to explain such a high

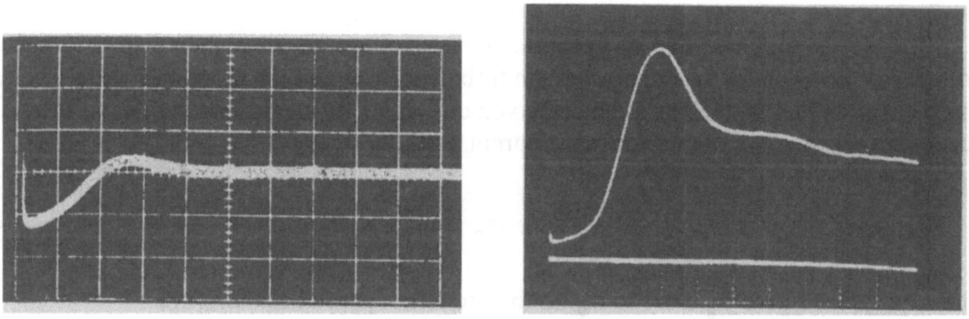

(a) (b)

Fig. 17-10 – Variations of current during the transient regime of space-charge-limited injection in pyralene 1460. a) without liquid motion (applied voltage $V < V_a$); b) in super-critical conditions ($V > V_a$); note the sharp current increase after the stability time and the long decrease toward the steady state value (*from Tobazeon* [14]).

efficiency of transient convection in transporting charge when compared with steady state convection : in transient conditions, the expansion of the charged zone occurs via filaments of charged liquid which invade the quasi motionless charge free zone and the backward filaments are roughly charge free [17]. In steady conditions the strong mixing due to convection results in a mean charge in the liquid bulk so that forward and backward filaments have only slightly differing charge densities [3,8].

5. SCL injection : interaction with a forced flow

A laminar forced flow with direction normal to the electric field does not alter the unstable character of the charge distribution [21]. We mentioned (§15-6) that for applied voltages V on the order of the non-linear critical voltage V_{nl}, the laminar forced flow is destabilised and convection is detectable beyond a certain distance from the edge of the injecting electrode. In the case of strong injection, for V high enough the resulting convection is chaotic even with forced flows of low Reynolds number Re.

The electrically induced turbulent agitation increases the friction on the walls [22]. Experiments of SCL injection in a rectangular duct in the presence of forced flow showed that the pressure difference required to maintain the same flow rate in the duct increases with the applied voltage (Fig. 11) by a factor which can exceed 10 [23,24]. Thus the

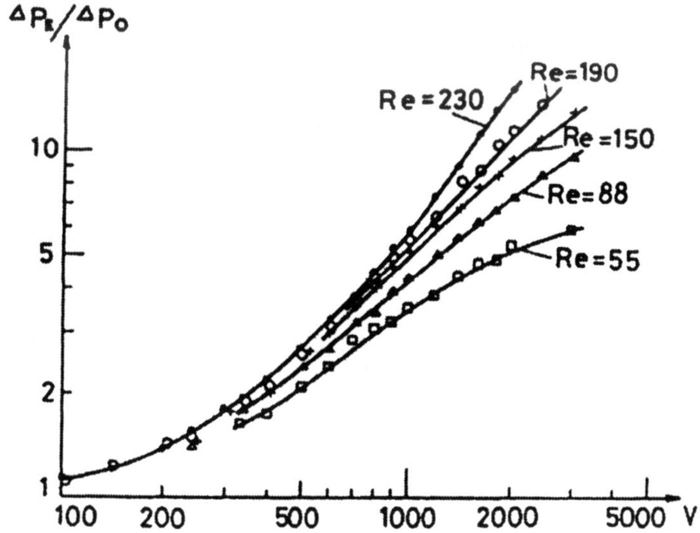

Fig. 17-11 – SCL unipolar injection in deionized nitrobenzene super-imposed on a forced flow between plane parallel electrodes (d = 0.84 mm). Variations of the ratio $\Delta p_E/\Delta p_o$ of pressure drops with and without applied voltage V for various flow rates *(from Atten & Honda* [23]).

injection induced secondary motion and turbulence is the basic phenomenon responsible for the improperly named "electroviscous effect" observed by numerous authors [23]. The forced flow superimposed on the injection induced convection very likely modifies the structure of electroconvection existing without forced flow. This complex problem has not been examined closely in the case of injection between plane parallel electrodes. We shall see that in co-axial wire-cylinder geometry, there is a clear effect of forced flow on the resulting motion structure and, more particularly, on the passage of current.

6. Heat transfer augmented by electroconvection

The analogy between convection induced by unipolar injection of charge carriers and thermoconvection in a horizontal layer of fluid (Rayleigh-Bénard problem) was emphasized all along these chapters. We consider now what occurs when superimposing an injection and a thermal gradient (horizontal plane parallel electrodes). Generally we can say that depending on the way of injecting and of heating (from above or from below), the effects of Coulomb and gravitational forces should add or counteract. In practice, when working with liquid layers of small thickness, the magnitude of Coulomb force is much higher than that of buoyancy and we expect the state of motion to be determined by the electrical forces alone. Then temperature can be considered as a passive contaminant transported by eddy diffusivity of the flow induced by charge injection. Formally, the situation is analogous to the case where turbulent motion is induced by the forced flow.

The dominance of electrical forces was proved by experiments performed with non polar liquids lying between copper electrodes. Injection of weak or moderate level was

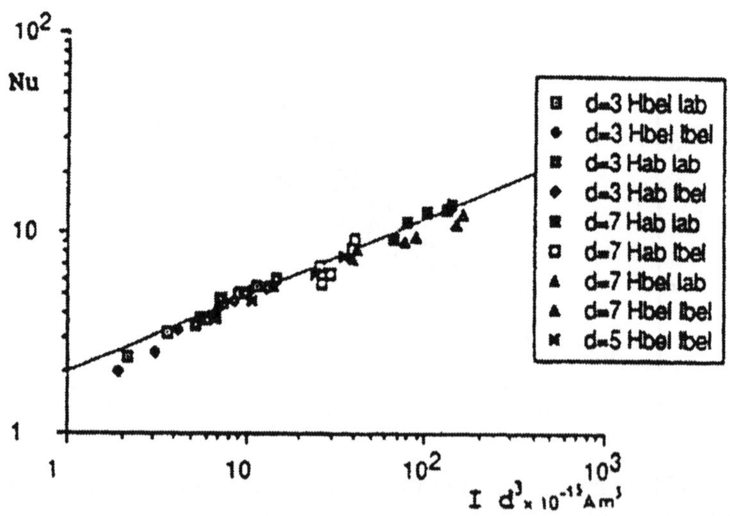

Fig. 17-12 – Nusselt number as a function of $I_{inj}d^3$ for different distances and experimental conditions in two-plate geometry (*from Perez et al.* [26]).

achieved by adding some salts [25]. Fig. 12 first shows that the Nusselt number Nu measuring the effect of convection on heat transfer does not depend on the fact that buoyancy and Coulomb force cooperate or counteract and this proves that buoyancy has but a negligible influence on convection [26,27]. The second result arising from these experiments is the good efficiency of electroconvection in transferring the heat because the heat flux can be increased by a factor of 10 (Fig. 12). A very rough analysis led to the determination of the parameters which control the heat flux depending on the convection regime and the injection level [10]. Fig. 12 shows that the product Id^3 is a relevant parameter and, further, strongly suggests that there is a power law relating the Nusselt number and the relevant non dimensional parameter. However the finite values of the injection parameter C and the fully turbulent character of the regime of convection prevented us from deriving a realistic power law by a simple analysis [27].

In the case of viscous dominated regime of convection induced by a weak injection ($C \ll 1$), nevertheless, a derivation is possible based on relation (17-6) giving the amplitude of velocity. A study of electrohydrodynamically enhanced heat transfer in silicone oils between copper plates provided an indirect test of the prediction [28]. In order to derive the variation law for the Nusselt number Nu in this particular case, we have to define first the thickness λ_T of the layer where thermal conduction is dominant. This is done by stating that at $z = \lambda_T$ the conduction heat flux H_k is equal to the convective heat flux H_w :

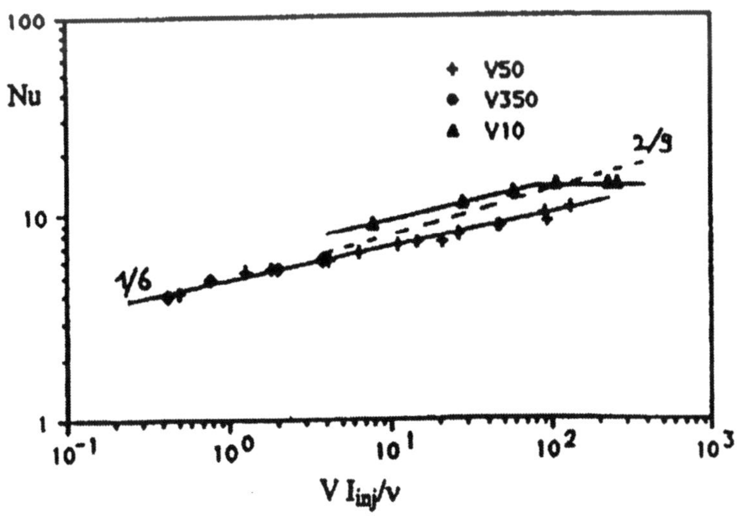

Fig. 17-13 – Variations of the Nusselt number Nu with the parameter VI_{inj}/ν for silicone oils of different viscosity : 10 mPa.s (V10), 50 mPa.s (V50), 350 mPa.s (V350). (*from Atten & Malraison* [28]).

$$H_k \sim k_T \frac{\Delta T}{\lambda_T} = H_w \sim \rho c_p \overline{\theta w} \tag{17-40}$$

where k_T is the thermal conductivity, ΔT the temperature difference between plates, c_p the specific heat and $\overline{\theta w}$ the correlation between temperature and velocity fluctuations. As the temperature fluctuation θ scales with ΔT, we have at $z = \lambda_T$:

$$w(\lambda_T)\,\lambda_T \sim \frac{k_T}{\rho c_p} = \kappa \tag{17-41}$$

κ being the thermal diffusivity. Assuming (as in §3-a) a quadratic variation for the normal velocity component in the layer $w(z) = \alpha\, w_{max}\,(z/d)^2$ and using (17-6), we finally obtain for the Nusselt number Nu [28] :

$$Nu = \frac{Hd}{k_T \Delta T} \sim \frac{d}{\lambda_T} \sim \left(\frac{\bar{j} V d^3}{\eta \kappa^2}\right)^{1/6} \tag{17-42}$$

The experiments performed in silicone oils in which an added salt promotes a weak injection give results (Fig. 13) in very good agreement with the prediction (17-42). This can be considered as an indirect successful test of the law (17-6) for typical velocity variations in the convection regime for which the viscous effects are dominant.

REFERENCES

[1] P. Atten, J.C. Lacroix and B. Malraison : Phys. Lett., **79A** (1980), 255-258.

[2] B. Malraison and P. Atten : Phys. Rev. Lett., **49** (1982), 723-726.

[3] J.C. Lacroix, P. Atten and E.J. Hopfinger : J. Fluid Mech., **69** (1975), 539-563.

[4] G.A. Ostroumov : Zh. Tekh. Fiz., **24** (1954), 1915-1919.

[5] O.M. Stuetzer : Phys. Fluids, **5** (1962), 534-544.

[6] N. Felici : Rev. Gén. Electr., **78** (1969), 717-734.

[7] P. Atten, F.M.J. McCluskey and A.C. Lahjomri : IEEE Trans. Ind. Appl., **IA23** (1987), 705-711.

[8] J.C. Lacroix : "*Instabilités hydrodynamiques et électroconvection lors d'injection d'ions dans les liquides isolants isotropes*", Doctoral Thesis, Grenoble University 1976.

[9] A. Castellanos : IEEE Trans. Electr. Ins., **IE26** (1991), 1201-1215.

[10] P. Atten, F.M.J. McCluskey and A.T. Perez : IEEE Trans. Electr. Ins., **EI23** (1988), 659-667.

[11] E.J. Hopfinger and J.P. Gosse : Phys. Fluids, **14** (1971), 1671-1682.

[12] R.H. Kraichnan : Phys. Fluids, **5** (1962), 1374-1389.

[13] D.R. Moore and N.O. Weiss : J. Fluid Mech., **58** (1973), 289-312.

[14] R. Tobazéon : J. Electrostatics, **15** (1984), 359-384.

[15] P. Atten and J.P. Gosse : J. Chem. Physics, **51** (1969), 2804-2811.

[16] J.C. Lacroix and R. Tobazeon : in Proceedings 4th Intern. Conf. Cond. Breakdown in Dielectric Liquids (Ed. T. Gallagher), Dublin, 1972, 93-96.
[17] N. Felici : Dir. Current, **2** (1972), 147-165.
[18] J.C. Lacroix, P. Atten and E.J. Hopfinger : Internal Report, 1979.
[19] P. Atten, B. Malraison and S. Ali Kani : J. Electrostatics, **12** (1982), 477-488.
[20] N. Felici and J.C. Lacroix : J. Electrostatics, **5** (1978), 135-144.
[21] A. Castellanos and N. Agräit : IEEE Trans. Ind. Appl., **IA28** (1992), 513-519.
[22] P. Atten, R. Baron and M. Goniche : J. Phys. Lettr., **41** (1980), L-1-L-4.
[23] P. Atten and T. Honda : J. Electrostatics, **11** (1982), 225-245.
[24] L. Elouadie : "*Electroconvection et augmentation des échanges thermiques produites par une injection unipolaire en géométrie fil-cylindre co-axiaux*", Ph. D. Thesis, Joseph Fourier University, Grenoble, 1991.
[25] A. Denat, B. Gosse and J.P. Gosse : J. Electrostatics; **7** (1979), 205-225.
[26] A.T. Perez, P. Atten, B. Malraison, L. Elouadie and F.M.J. McCluskey : in "*Experimental heat transfer, fluid mechanics and thermodynamics*" (Eds. R.K. Shah, E.N. Ganic and K.T. Yang), Elsevier 1988, 941-947.
[27] F.M.J. McCluskey, P. Atten and A.T. Perez : Int. J. Heat Mass Transfer, **34** (1991), 2237-2250.
[28] P. Atten and B. Malraison : in Conference Record 10th ICDL (Eds. P. Atten and R.Tobazeon), IEEE Cat. No 90CH2812-6, 1990, 323-327.

Chapter 18

INSTABILITY AND CONVECTION IN CYLINDRICAL GEOMETRY

1. Linear instability in cylindrical and spherical geometries

In chapter 15 we mentioned that the instability problem arises in the three electrode geometries of high symmetry when the injection is homogeneous on the injector. Chapter 15 was devoted to instabilities in the case of plane parallel electrodes. In this section we consider the two other configurations of coaxial cylinders and of concentric spheres. Given a constant value q_o of the injected space charge at the injecting electrode, the motionless state is a solution and the instability problem is to determine under what conditions this solution destabilises. Qualitatively the instability mechanism is the same than in the case of plane parallel electrodes. Due to coulombic repulsion, the charge density decreases when going from injector to collector along a radial line and, therefore, there is a positive coupling between velocity disturbances and the subsequent charge density perturbations (the argument of §15-1 is of general nature and is not specific to the case of parallel electrodes). Again the instability parameter is the nondimensional number T and the criterion depends on the injection parameter $C = q_o d^2/\varepsilon V$.

The linear instability analysis was developed for the two geometrical configurations, always with the assumption of the injection being autonomous, i.e. being characterised by a constant q_o value not depending on the field or the velocity perturbations. The cylindrical and spherical configurations result in a more intricate problem than the one of plane parallel plates because of two extra parameters. The first one is related with the question of which electrode is injecting, the inner one or the outer one. The two cases must be considered because it was shown, by using electrodialytic varnishes, that injection can be performed from an outer cylinder as well as from an inner one [1]. The second parameter is the ratio η of radii of the electrodes :

$$\eta = \frac{R_{inner}}{R_{outer}} < 1 \qquad (18\text{-}1)$$

It influences the criterion and plays a role in the selection of the convective cells which appear just above the criterion T_c.

The critical conditions have been determined for the cylindrical [2,3] and spherical [4] configurations as functions of C and η for the two cases of inner and outer electrode being the injector. For strong injection ($C >> 1$), injecting from the inner electrode results in a situation more unstable than when injecting from the outer electrode [3,4];

independently of the value of the radii ratio η we have :

$$(T_c)_{inner} < (T_c)_{planes} < (T_c)_{outer} \qquad (18\text{-}2)$$

This has been justified by qualitative arguments [3,4]. This inequality can be easily understood by the following remark : when injecting from the outer electrode, the space charge modified field increases strongly from outer (injecting) electrode to inner electrode, whereas it only moderately decreases from inner to outer electrode when the inner electrode is the injector; therefore the stationary charge density has a stronger gradient in the major part of the interval when the inner electrode is injecting than when injection is from the outer electrode. From the general argument of §15-1 showing the positive coupling between velocity and charge perturbations, it is clear that the charge density perturbation is proportional to the mean gradient of stationary charge density and, therefore, injecting from the inner electrode results in a more unstable situation than injecting from the outer electrode. An experimental study performed with a ratio $\eta = 1/6$ [1] gave results which confirm relations (18-2). Note that these relations (18-2) are reversed in the case of weak injection ($C << 1$) [3,4].

Fig. 1 shows the variations of the critical value T_c as a function of the radii ratio η for the configuration of coaxial cylinders. For $\eta = 1$ the criterion relative to parallel plates

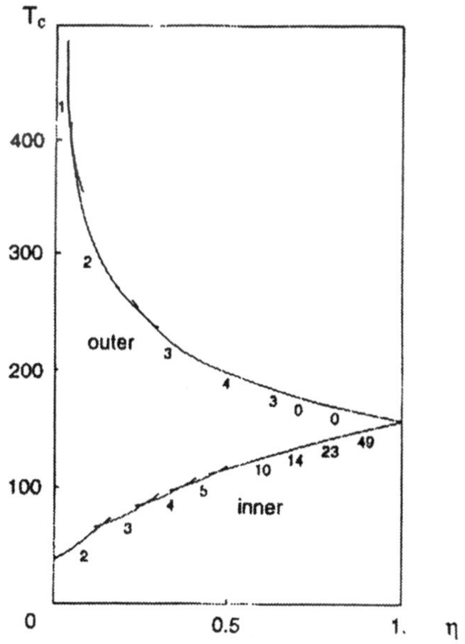

Fig. 18-1 – Instability criterion T_c for SCL injection between coaxial cylinders as a function of the ratio η of cylinders radii. Indices "inner" and "outer" are relative to the cylinder which is injecting and the integer numbers are the values of the azimuthal wave-number m corresponding to the part of curve (*from Agräit & Castellanos* [3]).

configuration is recovered. We see that for η close to 1, the two branches relative to inner and outer injectors are symmetrical. Note that these branches consist in numerous portions of curves corresponding to different values of the azimuthal wave number m and that there is discontinuity in the slope when passing from one portion to another one (at these points the m number is incremented or decremented by 1). One interesting result given by the analysis concerns the planform of the most unstable convective cells. For η smaller than about 0.7 the linear analysis shows that the three dimensional modes are preferred, which are characterised by non zero azimuthal (m) and axial (k) wave numbers [3]. The curvature of the liquid layer results in a clear selection of the most unstable pattern of convective cells in contrast to the case of plane parallel plates where the linear analysis determines the critical wave number k_c but gives no information on the planform of the convective cells which appear for $T > T_c$. In the case of an injection into a spherical shell, the linear instability analysis also determines unambiguously the pattern of convective cells [4].

Up to now there has been no study of non linear instability in the two configurations of coaxial cylinders and concentric spheres. As the problem tends asymptotically to that of plane parallel electrodes as $\eta \to 1$, it is sure that, at least in a given interval of η, there exists a non linear criterion lower than T_c and a finite amplitude convection occurs for $T > T_c$. The most interesting case to investigate would be the wire/cylinder configuration characterised by $\eta \ll 1$, the wire being the injector.

2. Electroconvection experiments in co-axial wire and cylinder

The analogy in instability phenomena between planar and cylindrical configurations suggests that in the wire/cylinder geometry, the convective transport of charge also plays an important role and can significantly increase the current. A preliminary study in the case of a moderate value of the ratio of radii of the cylinders ($\eta = 1/6$) indicated a strong analogy with the planar configuration [1]. But for very thin wires as inner cylinders ($\eta \ll 1$), the particular shape (quasi triangular cross section) of the convective cells can result in different properties of electroconvection. From a qualitative point of view, in liquids such that $M \gg 1$, we again expect a convective regime for which the viscous effects are dominant and a fully turbulent regime; but the transition between these two regimes is more complex than in the case of parallel plates and its characterisation is not easy because of a non unique Reynolds number associated with the convective flow.

Another difficulty arises from the largely uncontrolled injection strength. Generally, applying a high voltage V results in injection phenomena from the wire closely related with the high electric field at the wire surface. But the injection parameter C varies with V and, moreover, the injection is not space-charge-limited.

2.- *a.* Charge injection from a wire

In cylindrical geometry it is not possible to use the permselective membranes which made it easy to promote SCL injections from plane electrodes [5,6]. An electrodialytic varnish of properties similar to those of membranes could be coated on metallic cylinders and this led to strong injections from either the inner or the outer cylinder [1]. However such varnishes have a finite life time and, moreover, are not well suited for heat transfer experiments. Therefore it is convenient to work with the natural electrochemical reactions which take place at the metal/liquid interfaces.

In the co-axial wire/cylinder configuration it is easy to obtain currents increasing strongly with the applied voltage. Indeed the high electric field on the wire presumably results in strong injection [7,8]. But, superimposed on this injection, a field enhanced dissociation can occur in the liquid just around the wire (the latter phenomenon alone can induce intense liquid motion [9,10]). The importance of field enhanced dissociation depends on the polar or non polar nature of the liquid. In polar liquids the dominant process is the injection due to electrochemical reactions [8]. In weakly or non polar liquids, the field enhanced dissociation of electrolytic impurities (or added salt) is more important than in polar solvents. A study of the effect of field enhanced dissociation in highly diverging fields drew the conclusion that, in pure enough liquids, the injection process gives the major contribution to the current [11]. But when adding an appropriate salt to the liquid, there is a noticeable field enhanced dissociation in the liquid close to the thin wire and the ions of the wire polarity can be considered as being "injected" like those really created at the metal/liquid interface. Therefore this dissociation process can be viewed as reinforcing the injection as far as convection phenomena are considered.

A property of such injections from thin wires is their finite level character. The space charge q_o on the injector (and also the amount of charge created by field enhanced dissociation close to the wire) is an increasing function of the field E_o : $q_o = f(E_o)$. As E_o decreases when the space charge in the liquid bulk is increased, the steady state solution is such that the field E_o on the wire takes a value greater or much greater than the mean field V/d $(d = R_{outer} - R_{inner})$. Therefore the injection will never be space charge limited and the value of the injection parameter C generally will not be strictly constant.

2.- *b.* Steady state current-voltage characteristics

In conditions where there is strong injection from a thin wire, the critical conditions for hydrodynamic instability (see §1) give values of the critical voltage V_c lower than in the case of SCL injection between parallel plates (for non polar liquids, $V_c < 100$ V). The charge density again can be considered as playing both an active and a passive role and the convective motion transports charge and magnifies the current. The efficiency of the induced motion is illustrated on Fig. 2 which presents results obtained in mineral oil using a cell of following characteristics : length L = 50 cm, $R_{outer} = 1$ cm and $R_{inner} = 50\mu m$.

Three conduction regimes appear on Fig. 2. In a first zone, i.e. for applied voltage $V < V_1$ $(V_1 \sim 1$ kV), the linear variation of the current corresponds to the ohmic regime of conduction (the rather large value of interelectrode spacing $d \approx 1$ cm makes this conduction current to overcome the current due to space charge alone). In the third zone $V > V_2$ (with $V_2 = 10$ to 15 kV on Fig. 2) the current to a first approximation varies like V^2 and takes values which do not depend very much on temperature. From the further observation of an intense liquid motion with very small scale eddies [12], it is clear that this third regime is associated with a fully turbulent convection where the inertial effects are dominant. In the

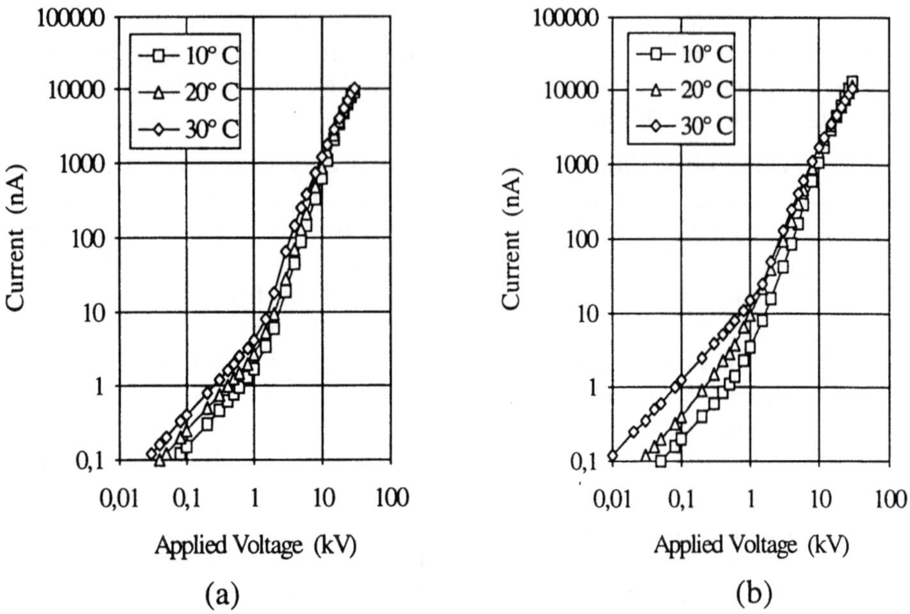

Fig. 18-2. – Current-voltage characteristics at various temperatures in a) pure transformer oil b) oil with TBAP additive lying between co-axial wire and cylinder. (*from Atten & Elouadie* [13]).

second conduction régime ($V_1 < V < V_2$) the current I varies like V^{α} with α ranging from 2.2 to 2.8. This regime is not so well characterised because the exponent α apparently depends on temperature. Under low applied voltage (1 or 2 kV) there exists a fairly well organised motion with often 6 more or less regular and slowly fluctuating convective cells [12]. This would suggest to extrapolate to the cylindrical case the conclusion drawn in the planar geometry and associating the $V^{5/2}$ variation of the current to the convection regime where the viscous effects are dominant (see §17-3-*a*). However such an extrapolation is not valid because the transition between the two convection regimes should occur at a much lower voltage V_t ($V_t \sim 2$ kV for the mineral oil tested in [12,13]). Remark finally that the addition of the salt TBAP (tetra-butyl-ammonium picrate) has a visible effect mainly at low applied voltages through the increase of conductivity [13].

Before developing a crude analysis, it is interesting to examine the results in a different way and to plot the ratio I/V and $I^{1/2}$ as functions of V. For the corona effect in gases in the wire/cylinder configuration, the variations of I/V lead to the determination of the threshold voltage V_{th} of corona discharge and allow one to recover the ion mobility. At greater applied voltages, the best representation is given by plotting $I^{1/2}$. The results of Fig. 3-*a* do not clearly characterise a threshold for the injection in the oil and the curves are not straight segments. The lack of linearity of these curves is presumably due to the contribution of convection to the passage of current and, moreover, possibly to the

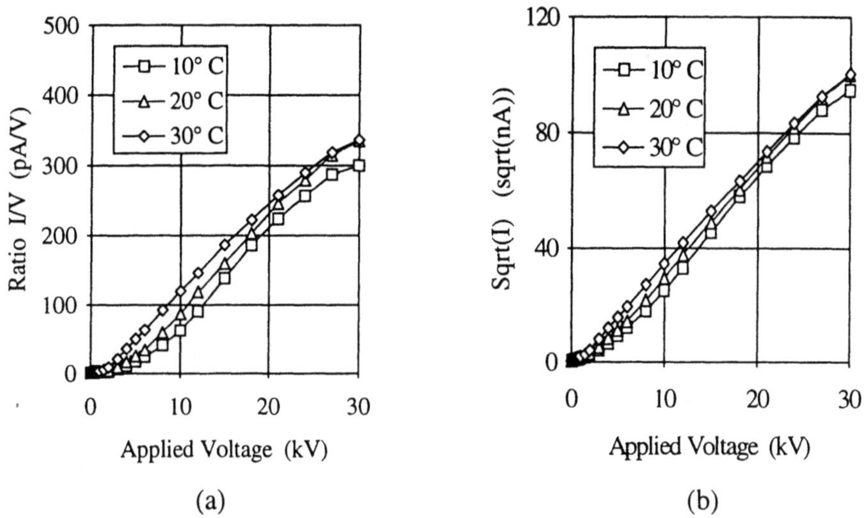

Fig. 18-3. – Current-voltage characteristics in pure transformer oil between co-axial wire and cylinder at various temperatures. a) Ratio I/V ; b) Values of $I^{1/2}$ (*from Atten & Elouadie* [13]).

injection parameter C not being constant. The curves of Fig. 3-*b* (plots of $I^{1/2}$) exhibit a less marked curvature. A mean apparent mobility can be deduced from the measured currents for $V \geq 20$ kV by applying relation (18-5). This gives $\mu_{app} \cong 2\ 10^{-8}\ m^2/V.s$ for mineral oil [12,13], a value much greater than the true mobility of ions $\mu \cong 1.3\ 10^{-8}\ m^2/V.s$ [12]. This μ_{app} value is also greater than the apparent mobility relative to fully turbulent convection between plane parallel electrodes which is $(\mu_{app})_{plane} = (\mu\ \mu_{ehd})^{1/2} \cong 8\ 10^{-9}\ m^2/V.s$ at 20°C. This suggests a more vigorous turbulence to exist in the wire/cylinder configuration.

3. Some characteristics of electroconvection

In the absence of fluid motion, there is an analytical solution of the electrical equations. For a pure injection from the inner cylinder, the electric field is :

$$E(r) = \left[E_o^2 \left(\frac{R_{inner}}{r}\right)^2 + \frac{J}{2\pi\mu\varepsilon}\left\{1 - \left(\frac{R_{inner}}{r}\right)^2\right\} \right]^{1/2} \qquad (18\text{-}3)$$

where E_o is the field on the wire and $J = 2\pi\mu rqE$ the steady current per unit length. E_o and J can be expressed as functions of the charge density q_o on the injector and of the applied voltage V. For a strong injection from a thin wire ($\eta << 1$) relation (18-3) simplifies into :

$$E(r) = \left[\frac{J}{2\pi\mu\varepsilon} \right]^{1/2} \tag{18-4}$$

valid over the major part of the interval $[R_{inner}, R_{outer}]$. The SCL current is therefore :

$$J_{SCL} = 2\pi\mu\varepsilon \frac{V^2}{d^2} \tag{18-5}$$

In the following, we assume that there is a strong enough injection and that the electric field is roughly constant. An order of magnitude analysis analogous to the one relative to a planar layer of liquid (§17-3) has not been developed until now. The derivation of estimates for the current is a difficult task and only very rough estimates are proposed for the amplitude of velocity perturbations.

3.- a. Velocity estimates

In the convective regime for which the viscous effects are dominant, when retaining the main terms in the Navier-Stokes equation, we have :

$$\eta \left\| \nabla^2 \mathbf{u} \right\| \sim \left\| q\mathbf{E} \right\| \tag{18-6}$$

The order of magnitude of the viscous term can be expressed taking into account the divergence of the streamlines in the convective cells. Retaining streamlines similar to the ones existing at critical conditions, it appears that at radius r, the width relative to the main contribution to viscous friction is proportional to r. Hence, we can write :

$$\eta \frac{u'}{r^2} \sim q'\overline{E} \tag{18-7}$$

where u' and q' are the typical values of the radial component of fluid velocity and of the charge density perturbation respectively and the overbar indicates an averaging over a cylinder of radius r. Assuming again that $q' \sim \overline{q}$ analogously to the case of plane parallel electrodes (see §17-3-a and [5]) and taking for q' the expression :

$$q' \sim \overline{q} \sim \frac{\varepsilon}{r} \frac{d}{dr}\left(r\overline{E}\right) \tag{18-8}$$

we obtain after elementary calculations :

$$u' \sim \frac{\varepsilon}{\eta} r \overline{E}^2 \tag{18-9}$$

In the strong injection case the field $\overline{E}$ should only slowly decrease with r. Relation (18-9) indicates that u' takes a maximum value for $r \sim R_{outer}$. As the mean field is $\overline{E} \cong V/d$ (with $d = R_{outer} - R_{inner} \approx R_{outer}$ for $\eta << 1$), we finally obtain :

$$\frac{u'_{max} d}{\mu V} \sim \frac{\varepsilon V}{\mu \eta} = T \tag{18-10}$$

a relation fully similar to (17-9).

In the other convection regime for which the inertial effects dominate, it is possible to derive an estimate of the radial velocity magnitude. Consider a blob of liquid leaving the immediate vicinity of the injecting wire. By stating that the work of the Coulomb force relative to this blob is converted into kinetic energy, we have [13] :

$$\frac{1}{2}\rho u'^2 \sim \int_{R_{inner}}^{r} q\,\bar{E}\,dr \qquad (18\text{-}11)$$

At any r value, the charge density in the blob is $\bar{q}+q'$ and it is assumed here that $q' \sim \bar{q}$. The order of magnitude of the mean charge density can be approximated from Poisson equation which takes here the expression :

$$\frac{1}{r}\frac{d}{dr}\left(r\bar{E}\right)=\frac{\bar{E}}{r}+\frac{d\bar{E}}{dr}=\frac{\bar{q}}{\varepsilon} \qquad (18\text{-}12)$$

The absolute value of the radial derivative of the mean field, $d\bar{E}/dr$, is generally lower than $\bar{E}/r$ because the mean charge cannot change in sign. In the fully turbulent regime it is presumably much lower and to a good approximation one has :

$$\bar{q} \sim \varepsilon\frac{\bar{E}}{r} \qquad (18\text{-}13)$$

Substituting the value V/d for the field $\bar{E}$ in (18-11) leads to :

$$\frac{1}{2}\rho u'^2 \sim \varepsilon\left(\frac{V}{d}\right)^2 \ln\left(\frac{r}{R_{inner}}\right) \qquad (18\text{-}14)$$

This relation shows that the velocity increase is initially very rapid and then becomes slower and slower (this corresponds to a velocity profile clearly different from the one expected in the viscous dominated regime). The maximum velocity is obtained for r close to R_{outer} and we deduce for the typical velocity scale (at $r \cong R_{outer}/2$) :

$$u' \sim \left[2\ln\left(\frac{R_{outer}}{2R_{inner}}\right)\right]^{1/2}\left(\frac{\varepsilon}{\rho}\right)^{1/2}\frac{V}{d}=\left[2\ln\left(\frac{1}{2\eta}\right)\right]^{1/2}\left(\frac{\varepsilon}{\rho}\right)^{1/2}\frac{V}{d} \qquad (18\text{-}15)$$

a relation to be compared with the similar relation (17-16) : $w' \sim (\varepsilon/\rho)^{1/2}\,(V/d)$ relative to the case of plane parallel electrodes. By making the hypothesis that the factor 1/3 found in the latter case (see §17-3-c and [5]) also applies to the wire/cylinder geometry, we can propose a more precise relation [13] :

$$\frac{u'd}{\mu V} \cong \left[2\ln\left(\frac{1}{2\eta}\right)\right]^{1/2}\frac{M}{3} \qquad (18\text{-}16)$$

It is clear from (18-16) that u' takes values higher than in the case of plane electrodes. This confirms that the electroconvection is more vigorous in the wire/cylinder configuration.

3.- *b.* Measurements in transient conditions

It was shown in §17-4 that the transient current technique used in the SCL injection between plane parallel electrodes allows one to recover the mobility μ of charge carriers in stationary conditions ($V < V_c$) [14] and to obtain transient mobility values $\mu_a > \mu$ in supercritical conditions ($V > V_c$) (μ_a is determined from the first stage of convection aided transfer of charge from injector to collector). For very high applied voltages ($V >> V_c$) a fully developed turbulence is induced; then μ_a takes a saturation value and the typical velocity $\mu_a V/d$ is very close to the steady characteristic velocity [14]. We expect a similar behaviour in the wire/cylinder configuration. Indeed the transient current technique applied to the case of strong injection between co-axial cylinders [1] leads to observations similar to the ones in the planar case : the transient current has a shape similar to the one exhibited by Fig. 17-10, with a well marked maximum defining the transit time t_T of first injected ions [1,12,13]. The transient mobility μ_a defined by :

$$\mu_a = \frac{d^2}{V\, t_T} \qquad \text{with} \qquad d = R_{outer} - R_{inner} \qquad (18\text{-}17)$$

increases with the applied voltage V up to a saturation value. Experiments performed with cylinders having a ratio η = 1/6 gave a saturation value of μ_a about 1.5 times the one obtained between plane parallel electrodes [1] (the factor $[2\ln(1/2\eta)]^{1/2}$ of relation 18-16 takes the value 1.48 which agrees very well with the experimental results). Other experiments performed in wire/cylinder geometry with the ratio η = 0.005 [12,13] gave the results shown in Fig. 4. The apparent mobility μ_a increases with the applied voltage V in

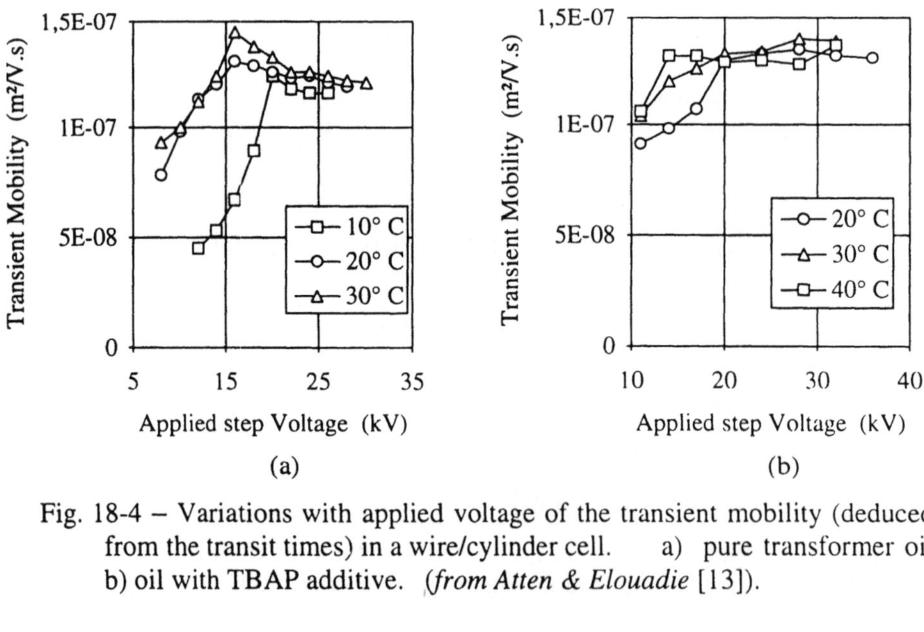

Fig. 18-4 – Variations with applied voltage of the transient mobility (deduced from the transit times) in a wire/cylinder cell. a) pure transformer oil b) oil with TBAP additive. (*from Atten & Elouadie* [13]).

the first zone ($V < V_{tr}$) and takes a constant value in a second zone ($V > V_{tr}$). The saturation values characterising the fully turbulent convection are $\mu_a \cong 1.25\ 10^{-7}\ m^2/V.s$ for pure oil and $\mu_a \cong 1.35\ 10^{-7}\ m^2/V.s$ for the oil with additive. They are close to the prediction derived from (18-16) : $(\mu_a)_{th} \cong 1.48\ 10^{-7}\ m^2/V.s$. The agreement appears to be good if we keep in mind the fact that the experiments do not correspond strictly to SCL conditions.

3.- c. Transition between the convective regimes

For the experimental results reported on in this section, the transition voltage V_{tr} as determined from the variations of steady and transient currents apparently ranges from 10 to 20 kV depending on the oil temperature [13]. But the transition is not so well defined as in the planar geometry (see §17-3-c) and V_{tr} would appear to be 4 or 5 times greater than for the same liquid subjected to SCL injection between parallel plates. This can be understood by taking into account the diverging character of the streamlines in cylindrical geometry. Instead of having a unique Reynolds number characterising the state of motion in planar geometry, the convective motion in wire/cylinder configuration must be accounted for by local Reynolds numbers :

$$Re(r) = \frac{r\,u'(r)}{\nu} \tag{18-18}$$

By assuming again that there is a transition Reynolds number $Re_{tr} \sim 10$ for which the inertial and viscous effects are of the same magnitude, it is possible to define a "transition radius" r_{tr} when the applied voltage V is between V_1 and V_2 , i.e. roughly for the second conduction regime of §2-b :

$$Re(r_{tr}) = Re_{tr} \tag{18-19}$$

Then we can picture the state of motion by distinguishing between an inner region ($r < r_{tr}$) where the viscous effects are dominant and an outer region where inertial effects dominate. Therefore we deal here with a range of applied voltage where there is a progressive transition between the two convection regimes for which the viscous effects and the inertial effects are dominant. Instead of defining a transition voltage, it appears more appropriate to delineate a mixed regime of convection and charge transport with a complex structure depending on the radial co-ordinate.

4. Influence of a forced flow

We mentioned in §15-6 that a Poiseuille flow between parallel plates does not modify the criterion for the instability induced by unipolar injection from one plate ($Re < Re_c$). The effect of the imposed flow is to select the pattern of secondary flow which appears to take the form of longitudinal rolls. In the case of co-axial cylinders the problem is different because in this configuration, without forced flow ($\overline{U} = 0$), generally there is no degeneracy of the planform of convective cells which can sustain at marginal state : the transverse (axisymmetric) rolls, the longitudinal (axial) rolls and the 3-D cells correspond to different T_c values and generally the 3-D pattern gives the most unstable convective cells. A superimposed laminar flow has a tendency to force longitudinal rolls but this occurs for high enough values of the Reynolds number Re. In their instability study Castellanos and Agräit showed that there are changes in the pattern of the most unstable convective cells at particular values of Re [15]. Note, however, that these studies are based on linear analysis and cannot fully account for the observed phenomena where, as was shown in chapters 15 and 16, non linear effects are predominant.

In the case of Poiseuille flow between parallel plates, in practice there is an entrance region beyond the abscissa delineating the non injecting and injecting parts of one of the plates (see §15-6). A similar entrance zone exists in the case of laminar flow between two co-axial cylinders. In subcritical conditions ($V < V_c$) the entrance length L_E depends on the mean velocity $\overline{U}$ of the forced flow and on the ionic drift velocity :

$$\frac{L_E}{R_{outer} - R_{inner}} \cong \frac{\overline{U}}{\mu \overline{E}} \tag{18-20}$$

For an applied voltage V above the critical value V_c, the forced laminar flow is destabilised and the secondary flow which arises induces a radial component of velocity u' which convects the charge more rapidly between the injecting and collecting cylinders. The entrance length then takes an expression similar to (18-20) by just replacing $\mu \overline{E}$ by u'. L_E can be estimated by taking for u' expressions like (18-10) or (18-16) obtained in §3-a for convection without forced flow. Such a derivation would give lower bounds for L_E because the mean flow interacts in a complex way with the injection induced secondary flow and very likely results in smaller values of u'.

Two global variables are affected by the resulting flow structure : the pressure drop Δp along the annular duct and the current I usually injected by the inner cylinder. As it can be seen on Fig. 17-11 for plane Poiseuille flow, the relative pressure drop $\Delta p_E / \Delta p_o$ (respectively with and without applied voltage) increases with the applied voltage and also with Re. This indicates that the forced flow plays a role in determining the resulting flow structure. A more systematic study performed in co-axial wire/cylinder configuration confirmed these observations and the fact that the electric variables V and I are not sufficient to fully characterise the problem. Indeed the increase in pressure drop is due to the friction on the cylinders which depends on the structure of the more or less turbulent flow. The particular convective flow which results from the interaction of forced flow and normal injection of charge has a structure which depends not only on the injection of charge (which gives rise to the secondary flow) but also on the extraction of kinetic energy from the mean flow by the Reynolds stresses. It is then natural to state that the degree of turbulence depends on the energy density ε' supplied to the liquid :

$$\varepsilon' \propto I\,V + \Delta p\, Q \tag{18-21}$$

Here the electrical power IV is assumed to be fully converted into kinetic energy (Joule heating negligible) and the forced flow also gives an input of kinetic energy from the

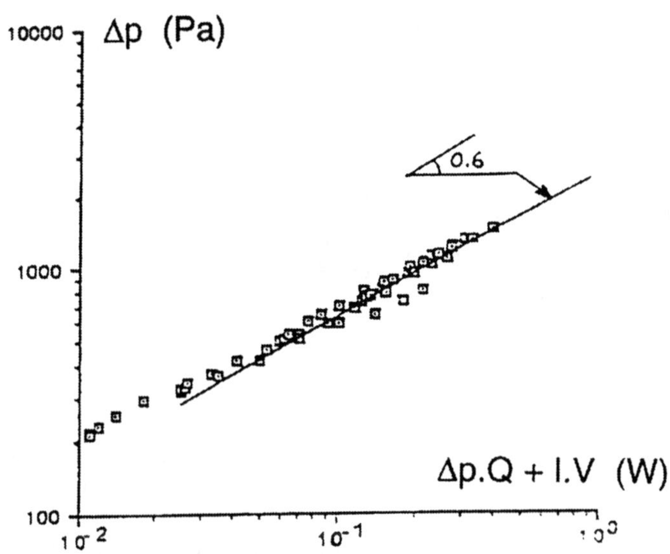

Fig. 18-5 – Variations of the pressure drop in mineral oil over the wire/cylinder duct as a function of the total energy input for various values of mean forced velocity and applied voltage (*from Elouadie* [12]).

mechanical energy $\Delta p\, Q$ where Q is the flow rate (this expression for ε' has been derived from the governing equations [12]). Fig. 5 shows that, to a good approximation, a unique curve is obtained when the pressure drop Δp is plotted as a function of ε'. For high enough energy input, the variations can be approximated by a power law $\Delta p \propto \varepsilon^{\xi}$ with $\xi \cong 0.6$. This exponent value is not very different from the one characterising a turbulent flow in a circular pipe : from the classical law $\Delta p \propto Re^{7/4}$ we deduce $\varepsilon' \propto \Delta p\, Q \propto Re^{11/4}$ and hence $\Delta p \propto \varepsilon^{\xi'}$ with $\xi' = 7/11 \approx 0.636$.

The second global variable affected by the structure of the resulting flow is the current I injected by the wire and convected by the complex flow. Of course the applied voltage remains the main physical parameter determining I and the current-voltage characteristic curves always exhibit the two zones found for $\overline{U} = 0$, where I varies like V^{α} and V^{β} with $\beta \cong 2$ and $\alpha > 2$. An interesting representation of the results obtained for different values of the forced flow velocity is given in Fig. 6 where the ratio I/V^2 is plotted as a function of V. We see that at low applied voltage (< 5 kV) the total current injected by the wire increases as $\overline{U}$ is increased, a phenomenon due to the important value of L_E and to subsequent complex space charge effects [12]. The most interesting fact concerns the variations of I for V > 20 kV. In these conditions the entrance length is small (< 2% of the total length of the duct) and its effect can be neglected. Fig. 6 then shows that the current slowly decreases as the mean velocity $\overline{U}$ is increased.

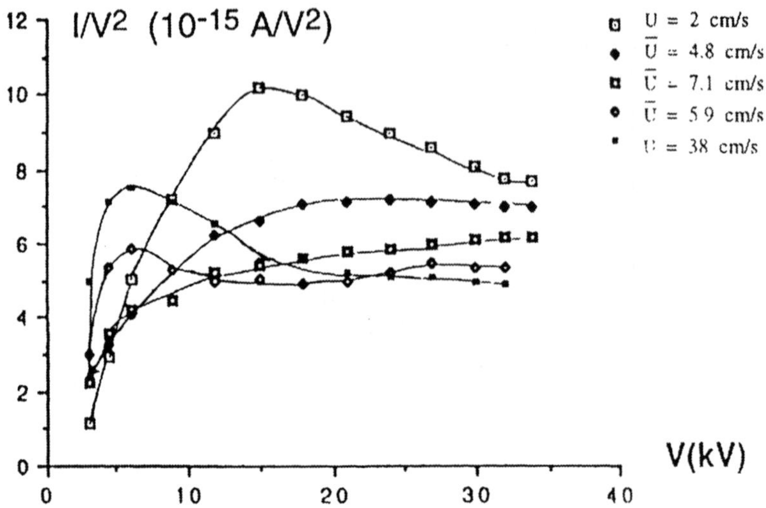

Fig. 18-6 – Injection from the wire into the mineral oil flowing in a wire/cylinder cell. Variations of the ratio I/V^2 of injected current and square of applied voltage V as a function of V for different values of mean velocity of forced flow (*from Elouadie* [12]).

The dependence on $\overline{U}$ is illustrated by Fig. 7. This figure shows the variations of the current normalised with its value for the pure convective case ($\overline{U} = 0$) as a function of the ratio of the mean ($\overline{U}$) and typical injection induced (u') velocities, where $u' = \mu_{ehd}\ V/d$ with μ_{ehd} defined from (18-16). A unique curve is obtained [12,16] and this implies that the change in the convective flow structure due to the forced flow depends only on the ratio of the typical radial and axial velocities (at least in a certain domain of variations of Re and of the ratio $\overline{U}/u'$). Clearly the convective transport of charge is reduced because very likely the radial velocity component is reduced. A very simple (and very naive) picture suggests that in the case $\overline{U} = 0$ the charge is convected by radial filaments [7] of liquid (grossly analogous to the thermals); but with a forced flow, we deal with rather helicoidal filaments which are more elongated and, therefore, experience a stronger viscous damping, leading to smaller radial velocity component and smaller convective transport of charge.

5. Effect on heat transfer

The strong analogy between transport of charge and of heat in an insulating liquid was underlined in §17-6 in the case of convection induced by unipolar injection between parallel plates. It was shown that the heat transfer can be significantly augmented by electroconvection. We expect a similar phenomenon to exist for the configuration of

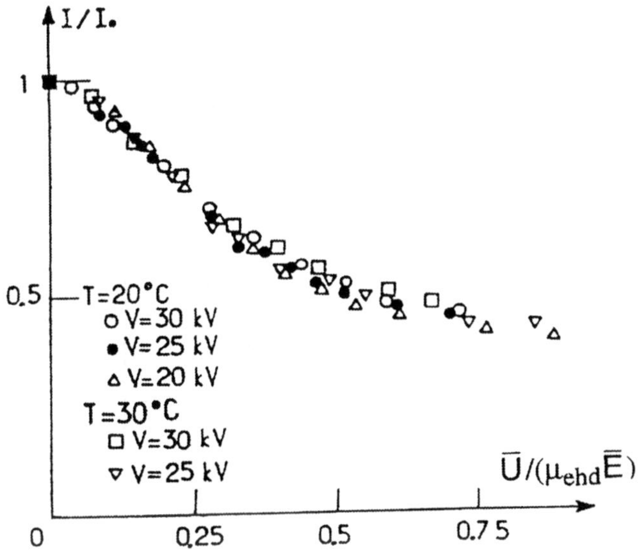

Fig. 18-7 – Injection into mineral oil from a wire co-axial with a cylinder, superimposed on a forced flow. Ratio I/I_0 of currents I with and I_0 without forced flow as a function of the ratio $\bar{U}d/\mu_{ehd}V$ of the mean flow velocity $\bar{U}$ and the typical injection induced convective velocity $\mu_{ehd}V/d$ *(from Elouadie* [12] *and Perez et al.* [16]).

co-axial cylinders which has the advantage of being very close to the basic element – pipe with forced flow – used in most heat exchangers. The other interest of examining the heat transfer in wire/cylinder configuration is to work with intense electroconvection. Unlike the planar case (§17-6), strong injections can be achieved from the wires, which leads to vigorous agitation of the liquid and, hence, to significant augmentation of the heat transfer at the outer cylindrical wall.

An experimental study was performed by Elouadie [12] in mineral oil using the wire/cylinder configuration already considered in §2 to §4 (R_{inner} = 50 μm, R_{outer} = 1 cm, L = 50 cm – see [12,13,16]). A uniform heat flux was imposed on the outer cylinder and the temperature distribution along the wall was measured to determine accurately the constant local Nusselt number Nu beyond the thermal entrance length [12]. The first basic observation was that, for a given potential difference V between wire and cylinder, the Nusselt number decreases when the Reynolds number of the forced flow is increased. This behaviour is similar to that exhibited by the current (Fig. 7). These decreases in I and H arise because the mean current density $\bar{j}$ and the mean heat flux $\bar{H}$ in the liquid bulk are strongly correlated with the typical radial component of liquid velocity u' which itself decreases with the forced flow velocity $\bar{U}$ (§4).

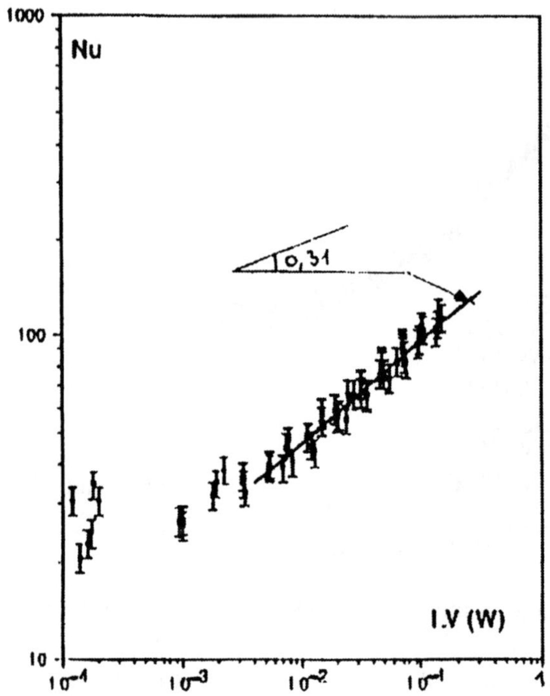

Fig. 18-8 – Variations of the Nusselt number Nu as a function of the injected electrical energy IV for various values of the mean velocity $\overline{U}$ of forced flow. Mineral oil (temperature : 20°C at entrance) in wire/cylinder configuration. (*from Elouadie* [12] *and Perez et al.* [16]).

The similar variations of the current and the heat flux suggest to examine more closely the correlation between the current and the Nusselt number characterising the heat transfer from the wall to the liquid. Indeed when plotting the Nu values as a function of I for various values of the applied voltage V and of the mean velocity $\overline{U}$, a well defined power law is obtained with rather low dispersion [12]. The most relevant electrical parameter, however, very likely is the product IV rather than I. Relation (17-42) for weak injection between parallel plates expresses Nu in terms of the product IV; moreover IV represents the electrical energy input which, for high enough IV values, is nearly fully converted into kinetic energy. Fig. 8 shows the variations of Nu with IV : the dispersion is low and for $IV > 7\ 10^{-2}$ W there is a well characterised power law :

$$Nu \propto (IV)^{\gamma} \qquad \text{with} \qquad \gamma = 0.31 \pm 0.02 \tag{18-22}$$

It is interesting to compare the influence on heat transfer of the electrically induced turbulence with that of the turbulence generated by the forced flow alone ($Re > Re_c$). In the latter case, an experiment performed in very similar conditions concerning the heat flux gave $Nu \propto Re^{\psi}$ with $\psi \approx 0.80$ [17]. Using the classical relation $u'/\overline{U} \sim Re^{-1/8}$ [18] leads to the following expression in terms of the turbulent Reynolds number $Re_t = 2Ru'/\nu$:

$$(Nu)_{flow} \propto (Re_t)^{\psi'} \qquad \text{with} \qquad \psi' = (8/7)\psi \approx 0.91 \tag{18-23}$$

For the injection induced turbulence, we easily obtain for the typical radial component u′ :

$$u' \propto \left(\frac{IV}{\rho}\right)^{1/3} \tag{18-24}$$

which finally gives, when using the associated turbulent Reynolds number :

$$(Nu)_{injection} \propto (Re_t)^{3\gamma} = (Re_t)^{0.93} \tag{18-25}$$

This exponent value does not significantly differ from $\psi' = 0.91$. Note that a slight difference is very plausible because the two turbulent flows have different anisotropies.

Finally a question arises when examining Figs. 5, 7 and 8 together. For the variations of the current I and of the Nusselt number Nu, only the electrical energy input IV is taken into account whereas the pressure drop Δp is expressed as a function of the total energy input $IV + \Delta p\, Q$. A possible reason for such a difference is the fact that I and Nu are determined mainly by the properties of radial transport of charge and heat in the liquid bulk whereas Δp is due to viscous friction on the outer cylinder and involves more particularly the region close to the wall. But there is also the possibility that the results presented in Figs. 7 and 8 concern rather low values of the mean velocity $\overline{U}$ which make the term $\Delta p\, Q$ nearly negligible compared to IV. Studies under much higher mean velocity are required to clarify this point.

REFERENCES

[1] P. Atten, J.C. Lacroix and R. Tobazeon : Proceed. 4th Intern. Conf. Cond. Breakdown in Dielectric Liquids (Ed. T. Gallagher), Dublin, 1972, 89-92.

[2] A.T. Richardson : J. Mech. Appl. Math., **33** (1980), 277-285

[3] N. Agräit and A. Castellanos : Phys. Fluids A, **2** (1990), 38-44.

[4] S. Oliveri and P. Atten : Phys. Fluids, **29** (1986), 1378-1385.

[5] J.C. Lacroix, P. Atten and E.J. Hopfinger : J. Fluid Mech., **69** (1975), 539-563.

[6] P. Atten and J.C. Lacroix : J. Mécanique, **18** (1979), 469-510.

[7] N. Felici : Dir. Current, **2** (1972), 147-165.

[8] N. Félici, J.P. Gosse and B. Gosse : Rev. Gén. Electr., **85** (1976), 861-874.

[9] M.S. Apfel'baum : Magnitnaya Gidrodinamika, **3** (1978), 57.

[10] Y.K. Stishkov and A.A. Ostapenko : Magnitnaya Gidrodinamika, **4** (1979), 46.

[11] P. Atten and M. Haidara : IEEE Trans. Electr. Ins., **EI-20** (1985), 187-198.

[12] L. Elouadie : "*Electroconvection et augmentation des échanges thermiques produites par une injection unipolaire en géométrie fil-cylindre co-axiaux*", Doctoral Thesis, Joseph Fourier University, Grenoble, 1991.

[13] P. Atten and L: Elouadie : J. Electrostatics, **34** (1995), 279-297.

[14] R. Tobazéon : J. Electrostatics, **15** (1984), 359-384.

[15] A. Castellanos and N. Agräit : in Conference Record 10th ICDL (Eds. P. Atten and R.Tobazeon), IEEE Cat. No 90CH2812-6, 1990, 311-315.

[16] A.T. Perez, P. Atten, B. Malraison, L. Elouadie and F.M.J. McCluskey : in *Experimental heat transfer, fluid mechanics and thermodynamics* (Eds. R.K. Shah, E.N. Ganic and K.T. Yang), Elsevier, 1988, 941-947.

[17] R.C. Martinelli : Trans. ASME, **69** (1947), 947-959.

[18] H. Schlichting : "*Boundary layer theory*", McGraw-Hill, New-York, 1981.

PART V

TWO-PHASE FLOW IN ELECTROHYDRODYNAMICS

J.-S. Chang
McMaster University, Hamilton, Ontario, Canada

Chapter 19

CONSERVATION LAWS FOR ELECTROHYDRODYNAMICS TWO-PHASE FLOW

19.1 The Phenomenological Approach for Two-Phase Flow

Two-phase flow phenomena have been a topic of discussion and controversy for more than a half century. In the past decade, a serious examination has taken place and some insights are beginning to emerge. Nevertheless, most heat and mass transfer processes in a two-phase flow system are not yet well understood. The research approach normally conducted for the two-phase flow is the phenomenological approach.

In the phenomenological approach for two-phase flow the following stages are commonly used (Hewitt, 1981):

(1) Identification of the flow patterns (or flow regimes)

(2) Observation of the detailed phenomena and the performance of appropriate measurements such as temperatures, pressures, void fractions, quality, flow rates or velocities, etc.

(3) Construction of physical models with theoretical links to describe the local phenomena.

(4) Integration of the local models to achieve a complete system description.

(5) Use of the integrated model for prediction and design.

19.2 Flow Regime Maps

In gas-liquid two-phase flow in pipes or containers, a number of flow patterns, normally called regimes, are found to exist. This results from the particular manner in which the gas-liquid flow is distributed in the pipe. Even though many authors define flow regimes somewhat differently, most agree with the classification shown in Figure 19.1:

(1) Stratified smooth flow: This regime often occurs in near horizontal pipes with the liquid at the bottom of the pipe and the gas flow along the top. The surface of the liquid is smooth.

(2) Stratified wavy flow: This flow regime is similar to stratified smooth flow, except the gas-liquid interface is wavy instead of smooth.

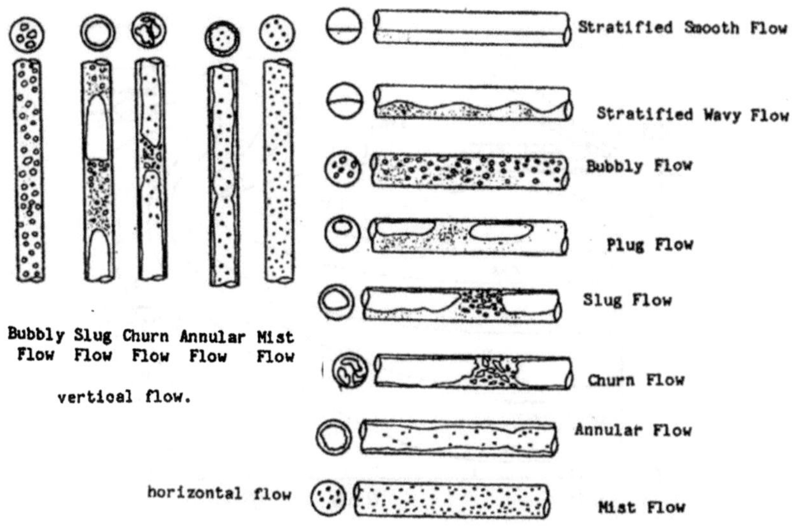

Fig. 19.1 Flow patterns for vertical and horizontal gas liquid two-phase flow

(3) Bubbly flow: This flow regime consists of small, nearly spherical bubbles (gas phase) inside the liquid flow.

(4) Elongated bubble (or plug) and slug flow (intermittent flow, Taitel and Dukler, 1976). These flow regimes are characterized by liquid bridging the gap between the gas-fluid interface and the top of the pipe. The difference between slug and plug flow depends on the degree of agitation case of slug flow where no entrained bubbles exit in the fluid slug.

(5) Churn flow: This flow regime is similar to that of slug flow except that the gas-liquid interface is wavy and often changing in short times.

(6) Annular flow: This regime occurs when the walls are wetted by a thin film of fluid while gas at high velocity flows through the center of the pipe. Fluid droplets are usually entrained in this gas phase. When the upper walls are intermittently wetted by large aerated waves sweeping through the pipe, it is not included with slug flow, which requires a complete fluid bridge, nor annular flow which requires a stable film. Again, following Barnea et al (1980), this has been designed as wavy annular flow.

(7) Mist and droplet flow: The gas completely fills the pipe while small liquid drops or mists are distributed through the body of the pipe. The difference between mist and droplet flow depends on the particle size of the liquid phase. Usually, if the effect of gravitational motion on the particles is much smaller than diffusion, we call it a "mist".

Identification of flow regimes is not easy in the case of metallic pipes and containers (Harvel and Chang, 1995). Some typical flow regime maps based on superficial velocities of gas and

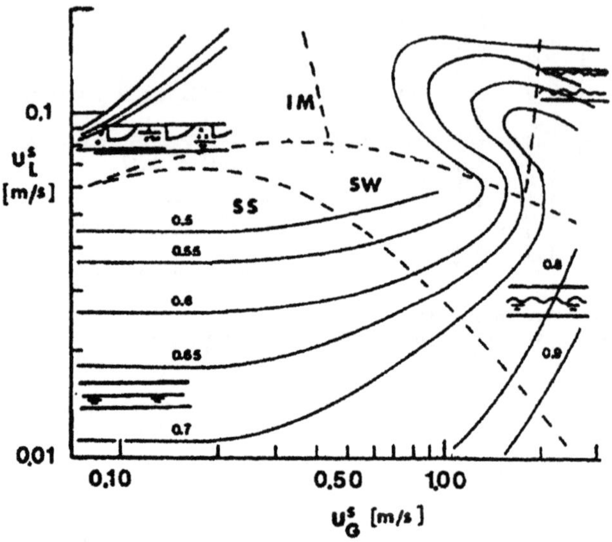

Fig. 19 2 Typical horizontal pipe gas-liquid two-phase flow regime map

liquid phases at the fully developed condition in horizontal pipes are shown in Figure 19.2, for an air-water two-phase flow at room temperature (Taitel and Dukler, 1976; Barnea et al, 1980). The descriptions of the transition mechanisms between these flow regimes can be obtained from various references such as Taitel and Dukler (1976).

19.3 Flow Quality and Void Fraction

If the mass rates of flow [Kg/s] of the gas and liquid phases are written in M_g and M_ℓ, respectively, the "flow quality" x can be defined as

$$x = M_g/(M_g + M_\ell) \tag{19.1}$$

and the total gas and liquid mass fluxes per unit cross-sectional area, m_t, m_g and m_ℓ, respectively, are given by:

$$m_t = (M_g + M_\ell)/A$$

$$m_g = M_g/A = m_t x$$

$$m_\ell = M_\ell/A = m_t(1 - x)$$

The value of the flow quality is not usually known locally along a given pipe so it is often calculated from the two-phase flow enthalpy h_ϕ. Here the two-phase enthalpy is calculated

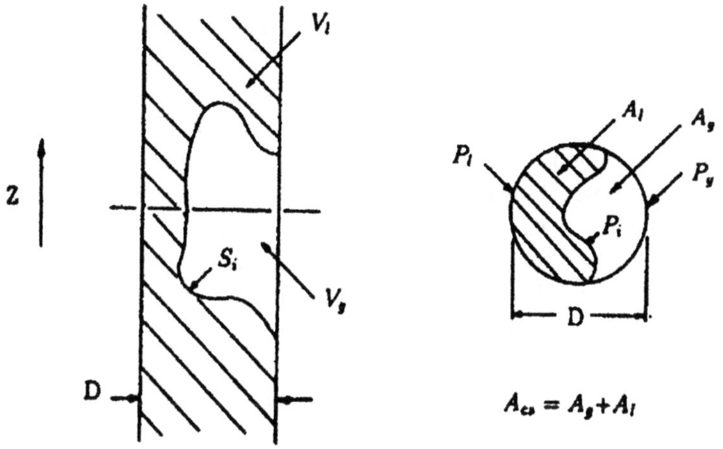

Fig. 19.3 Definitions of two-phase flow parameters in pipe flow: A is the area, V is the volume, p is the perimeter, ℓ denotes liquid, g denotes gas, i denotes interface, s_i is the interfacial area, D is the diameter and z is the axial position.

from the sum of the inlet enthalpy and the heat added up to the location at which flow quality is to be calculated per unit mass of material. Therefore, if h_ℓ and h_g are the liquid and gas phases (saturation enthalpies), respectively, we obtain

$$x = \frac{h_\phi - h_\ell}{h_g - h_\ell} \tag{19.2}$$

The fraction of pipe occupied by the gas phase is called the "void fraction", $\varepsilon_g (= V_g/V)$, and the fraction of volume occupied by the liquid is called the "hold-up", ε_ℓ, i.e. $\varepsilon_\ell = 1 - \varepsilon_g$ $(= V_\ell/V)$ (see Fig. 19.3). Now local phase velocities can be defined as follows:

$$u_g = u_{gs}/\varepsilon_g \; ; \qquad u_\ell = u_{g\ell}/\varepsilon_\ell \tag{19.3}$$

where u_{gs} and $u_{\ell s}$ are the superfacial gas and liquid velocities, respectively (volumetric flow rate/cross sectional area), V_g and V_ℓ are the volume occupied by gas and liquid phases, respectively, and V is the total volume $(= V_g + V_\ell)$.

19.4 Conservation Equations For Volume-Averaged Forms

19.4.1 Averaging Procedures

From the engineering viewpoint, averaging should preserve the signal and eliminate noise. This implies that in two-phase flow systems, the noise can be clearly separated from the signal. This is not always true as we shall discuss later, especially for flow transients. Furthermore, if the fluctuations are relatively large compared to the signal, they may well enter into design and operational considerations. Nonetheless, the primary objective of two-phase flow models at present is the prediction of meaningful averages of flow parameters.

In addition, the averaging procedure should preferably lead to flow parameters that are experimentally measurable. This is necessary because predictions from a model should be checked in order to determine its validity. Therefore, certain types of mathematically elegant multiple averages may be of little use because the quantities cannot be measured, or can only be measured with very great difficulty. Since two-phase flow models are often phrased as first order partial differential equations, it is desirable that an averaged flow parameter at least be continuous with continuous first derivatives.

With these desirable properties in mind, we can proceed to review the commonly used averaging procedures which are:

(i) volume or area averaging, with no averaging in time
(ii) time averaging, with no averaging in space
(iii) ensemble averaging, with no averaging in space
(iv) ensemble/space averaging or time/space averaging.

Reviews of these averaging operators from a mathematical viewpoint have been presented by Ishii (1975) and Banerjee (1979).

Consider first an average of any flow parameter (such as void fraction, velocity, pressure and enthalpy) taken over the cross-sectional flow area occupied by a phase at any instant. It is intuitively evident that such an area average is continuous with regard to the flow direction and time as long as the interface and the cross-sectional plane do not share a common area. (For example, a flat interface sitting at the cross-section plane would occupy an area and would lead to discontinuities in area averages.) This is not a particularly restrictive condition since it seldom occurs, so area averaged quantities may be considered to be continuous for practical purposes. On the other hand, the derivatives with regard to time and space of the averaged quantities will become discontinuous each time an interface becomes tangent to the cross-sectional plane, which could be expected to occur frequently.

Thus area averaging alone can be expected to lead to discontinuities in the derivatives of the averaged dependent variables. This problem, however, disappears if one considers a thin slice with a finite volume rather than an area. In this case, discontinuities do not arise. However, if volume averaging alone is done, the volume must be chosen such that the instantaneous value of the fluctuations (or noise) in the flow parameters is never large compared to their average. From a physical viewpoint, the average volume has to be large compared to the dimensions characterizing the flow regime involved.

19.4.2 Averaged Conservation Equations for the Multifluid Model

We will take an approach to deriving two-phase flow models that is becoming fairly standard at present. The approach is somewhat mathematical at the expense perhaps of the physical insight introduced by a control volume approach, but it is also very powerful and therefore worth discussing.

As mentioned before, one of the main approaches to two-phase flow modelling has been to develop separate sets of averaged conservation equations for each phase starting with the local instantaneous formulation. The resulting field equations and interfacial jump conditions are much simpler than the original formulation, but information is lost in the averaging process and must now be supplied in the form of auxiliary relationships (Ishii, 1975).

These auxiliary relationships are of two types - the first for interfacial and wall transfer of mass, heat and momentum; the second for intraphase distributions of the dependent variables. To illustrate this, consider the cross-sectionally averaged single phase equation for momentum conservation in a duct of uniform area:

$$\frac{\partial}{\partial t}<\rho u>_2 + \frac{\partial}{\partial z}<\rho u^2>_2 + \frac{\partial p}{\partial z} = <\rho F_z>_2 + <(\vec{n}\cdot\bar{\bar{\tau}}_z)>_{2w} + \frac{\partial}{\partial z}<\bar{\bar{\tau}}_{zz}>_2 \qquad (19.4)$$

where ρ is density, u is velocity, p is pressure, $\vec{F}$ is the external force, $\bar{\bar{\tau}}$ is the stress tensor, n is the unit normal vector and subscripts z and w denote components related to the z-direction and at the wall. $< >_2$ denotes the cross-section average $(1/A)\int_A dA$. The term $<(\vec{n}\cdot\bar{\bar{\tau}}_z)>_{2w}$ is then the component of the averaged wall shear stress tensor in the z-direction. By analogy with steady single phase flow, it is often approximated by

$$<(\vec{n}\cdot\tau)_z>_{2w} = \frac{2f\rho u^2}{d_h} \qquad (19.5)$$

where d_h is the hydraulic diameter, and f is the friction factor. This then is an auxiliary relationship of the first type.

On the other hand, (19.4) contains cross-section averages of products of the dependent variables ρ and u, e.g. $<\rho u>_2$ and $<\rho u^2>_2$. We generally wish to solve the equation so that the cross-section averages of each dependent variable ($<\rho>$ and $<u>$) is obtained. Therefore we need relationships between $<\rho u^2>_2$ and $<\rho>_2<u>_2^2$. These are auxiliary relationships of the second type - intraphase distribution relationships. These auxiliary relationships must be obtained from empirical and theoretical considerations outside the averaged equations. However, some information on their form can be obtained, by requiring that the more complex models reduce smoothly to simpler models.

19.4.3 Phase Conservation Equations

First we will derive the averaged conservation equations for each phase and then for the two-phase mixture. To illustrate the mathematical procedures used in averaging, we will derive the volume averaged one-dimensional conservation equations and jump conditions for transient multi-phase flow in a duct of uniform cross-sectional area. Other types of averages can be derived in exactly the same way and do not change the form of the equation. In particular, time or ensemble averaging could be done first followed by volume averaging. Extensions to multidimensional flows and variable area ducts are straightforward.

Consider the flow situation in Figure 19.3. Let V_k be the volume of phase k enclosed between the walls and the cross-sectional planes spaced a distance Z apart (Z can be arbitrarily small). To derive the volume averaged form of the conservation equations, we will use Gauss' theorem and Leibnitz rule (Bird et al. (1960)). The particular forms applying to Figure 19.3 are given below.

As mentioned previously, the equations are derived in volume averaged form rather than cross-sectional averaged form because some difficulties arise in the latter form when the interface and the cross-section coincides as, for example, when the whole interface occupies the cross-section in a refilling problem. The distance Z shown in Figure 19.3 can be made as small as desired, so we essentially average over a slice. The theorems we will use are given below:

Leibnitz Rule

$$\frac{\partial}{\partial t}\int_{V_k(z,t)} f(x,y,z,t)\,dV = \int_{V_k(z,t)} \frac{\partial f}{\partial t}\,dV + \int_{a_i} f(\vec{v}_i\cdot\vec{n}_k)\,dS \tag{19.6}$$

Gauss' Theorem

$$\int_{V_k(z,t)} \nabla\cdot\vec{a}\,dV = \frac{\partial}{\partial z}\int_{V_k(z,t)} \vec{n}_z\cdot\vec{a}\,dV + \int_{a_i} \vec{n}_k\cdot\vec{a}\,dS \tag{19.7}$$

We will define averages by the following symbols

$$<f_k> = \frac{1}{V_k}\int_{V_k} f_k\,dV \tag{19.8}$$

$$<f_k>_i = \frac{1}{V}\int_{a_i} f_k\,dS \tag{19.9}$$

where $V = \Sigma V_k$

If $\rho_k\psi_k$ is the quantity being conserved in the kth phase, and $\vec{j}_k$ and $\hat{S}_k$ are the flux and source of ψ_k, then

$$\frac{\partial \rho_k\psi_k}{\partial t} + \nabla\cdot\rho_k\psi_k\vec{v}_k + \nabla\cdot\vec{j}_k - \rho_k\hat{S}_k = 0 \tag{19.10}$$

The general conservation equation may be volume averaged using Equations (19.6) and (19.7) as follows:

$$\frac{\partial}{\partial t}\int_{V_k} \rho_k \psi_k \, dV + \frac{\partial}{\partial z}\int_{V_k} \vec{n}_k \cdot (\rho_k \psi_k \vec{v}_k + j_k)\, dV - \int_{V_k} \rho_k \hat{S}_k \, dV$$

$$= \int_{a_i} \rho_k \psi_k \vec{n}_k \cdot (\vec{v}_i - \vec{v}_k)\, dS - \int_{a_i + a_{kw}} \vec{n}_{kw} \cdot \vec{j}_k \, dS$$

where the usual assumption that $\vec{v}_k = 0$ at the wall (a_{kw}) has been used. Let $\alpha = V_k/V$ = phase volume fraction. Then using Equation (18.9) (the definition of <>), we have

$$\frac{\partial}{\partial t} \alpha_k <\rho_k \psi_k> + \frac{\partial}{\partial z} \alpha_k <\vec{n}_k \cdot (\rho_k \psi_k \vec{v}_k + \vec{j}_k)> + \alpha_k <\rho_k \hat{S}_k>$$

$$= -\frac{1}{V}\int_{a_i} (\dot{m}_i \psi_k + \vec{j}_k \vec{n}_k)\, dS - \frac{1}{V}\int_{a_{kw}} n_{kw} \cdot \vec{j}_k \, dS \qquad (19.11)$$

where we have written the interphase mass transfer rate as $\dot{m}_k = \rho_k \vec{n}_k \cdot (\vec{v}_k - \vec{v}_i)$

The forms of the conservation equations for each quantity (mass, momentum in the z direction, and energy) for each phase may now be derived.

Mass

In this case, $\psi_k = 1$, $\vec{j}_k = 0$, $\hat{S}_k = 0$. We have

$$\frac{\partial}{\partial t} \alpha_k <\rho_k> + \frac{\partial}{\partial z} \alpha_k <\rho_k u_k> = -<\dot{m}_k>_i = \Gamma_{mk} \qquad (19.12)$$

where equation (19.9) has been used to write the right-hand side. In general, the volume averaged interfacial mass transfer rate is not known 'a priori' and a correlation must be supplied.

Linear Momentum

In this case, $\psi_k = \vec{v}_k$, $\vec{j}_k = p_k \bar{\bar{I}} - \bar{\bar{\tau}}_k$, $\hat{S}_k = \vec{F}_k$. Taking the dot product of the conservation equation (19.11) with the unit z direction vector $\vec{n}_z$, we obtain the equation for conservation of z direction momentum as:

$$\frac{\partial}{\partial t}\alpha_k<\rho_k u_k> + \frac{\partial}{\partial z}\alpha_k<\rho_k u_k^2> + \frac{\partial}{\partial z}\alpha_k<p_k> + \frac{\partial}{\partial z}\alpha_k<\vec{n}_z\cdot(\bar{\bar{\tau}}_k\cdot\vec{n}_z)> - \alpha_k<f_k F_{z,k}>$$

$$= \frac{1}{V}\int_{a_i}[\dot{m}_k u_k + \vec{n}_z\cdot\vec{n}_z p_k - \vec{n}_z\cdot\vec{n}_k\cdot\bar{\bar{\tau}}_k)]\,dS + \frac{1}{V}\int_{a_{kw}}\vec{n}_z\cdot(\vec{n}_{kw}\cdot\bar{\bar{\tau}}_k)\,dS$$

(19.13)

To derive equation (19.13), we must have used the relationship $\vec{n}_{kw}\cdot\vec{n}_z p_z = 0$ (since $\vec{n}_{kw}\cdot\vec{n}_z = 0$ if there is no area change).

Energy

In this case,

$$\psi_k = E_k = \left(e_k + \frac{u_k^2}{2}\right);\quad \vec{j}_k = \vec{q}_k + (p_k\bar{\bar{I}} - \bar{\bar{\tau}}_k)\cdot\vec{v}_k;\quad \hat{S}_k = \vec{F}_k\cdot\vec{v}_k + Q_k \qquad (19.14)$$

We have

$$\frac{\partial}{\partial t}\alpha_k<\rho_k E_k> + \frac{\partial}{\partial z}\alpha_k<\rho_k E_k u_k> + \frac{\partial}{\partial z}\alpha_k<q_{z,k}> + \frac{\partial}{\partial z}\alpha_k<p_k u_k>$$

$$-\frac{\partial}{\partial z}\alpha_k<\vec{n}_z\cdot(\bar{\bar{\tau}}_k\cdot\vec{v}_k)> = \alpha_k<\rho_k(\vec{F}_k\cdot\vec{v}_k + Q_k)> \qquad (19.15)$$

$$-\frac{1}{V}\int_{a_i}[\dot{m}_k E_k + \vec{n}_k(\vec{q}_k + p_k\vec{v}_k - \bar{\bar{\tau}}_k\cdot\vec{v}_k)]\,dS + \frac{1}{V}\int_{a_{kw}}\vec{n}_{kw}\cdot\vec{q}_k\,dS$$

The equation can be written in enthalpy form as

$$E_k = h_k + \frac{u_k^2}{2} - \frac{p_k}{\rho_k} \qquad (19.16)$$

Now $\dot{m} = \rho_k(\vec{v}_k - \vec{v}_i)\cdot\vec{n}_k$ by definition

$$\therefore\ \vec{n}_k\cdot p_k\vec{v}_k = \frac{p_k}{\rho_k}\{\vec{n}_k\cdot\vec{v})_i)\,\rho_k\} + p_k\vec{n}_k\cdot\vec{v}_i = \dot{m}_k\frac{p_k}{\rho_k} + p_k\vec{n}_k\cdot\vec{v}_i$$

or

$$-\frac{1}{V}\int_{a_i}(\dot{m}_k E_k + \vec{n}\cdot p_k \vec{v}_k)\, dS = -\frac{1}{V}\int_{a_i}\left[\dot{m}\left(h_k + \frac{u_k 2}{2}\right) + p_k(\vec{n}_k\cdot\vec{v}_i)\right] dS \quad (19.17)$$

19.4.4 Interface Jump Conditions

If the interface is treated as a contact discontinuity the local instantaneous form of the generalized conservation equation across the interface (i.e, jump condition) is

$$\sum_{k=1}^{2}[\rho_k \psi_k(\vec{v}_k - \vec{v}_i) + \vec{j}_k]\cdot\vec{n}_k = 0$$

or

$$\sum_{k=1}^{2}(\dot{m}_k \psi_k + \vec{j}_k\cdot\vec{n}_k) = 0 \quad (19.18)$$

These local instantaneous equations may be volume averaged in the same way as the phase conservation equations.

Interface Mass

$$\sum_{k=1}^{2}\frac{1}{V}\int_{a_i}\dot{m}_k\, dV = 0 \qquad \text{or} \qquad \langle\dot{m}_1\rangle_i = -\langle\dot{m}_2\rangle_i \quad (19.19)$$

Interface Linear Momentum

$$\sum_{k=1}^{2}\frac{1}{V}\int_{a_i}[\dot{m}_k u_k + \vec{n}_z\cdot(\vec{n}_k p_k) - \vec{n}_k\cdot(\vec{n}_k\cdot\bar{\bar{\tau}}_k)]\, dS = 0 \quad (19.20)$$

or

$$\sum_{k=1}^{2}\left[-\langle\dot{m}_k u_k\rangle_i - \langle p_{ki} - \langle p_{ki}\rangle\rangle_i + \langle(\vec{n}\cdot\vec{\tau}_z)\rangle_i\right] = 0$$

Interface Energy

$$\sum_{k=1}^{2}\frac{1}{V}\int_{a_i}[\dot{m}_k E_k + \vec{n}_k(\vec{q}_k + p_k\vec{v}_k - \bar{\bar{\tau}}_k\cdot\vec{v}_k)]\, dS = 0 \quad (19.21)$$

or $$\sum_{k=1}^{2} - \left\langle \left[\dot{m}_k \left(\frac{u_k^2}{2} \right) + \vec{n}_k \cdot \vec{v}_i (p_{ki} - \langle p_{ki} \rangle) + \vec{n}_k \cdot \vec{q}_k - \vec{n}_k \cdot \vec{v}_k \cdot \overline{\overline{\tau}}_k \right] \right\rangle_i = 0$$

All equations in this section have been derived instantaneous volume averaged form. They can now be time or ensemble averaged and the form remains exactly the same. Terms like $\frac{\partial \alpha_k \langle \rho_k u_k \rangle}{\partial T}$ become $\overline{\frac{\partial \alpha_k \langle \rho_k u_k \rangle}{\partial t}}$ where the overbar denotes time or ensemble averaging. The averaging operators are cumulative.

19.5 Mixture Models

19.5.1 Derivation of the Mixture Equations

For many practical purposes, the level of sophistication embodied in such models is not needed. The approach is then to combine the separate phase conservation equations into mixture equations. Some aspects of the multifluid model can be retained by using empirical correlations relating the phase velocities to each other, or the cross-sectional distribution of velocities and local void fractions. The mixture equations then are:

Mass

When the separate phase conservation equations for mass given in equation (19.12) are summed and the interface jump condition (19.19) is applied, we obtain

$$\frac{\partial \rho}{\partial t} + \frac{\partial G}{\partial z} = 0 \tag{19.22}$$

where the mixture density, $\rho = \Sigma \alpha_k \rho_k$; and the mixture mass velocity, $G = \Sigma \alpha_k \rho_k u_k$. The averaging brackets have been dropped and by u_k we mean $\langle \overline{u_k} \rangle_k$. The mass transfer terms on the right hand of equation (19.12) does not appear in (19.23) because they cancel due to (19.19). Note that at steady state, the mass flow rate, G, is a constant for a constant area duct.

Momentum

The separate phase conservation equations for momentum given in equation (19.13) may be summed and the jump condition (19.20) applied to yield

$$\frac{\partial G}{\partial t} + \frac{\partial (G^2 v')}{\partial z} + \frac{\partial p}{\partial z} = \Sigma \frac{1}{V} \int_{a_{kw}} \vec{n}_z \cdot (\vec{n}_w \cdot \tau_k)\, dS = \tau_w \tag{19.23}$$

where $v' = \Sigma \dfrac{x_k^2}{\alpha_k \rho_k}$ = momentum weighted mixture specific volume and x_k = flow quality of phase $k = \dfrac{\alpha_k \rho_k u_k}{\Sigma \alpha_k \rho_k u_k}$, and the right hand side of equation (19.23) is the total wall shear stress.

Energy

If we sum the separate conservation equations for energy given in (19.16), and apply the jump condition (19.21), then

$$\frac{\partial}{\partial t} \rho \left(e + \frac{G^2 v'}{2\rho} \right) + \frac{\partial}{\partial z} \left[G \left\{ h + \frac{(G v'')^2}{2} \right\} \right] = - \Sigma \frac{1}{V} \int_{a_{kw}} \vec{n}_{kw} \cdot \vec{q}_k \, dS = q_w \quad (19.24)$$

where q_w = total wall flux heat,

$$v_k'' = \Sigma \frac{x_k^3}{(\alpha_k \rho_k)^2},$$

h_k = flow enthalpy = $\Sigma x_k h_k$,

e_k = internal energy = $\Sigma \hat{x}_k e_k$, and

$\hat{x}_k$ = thermodynamic quality of phase k = $\alpha_k \rho_k / (\Sigma \alpha_k \rho_k)$.

The quantity v_k'' again does not have a clear physical significance and therefore various other ways of writing these equations are also used.

In general, the phases are not at thermal or mechanical equilibrium, so empirical correlations have to be used to close equations (19.22), (19.23) and (19.24), in addition to the usual ones for wall shear stress and heat flux. The equations of state for each phase $h_k = h_k (p, \rho_k)$ are also assumed to be known.

19.5.2 The Homogeneous Equilibrium Model (HEM)

If the phases are assumed to be in equilibrium with each other, then a particularly simple and quite useful set of mixture equations is obtained. It is called the homogeneous equilibrium (or equal velocities-equal temperatures) model and essentially ignores all formulation and the equations are similar to those for compressible gas dynamics.

The homogeneous equilibrium model only requires explicit numerical information for wall shear stress and heat flux.

The model implies that

(i) $\hat{x}_k = x_k$ since $u_1 = u_2$

and (ii) $v' = v'' = 1/\rho$.

In this case, the quantities v′ and v″ have a clear physical significance as the reciprocal of the mixture density, ρ.

For steady flow, the mixture momentum equation (19.23) may be phrased in terms of total pressure drop as

$$-\frac{dp}{dz} = \tau_w + \rho u \frac{\partial u}{\partial z} + \rho g \sin \alpha \tag{19.25}$$

frictional accelerational gravitational

where the significance of each of the components is identified above.

19.6 Conservation Equations Under Electromagnetic Field

For a dielectric fluid, electromagnetically induced flow velocity normally does not exceed Mach number M = 0.2, hence subsonic incompressible flow is only considered in this work. We also assumed that the mean free path of the fluid particle is much less than the characteristics length of the flow tubes or bodies, hence, the continuum fluid flow mechanics is considered. We limited discussions on the Newtonian fluid without the viscous dissipation heating of the fluid. We also assumed the Boussineq approximation (Chang and Watson, 1994) for natural convection flow with small density changes.

Under these conditions with an electromagnetic field, the mass, momentum and energy conservation equations in fluid flow for gas and liquid phases can be expressed as follows

(i) **Mass conservation**

$$\frac{\partial \rho}{\partial t} + \nabla \cdot (\rho \mathbf{U}) = 0 \tag{19.26}$$

(ii) **Momentum conservation**

$$\rho \frac{\partial \mathbf{U}}{\partial t} + \rho(\mathbf{U} \cdot \nabla)\mathbf{U} = -\rho_g \beta(T - T_s) - \nabla P + \mathbf{F}_{EB} + \nabla \cdot [(\mu_D + \varepsilon_D)\nabla \mathbf{U}] \tag{19.27}$$

(iii) **Energy conservation**

$$\frac{\partial T}{\partial t} + \mathbf{U} \cdot \nabla T = \frac{k}{\rho C_p} \nabla^2 T + Q_{EB} \tag{19.28}$$

where ρ is the density, **U** is the flow velocity, β is the coefficient of thermal expansion of the fluid, k is the thermal conductivity, T is the temperature, P is the pressure, μ_D and ε_D are viscosity and eddy viscosity, respectively. T_s is the reference temperature, and C_p is the specific heat. $\mathbf{F}_{EB}$ and Q_{EB} are the momentum and energy change due to the presence of electric and magnetic fields, respectively as follows:

$$\mathbf{F}_{EB} = \rho_{ie}\mathbf{E} + \mathbf{J}\mathrm{x}\mathbf{B} - \frac{1}{2}E^2\nabla\varepsilon - \frac{1}{2}H^2\nabla\mu + \nabla\left[\frac{1}{2}\rho E^2\cdot\left(\frac{\partial\varepsilon}{\partial\rho_T}\right)_T + \frac{1}{2}\rho H^2\left(\frac{\partial\mu}{\partial\rho_T}\right)_T\right] \tag{19.29}$$

where the term $\rho_{ie}\mathbf{E}$ is the force density due to the space charges, the term $\mathbf{J}\times\mathbf{B}$ is the force density due to the charged particle motion, the term $1/2E^2\nabla_\varepsilon$ is the force density due to the dielectric property changes, the term $1/2H^2\nabla\mu$ is the force density change due to the fluid permeability changes, and the term [ ] is the force density change due to the electrostriction and magnetostriction effects.

$$Q_{EB} = (\mathbf{J} - \rho_{ie}\mathbf{U})\cdot(\mathbf{E} + \mathbf{U}\mathrm{x}\mathbf{B}) + \nabla\cdot[(\mathbf{E} + \mathbf{U}\mathrm{x}\mathbf{B})\,\mathrm{x}(\mathbf{H} - \mathbf{U}\mathrm{x}\mathbf{D})]$$

$$+\left[\mathbf{E}\cdot\frac{d}{dt}\left(\frac{\mathbf{D}}{\rho}\right) + \mathbf{H}\cdot\frac{d}{dt}\left(\frac{\mathbf{B}}{\rho}\right)\right]\rho \tag{19.30}$$

where the first term is the heat generation due to flow of charged particles such as ohmic heatings, the second term is the energy due to the polarizations such as electromagnetic hysteresis loss, the third term is energy due to the displacement current and time varying magnetic fields such as energy storage in an electromagnetic system. The relative importance of each term in equations (19.29) and (19.30) depends on working fluid and is summarized in Table 19.1.

Electron, positive and negative ion transport equations are:

$$\mathbf{U}N_c \pm \mu_c N_c \mathbf{E} - D_c\nabla N_c = \mathbf{J}_c \tag{19.31}$$

$$\nabla\cdot\mathbf{J}_c = \frac{\partial N_c}{\partial t} + (\text{source})_c - (\text{sink})_c \tag{19.32}$$

$$\nabla\cdot\mathbf{D} = -e(N_+ - N_e - N_-) \tag{19.33}$$

$$\mathbf{E} = -\nabla V \tag{19.34}$$

where N is the number density, V is the electric potential, and subscripts +, −, e and c are positive ion, negative ion, electron and charged particles (+, −, or e).

The dimensionless equations now become

$$\tilde{\nabla}\cdot\mathbf{u} = 0 \tag{19.35}$$

$$\frac{\partial\mathbf{u}}{\partial\tau} + (\mathbf{u}\cdot\tilde{\nabla})\mathbf{u} = -\frac{Ra\,\mathbf{k}}{Pr\,Re^2}(\theta - 1) - \tilde{\nabla}p + \tilde{\nabla}\cdot\left[\left(\frac{1 + \varepsilon_D'}{Re}\right)\tilde{\nabla}\mathbf{u}\right] + \mathbf{F}_E'$$

Table 19.1

The momentum and energy terms due to the existence of an electromagnetic field in a conservation equation for various fluids. D': diffusion coefficient, μ': mobility, I: current, σ': electrical conductivity, a_g: void fraction.

Type of Fluid		Momentum terms due to EMF: F_{EB}	Energy terms due to EMF: Q_{EB}	Comments
Non-conducting Fluids (I → small)	Gas Single Phase	$\frac{1}{2}\varepsilon_o(\varepsilon_s-1)\nabla E^2 + \frac{1}{2}\mu_o(\mu_s-1)\nabla H^2$	$\nabla\cdot(\underline{E}^*X\underline{H}^*) + [\underline{E}\cdot d(\underline{D}/\rho)/dt + \underline{H}\cdot d(\underline{B}/\rho)/dt]\rho$	$\mu_s \doteq 1$, $\varepsilon_s \doteq 1$, $\nabla\varepsilon, \nabla\mu \rightarrow$ small
	Liquid Single Phase	$\frac{1}{2}E^2(\partial\varepsilon/\partial T)_\rho \nabla T$	(same as for gas)	$\rho\,\partial\varepsilon/\partial\rho$, $\rho\,\partial\mu/\partial\rho$, $\nabla\mu \rightarrow$ small
	Gas-Liquid Two-Phase	$-\frac{1}{2}E^2\nabla\varepsilon - \frac{1}{2}H^2\nabla\mu$ (for each phase) $+\nabla(\frac{1}{2}E^2\rho\,\partial\varepsilon/\partial\rho + \frac{1}{2}\rho H^2\,\partial\mu/\partial\rho)$	(same as for gas but for each phase)	$\rho_\phi = \alpha_g\rho_g + (1-\alpha_g)\rho_l$ $\varepsilon_l, \mu_l >> \varepsilon_g, \mu_g$
Conducting Fluids (I ≠ 0)	Gas Single Phase	$\rho_{le}\underline{E} + \underline{J}\,X\,\underline{B} + \nabla(\frac{1}{2}\rho E^2\,\partial\varepsilon/\partial\rho + \frac{1}{2}\rho H^2\,\partial\mu/\partial\rho)$	$\underline{J}\cdot\underline{E}^* + \nabla\cdot(\underline{E}^*X\underline{H}^*) + [\underline{E}\cdot d(\underline{D}/\rho)dt + \underline{H}\cdot d(\underline{B}/\rho)/dt]\rho$	$\rho_{le} \neq 0$, $\nabla\varepsilon$, $\nabla\mu \rightarrow 0$ $\underline{J} = \underline{U}\,N\pm\mu'\,N\underline{E} - D'\nabla N$
	Liquid Metal Single Phase	$\underline{J}\,X\,\underline{B}$	$\underline{J}^{*2}/\sigma' + \nabla\cdot(\underline{E}^*X\underline{H}^*) + \rho\underline{H}\cdot d(\underline{B}/\rho)/dt$	$\rho\,\partial\varepsilon/\partial\rho$, $\rho\,\partial\mu/\partial\rho \rightarrow 0$ $\underline{J} = \mu'\,\underline{E}$, $\rho_{le} = 0$, $\nabla\varepsilon$, $\nabla\mu \rightarrow 0$
	Liquid Single Phase	$\rho_{le}\underline{E} + \underline{J}\,X\,\underline{B} + \frac{1}{2}E^2(\partial\varepsilon/\partial T)_\rho\nabla T$	(same as for gas)	$\nabla\mu$, $\rho\,\partial\varepsilon/\partial\rho$, $\rho\,\partial\mu/\partial\rho \rightarrow 0$ $\rho_{le} = \rho_+ - \rho_-$
	Gas-Liquid Two-Phase	$-\frac{1}{2}E^2\nabla\varepsilon - \frac{1}{2}H^2\nabla\mu + \nabla(\frac{1}{2}\rho E^2\partial\varepsilon/\partial\rho + \frac{1}{2}\rho H^2\partial\mu/\partial\rho) + \underline{J}X\underline{B} + \rho_{le}\underline{E}$ (for each phase)	(same as for gas but for each phase)	$\underline{J}^* = \sigma'\underline{E}^*$ $\underline{E}^* = \underline{E} + \underline{U}X\underline{B}$ $\underline{H}^* = \underline{H} - \underline{U}X\underline{D}$

$$\frac{\partial \theta}{\partial \tau} + \mathrm{Re}\,\mathrm{Pr}\,\mathbf{u}\cdot\tilde{\nabla}\theta = \tilde{\nabla}\cdot(f(\theta)\tilde{\nabla}\theta$$

by introducing the following nondimensional variables:

$$u = U/U_s,\quad p = P/\rho_s U_s^2,\quad \tau = tU_s/L,\quad \tilde{\nabla} = L\nabla,$$

$$\mathrm{Pr} = \nu/\alpha,\quad \mathrm{Ra} = g\beta\, T_s L^3/\nu\alpha,\quad \theta = T/T_s,\quad \mathrm{Re} = U_s L/\nu,$$

$$\varepsilon_D' = \varepsilon_D/\mu_D,\quad f(\theta) = k(T)\,\rho_s/k(T_s)\rho,\quad F_E' = L\,F_E/\rho_s\,U_s^2$$

where subscripts s refer to reference values, Pr, Re and Ra are Prandtl, Reynolds and Rayleigh numbers, ν is the kinematic viscosity, α is the thermal diffusivity, p is the pressure, ε_D is the Eddy viscosity, and L is a characteristic length. The nondimensional form of equation with the Poisson's equation and the charged particle transport equations (19.35) then become:

$$\mathbf{F}_E' = \Big\{E_{1\sigma}(\eta\, n_{ie} + \mathbf{j}\times\zeta\, H_{em}\,\mu_s) - E_{1\varepsilon}\,\eta^2\tilde{\nabla}\varepsilon_s - H_{mg}\,\zeta^2\tilde{\nabla}\mu_s$$

$$+\tilde{\nabla}\left[E_{1\varepsilon}\,\eta^2\left(\frac{\partial\varepsilon_s}{\partial\rho'}\right)_\theta + H_{mg}\,\zeta^2\left(\frac{\partial\mu_s}{\partial\rho'}\right)_\theta\right]\Big\}/\mathrm{Re}^2$$

$$\tilde{\nabla}^2\phi = -(L/\lambda_D)^2(n_+ - n_-) = \tilde{\nabla}\cdot\eta \tag{19.36}$$

$$\mathrm{ReSc}_c\,\mathbf{u}\cdot\tilde{\nabla}n \pm \tilde{\nabla}\cdot(n\tilde{\nabla}\phi) - \tilde{\nabla}^2 n = 0$$

where the nondimensional variables are

$$\zeta = H/H_s;\quad n = N/N_S;\quad \phi = eV/kT;\quad Sc = \nu/D;$$

$$\eta = E/E_s,\ j = J/J_s;\quad H_{mg} = \mu_o H_s^2 L^2/2\rho_s\nu^2;$$

$$E_{\ell\varepsilon} = \varepsilon_o L^2 T_S\, f(T)\, E_s^2/2\rho_s\nu^2;\quad H_{em} = \mu_c B_s;$$

$$E_{\ell\sigma} = eN_sE_s\,L^3/\rho_s\,\nu^2 = \sigma_s E_s L^3/\rho_s\nu^2\mu_c = J_s L^3/\rho_s\nu^2\mu_c = I_s\,L^3/\rho_s\,\nu^2\,\mu_c\,A\,.$$

Here V is the electric potential, D is the diffusion coefficient, μ is the mobility, σ is the electric conductivity, J_s is the current density value at the electrodes, I_s is the total current, N is the charged particle number density, A is the electrode surface area, k is Boltzmann's constant, λ_D is the Debye length $(\varepsilon\, k\, T/e^2 N_s)^{1/2}$, Sc is the Schmidt number, subscript c, + and - refer to charged particles, positive and negative ions respectively. $E_{\ell\varepsilon}$ and $E_{\ell\sigma}$ are the dielectric and conductive component of EHD forces, and called "the Dielectric Electric Rayleigh or EHD and the Conductive Electric Rayleigh or EHD Numbers" in this text. Equations (19.35)-(19.36) show that EHD forces become significant when $E_{\ell\varepsilon}$ or $E_{\ell\sigma}$ > RaPr and Re^2 for natural and forced convections, respectively. Here, the EHD number used by other authors (Chang et al., 1995) corresponds to $E_{\ell\sigma}{}^{1/2}$. Classical magnetic pressure number $P_B = H_{mg}/Re^2$ shows more physics of EMHD flow than the Hartmann or magnetic Reynolds numbers.

References

Banajee, S. and A.M.C. Chan (1983), Int. J. Multi-phase Flow, 9, 117.

Barnea, D., O. Shoham, Y. Taitel and A.G. Dukler (1980), Int. J. Multiphase Flow, 6, 217-225.

Bird, R.B., Stewart, W.E., Lightfoot, E.N. (1960), "Transport Phenomena", Wiley.

Chang, J.S. and A. Watson (1994), IEEE Trans. Dielectrics & Elect. Insul., 1, 871-895.

Chang, J.S. A. Kelley and J. Crowley, Eds. (1995), "Handbook of Electrostatic Processes", Marcel Dekker Press, New York.

Chu, B.T. (1959), Phys. Fluids 5, 473-484.

Harvel, G.D. and J.S. Chang (1995), Chapter 13 of "Handbook of Electrostatic Processes", J.S. Chang et al., Eds., Marcel Dekker Inc.

Hewitt, G.F. (1981), ASEM/AIChE 20th Heat Transfer Conference, D.Q. Kemn Lecture.

Hewitt, G.F. (1978), "Two-Phase Flow and Heat Transfer With Application to Nuclear Reactor Design Problem", J.J. Giboux, Ed., Hemisphere Press.

Ishii, M. (1975), "Thermo-Fluid Dynamic Theory of Two-Phase Flow", Eyrolles, Paris.

Mandhane, J.M., G.A. Gregory and K. Aziz (1974), Int. J. Multiphase Flow, 1, 537-537.

Taitel, Y. and A.E. Dukler (1976), AIChE J., 22, 47-55.

Vernier, P. and J.M. Delhaye (1968), Energie Primaire, 4, No. 1-2.

Wallis, G.B. (1969), "One-Dimensional Two-Phase Flow", McGraw-Hill.

Chapter 20

EHD GAS-LIQUID TWO-PHASE FLOW

20.1 Introduction

Flow and heat transfer enhancement due to an electromagnetic field (EMF) and pulses (EMP) on single and two phase fluid flow in a heat exchanging system has increased in importance as a result of recent developments in high power transformers and transmission line cable cooling systems (Chang and Watson 1994). In the boiling water and pressurized water type nuclear power plants (BWR and PWR), the boiling phenomenon becomes important to estimate heat flow at the nuclear fuel bundle surface when the reactor or turbine is being tripped, since the heat transport system is under EMF or EMP, and quite often "lightning" strikes or other types of EMP are the cause of these trips. In a nuclear fusion reactor primary heat transport system, EMP and EMF effects would become a more common problem, since relatively large magnetic field and pulse currents must be used to confine or generate plasmas. However, the effects of the EMF or EMP on the quantities of engineering interest such as the critical heat flux, heat transfer rate and pressure drops are poorly understood at the present stage. The direction of the electric or magnetic field to the gas-liquid interface may be complex, in a two-phase flow and in some case EMF or EMP may also be generated as a result of fluid flow (see Fig. 20.1). Here we must note that the effect of leak EMF becomes significantly important as a result of recent development on the new materials such as less conductive ceramics and insulative polymers. Modelling of electrohydrodynamically induced flow has been studied for a simple geometry, such as pipe, annulus and parallel plates etc. by numerical simulation techniques. However, it is difficult to calculate more complex geometries, since the charged particle transport and Poisson's equations are required to solve in additional to the conservation of mass, momentum and energy equations. On the other hand, the volume averaged conservation equations are normally used to simulate multi-phase flow phenomena due to the additional complexity between two-phase interfaces. In this case, the volume averaged conservation equations, $F(z', t)$ for each phase were introduced from the averaging of the three-dimensional conservation equations, $f(x, y, z, t)$, and hence, the complex three dimensional flow piping system with many different flow components, such as pumps, heat exchangers etc., can be included in calculations. For EHD two-phase flow model, six conservation equations, three

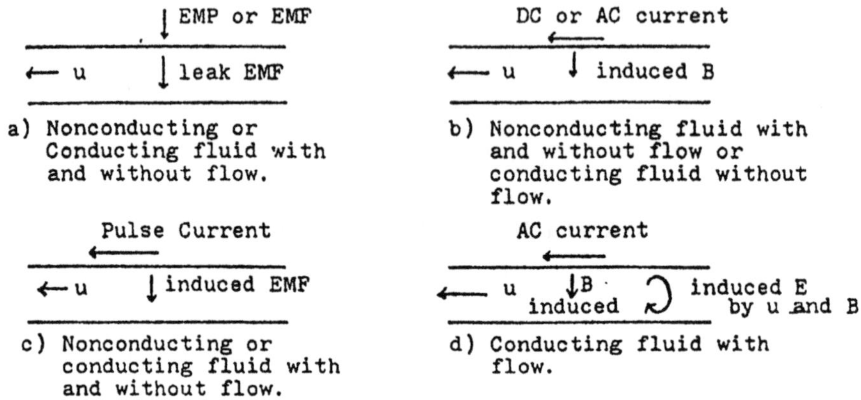

Fig. 20.1 Schematics of the electric and magnetic field directions under electromagnetic field and electromagnetic pulse impacts.

interfacial jump conditions and several constitutive equations are required. In this chapter, EHD equations for two phase pipe flow is outlined.

20.2 Separated Flow Model of Stratified Gas-Liquid Flow Under Electromagnetic Fields

In the present, the primary property of this regime is the liquid level. This is found by equating the pressure drops in air and water and considering the momentum balance. Consider the cross-sectional planes shown in Figure 20.2. D is the pipe diameter, h_L is the height of the liquid in the pipe, A_G and A_L are the gas and liquid cross-sectional areas respectively, S_G is the gas perimeter, S_L is the liquid perimeter and S_i is the length of the interface between the two. In the second diagram U_G and U_L are the gas and liquid velocities, α is the angle of the pipe with the horizontal, and γ is the angle of the of the electric or magnetic field with the pipe (see Fig. 20.2).

Since the gas-fluid system is in equilibrium the total change in momentum for each system must be zero. This gives the following equations for the fluid and gas phases respectively,

$$-A_L \frac{dp}{dx} - \tau_{WL} S_L + \tau_i S_i + \rho_L A_L g \sin\alpha - S_i F_{EB} \cos\gamma = 0 \tag{20.1}$$

$$-A_G \frac{dp}{dx} - \tau_{WG} S_G + \tau_i S_i + \rho_G A_G g \sin\alpha + S_i F_{EB} \cos\gamma = 0 \tag{20.2}$$

where dp/dx is the change in pressure along the pipe, τ_{WL} is the shear stress of the wall on the liquid, τ_{WG} is the shear stress of the wall on the gas, τ_i is the interfacial shear stress, ρ_L

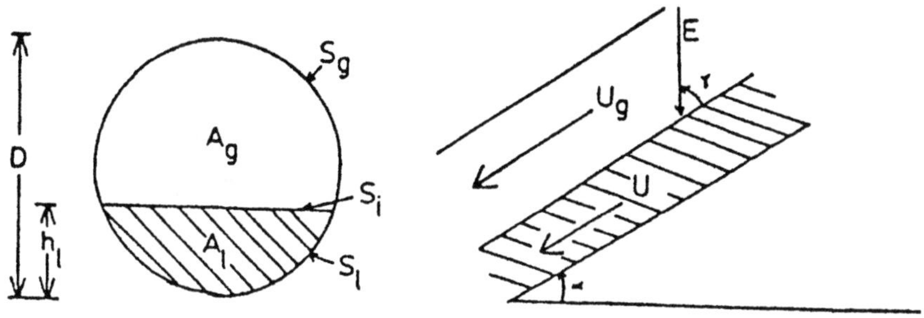

Fig. 20.2 Gas-liquid system in equilibrium stratified flow.

and ρ_G are the liquid and gas densities, F_{EB} is the electromagnetic force from (19.29) and g is the gravitational acceleration.

The shear stresses are evaluated in the usual manner, which is

$$\tau_{WL} = \frac{f_L \rho_L U_L^2 L}{2}; \qquad \tau_{WG} = f_G \frac{\rho_G U_G^2}{2};$$

$$\tau_i = f_i \frac{\rho_i (U_G - U_L)^2}{2}; \qquad f_i = f_G; \tag{20.3}$$

$$f_L = C_L \left(\frac{D_L U_L}{\nu_L} \right)^{-n}; \qquad f_G = C_G \left(\frac{D_G U_G}{\nu_G} \right)^{-m}$$

where D_L, D_G are the hydraulic diameters, and ν_L, ν_G are the liquid and gas viscosities, respectively. C_G, C_L, n and m are constants whose value depends on whether the flow is laminar or turbulent, where hydraulic diameters are

$$D_L = \frac{4 A_L}{S_L}; \qquad D_G = \frac{4 A_G}{S_i + S_G}$$

It is useful to transform the equations to dimensionless form. The normalizing variables are: D for length, D^2 for area, and the superficial velocities U_G^S and U_L^S for the gas and liquid velocities. By combining equations (20.1) to (20.3), we obtain

$$X^2 \left[(\tilde{U}_L \tilde{D}_L)^{-n} \tilde{U}_L^2 \frac{\tilde{S}_L}{\tilde{A}_L} \right] - \left[(\tilde{U}_G \tilde{D}_G)^{-m} \tilde{U}_G^2 \left(\frac{\tilde{S}_G}{\tilde{A}_G} + \frac{\tilde{S}_i}{\tilde{A}_L} + \frac{\tilde{S}_i}{\tilde{A}_G} \right) \right]$$

$$- 4 Y - 4 \tilde{S}_i \left(\frac{1}{\tilde{A}_L} + \frac{1}{\tilde{A}_G} \right) Z = 0 \tag{20.4}$$

$$X^2 = \frac{\frac{4C_L}{D}\left(\frac{U_L^S D}{\nu_L}\right)^{-m}\frac{\rho_L (U_L^S)^2}{2}}{\frac{4C_G}{D}\left(\frac{U_G^S D}{\nu_G}\right)^{-m}\frac{\rho_G (U_G^S)^2}{2}} = \frac{|(dp/dx)_L^S|}{|(dp/dx)_G^S|} \tag{20.5}$$

$$Y = \frac{(\rho_L - \rho_G)\, g \sin\alpha}{\frac{4C_G}{D}\left(\frac{U_G^S D}{\nu_G}\right)^{-m}\frac{\rho_G (U_G^S)^2}{2}} = \frac{(\rho_L - \rho_G)\, g \sin\alpha}{|(dp/dx)_G^S|} \tag{20.6}$$

$$Z = \frac{2\, D\, F_{EB} \cos\gamma}{\frac{4C_G}{D}\left(\frac{U_G^S D}{\nu_G}\right)^{-m}\frac{\rho_G (U_G^S)^2}{2}} = \frac{2\, D\, F_{EB} \cos\gamma}{|(dp/dx)_G^S|} \tag{20.7}$$

The dimensionless quantities are denoted by a tilde (~). The dimensionless variables in equation (20.4) can be calculated with the dimensionless water level $\tilde{h}_L = h_l/D$ only, as follows,

$$\begin{aligned}
&\tilde{A}_L = 0.25\left[\pi - \cos^{-1}(2\tilde{h}_L - 1) + (2\tilde{h}_L - 1)\sqrt{1 - (2\tilde{h}_i - 1)^2}\right]; \qquad \tilde{U}_L = A/A_L \\
&\tilde{A}_G = 0.25\left[\cos^{-1}(2\tilde{h}_L - 1) - (2\tilde{h}_L - 1)\sqrt{1 - (2\tilde{h}_L - 1)}\right]; \\
&\tilde{S}_i = \sqrt{1 - (2\tilde{h}_L - 1)^2}; \qquad \tilde{U}_G = A/A_G \\
&\tilde{S}_L = \pi - \cos^{-1}(2\tilde{h}_L - 1); \qquad \tilde{S}_G = \cos^{-1}(2\tilde{h}_L - 1)
\end{aligned} \tag{20.8}$$

20.3 Flow Regime Transition Criteria

Flow regime transition in gas-liquid two-phase flow also influenced by electromagnetic field as follows:

(i) Transition from the stratified to the intermittent or annular flow regimes

The intermittent flow regime covers the plug (elongated bubble), and slug flow regimes (see Chapter 19). Flow in these regimes starts as stratified flow. As the liquid flow rate is increased, the waves will tend to reach the top of the pipe forming a bridge. For plug and slug flow, a complete bridge is formed (see Fig. 20.3).

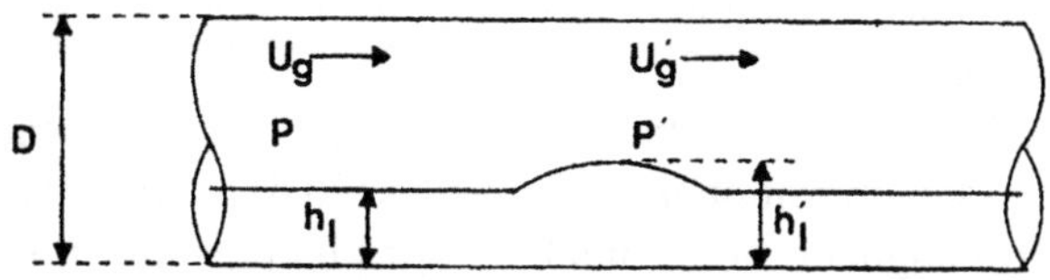

Fig. 20.3 Schematics of slug flow formations.

Transition criteria from the stratified to the intermittent flow regimes can be modified from the work of Brunner and Chang (1980), which is based on the Kelvin–Helmholtz instability analysis, and the results can be expressed as follows:

$$f_{EB}^2 \left[\frac{1}{2} \frac{\tilde{U}_G \, d\tilde{A}_L / d\tilde{h}_L}{\tilde{A}_G} \right] \geq 1 \; ; \qquad f_{EB} = \frac{U_G^S}{\sqrt{D}} \left[\frac{\rho_G}{(\rho_L - \rho_G) g - F_{EB}} \right]^{1/2} \tag{20.9}$$

where f_{EB} is an electromagnetic Froude number of the form, and dA_L/dh_L can easily be found from the expression for A_L. It should be noted that if the electric field goes to zero, then the expression reduces to the formula obtained by Taitel and Dukler (1976).

(ii) Transition from the stratified smooth to stratified wavy flow

Stratified wavy flow occurs when the gas velocity is great enough to cause waves, but lower than that needed to cause the rapid wave growth characteristic of intermittent or annular flow. This happens when the force of the wind is great enough to overcome the viscous dissipation of the wave. By modification of the model presented by Taitel and Dukler (1976), we obtain (Chang, 1989)

$$\tilde{U}_G > \left[\frac{4\nu \, (\rho_L - \rho_G) g^2 - 2 F_{EB} \tilde{U}_L^3 (U_L^S)^3}{\beta \, \rho_G \tilde{U}_L (U_L^S) (U_G^S)^2} \right]^{1/2} \tag{20.10}$$

(iii) Interfacial electromagnetic forces

The cause of above interfacial phenomena may be due to the differences of the dielectric and magnetic property of gas and liquids for present magnetic and electric fields direction, and the most important forces in the interfaces due to the electromagnetic fields are

$$\begin{aligned} F_{EBI} &= \frac{\varepsilon_g (\varepsilon_\ell - \varepsilon_g)^2}{\varepsilon_\ell (\varepsilon_\ell + \varepsilon_g)} E_g^2 + \frac{\mu_g (\mu_\ell - \mu_g)^2}{\mu_\ell (\mu_\ell + \mu_g)} H_g^2 \\ &= \frac{\varepsilon_\ell (\varepsilon_\ell - \varepsilon_g)^2}{\varepsilon_g (\varepsilon_\ell + \varepsilon_g)} E_g^2 + \frac{\mu_g (\mu_\ell - \mu_g)^2}{\mu_g (\mu_\ell + \mu_g)} H_\ell^2 \end{aligned} \tag{20.11}$$

where E_g, E_ℓ, H_g, H_ℓ are electromagnetic field near interfaces for E and H perpendicular to the gas-liquid interfaces.

By using the method employed by Chang et al. (1983, 1989), a typical numerical result for the stratified smooth to slug transition using equations (20.4) to (20.10) to predict the flow transition under an electromagnetic field for a 0.019 [m] ID pipe are shown in Figure 20.4. It should be noted that the result without applied field is equivalent to the results obtained by Taitel and Dukler (1976). The applied voltage causes the transition from stratified to slug flow to occur at lower gas and water flows. The effect increases as the voltage is increased, being more pronounced at low gas flows. This occurs because at low gas flow, the liquid level is near the top electrode, thus giving a large force, since, $F_{EBI} \propto [D-h_\ell]^{-2}$ for the conductive tap waters and gas. However, if we calculated the effect of magnetic field in equation (19.11), the order of magnitude of magnetic field component is much less than the both electric field and the gravity terms.

The effect of applied voltage on the stratified smooth to wavy flow regime transitions for 1.9×10^{-2} [m] tubes are shown in Figure 20.5, respectively. Figure 20.5 show that the non-monotonic effect of applied electric field on the stratified smooth to wavy flow boundary. The flow regime boundary shifted towards smaller superfacial velocities of gas and liquid areas when small electric field are applied and then shifted back towards larger superfacial velocity area when electric field increases. Since time averaged void fraction or time averaged liquid level in this case, is not influenced by applied field we may conclude that this nonmonotonic dependence origin from the surface wave.

Figures 20.4 and 20.5 show that present results agree both qualitatively and quantitatively with experimental results, except that the effect of applied field by the current prediction method overestimated EHD force effects. This may be due to the building up of space charges or the inadequate estimations of the constitutive relationships used in the present models. Nevertheless, the effect of applied electric field in a flow regime map was qualitatively confirmed.

20.4 Two-Component Two-Phase EHD Flow

From various fundamental studies, the effect of EMHD forces on the fluid flow can be categorized as follows (see Table 20.1, Chang and Watson, 1995):

1. Micro and bulk conversions
2. Multi-phase interfacial instabilities
3. Enhancement of entrainments
4. Electrostatic adhesive forces in between solid particles
5. Onset of turbulence.

Therefore, the most of EHD technology applies one or two of these driving forces. In the past, an application of the momentum charges due to the charged particle injections (term 1 in Eq. (19.29) is called "electrohydrodynamics", and an application of the momentum exchanges due to the different phases, such as gas bubbles or solid particles in liquid (term 3 in Eq. (19.29)), is called "dielectrophoresis". In this work, we define electromagnetic hydrodynamics by using all the driving forces in Eq. (19.29) instead of classical definitions,

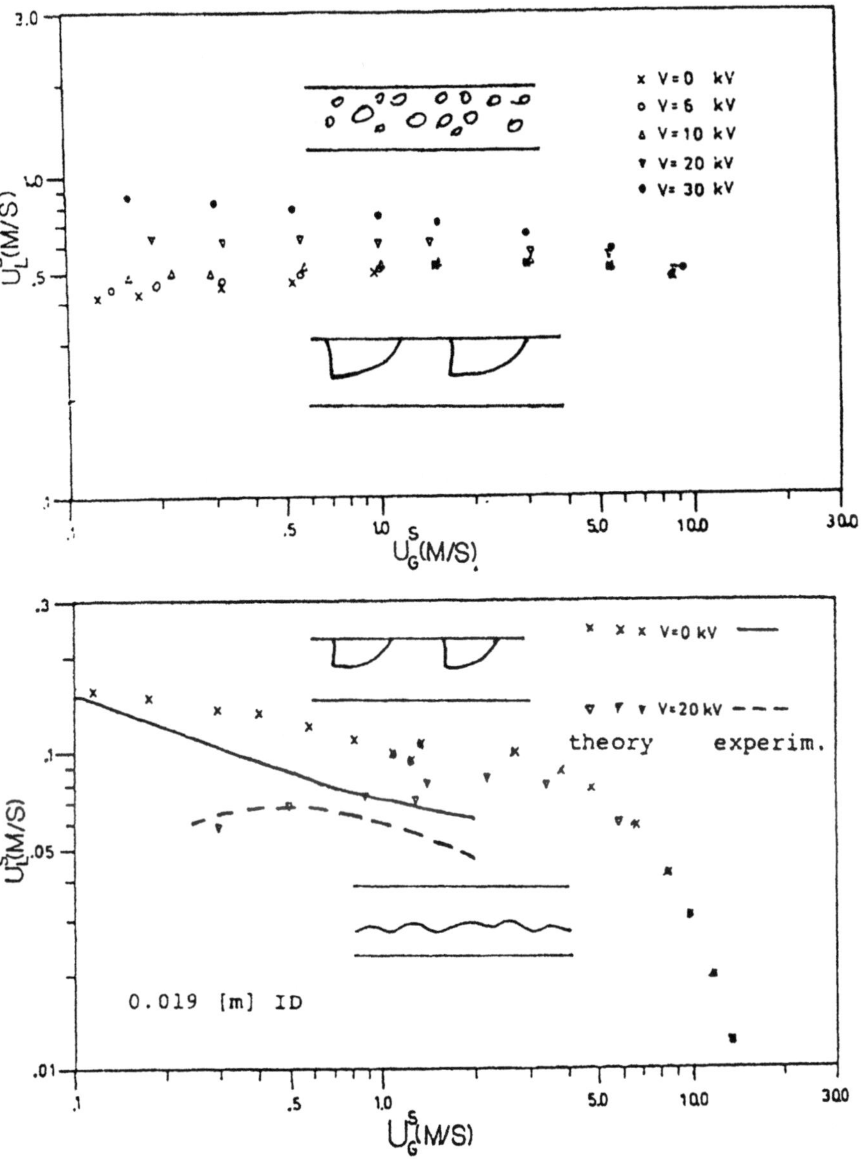

Fig. 20.4 The effect of EHD forces on the two-phase flow transition from stratified to intermittent flow.

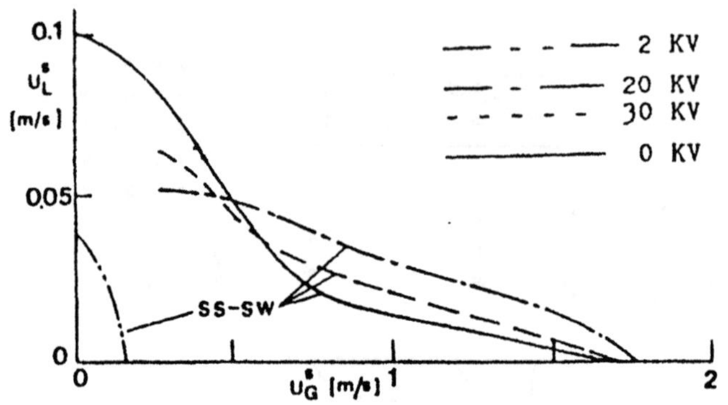

Fig. 20.5 The effect of EHD forces on the two-phase flow transition from stratified smooth to wavy flow.

and two-phase EMHD flow in this chapter (see Table 19.1). This phenomenon is particularly important to heat pipe and pump applications (Yabe et al., 1985; Jones and Perry, 1974).

Table 20.1

Effects of electrohydrodynamic forces on the fluid flow

Type of Flow	Mechanisms
Gas Single Phase Flow	Micro and Bulk Convections On-set of Turbulence
Liquid Single Phase Flow	Micro and Bulk Convections On-set of Turbulence
Gas-Liquid Two Phase Flow	Phase Micro and Bulk Convections On-set of Turbulence Interfacial Instability Entrainment Enhancement
Gas-Solid Two-phase Flow	Electrostatic Adhesive Forces Entrainment Enhancement Phase Bulk Convection

20.4.1 Bubble Column

It is well known that bubble column reactors have a wide range of applications in many components of chemical industrial process such as absorption, catalytic slurry reactions, bioreactions, coal liquefaction, etc. These reactors are preferred because of simplicity of operation, and low operating costs. The review of bubble column reactors was recently done by Shah et al. (1982). Main request to design better mass transfer bubble column reactors other than energy efficiency is for better insight of the residence time, the uniformity, and the gas-liquid interfacial area. Currently existing bubble dispersion technique based on applied electric field method and dispersion fluid is shown in Figure 20.6 and the basic bubble disruption and dispersion mechanisms as well as system efficiency enhancement by applied electric field may differ depending on electrode geometry and gas injection system (Ogata et al., 1985). Typical bubble size distribution observed by Sato (1979) is shown in Figure 20.7. The mechanism of the bubble disruption may be due ot the three steps as schematically shown in Figure 20.8 and as follows:.

(1) Bubble Disruption Via Bubble Elongation Due to the Applied Electric Field. This process has been analyzed by Garton and Krasucki (1984). The elongation of bubble due to the electric field in a gas-liquid system can be expressed by:

$$E = \left(\frac{\sigma}{\varepsilon_0 \varepsilon_1 \varepsilon_2}\right)^{1/2} \left\{\frac{\varepsilon_1}{\varepsilon_1 - \varepsilon_2} - G\right\} \left\{\frac{8 A^2 B}{3 \gamma (\varepsilon_1 - \varepsilon_2)}\right\}^{1/4} \cosh^{1/2}\theta \qquad (20.12)$$

where

$$3\theta = \cosh^{-1}\left[\frac{P r_0}{\sigma}\left\{\frac{27 \gamma^5}{2(1+\gamma^2)^3}\right\}\right]^{1/2} \; ; \quad A = 1\gamma - 1 - 1/\gamma^2$$

$$B = 2\varepsilon_1 \gamma^3 - \varepsilon_2 (1+\gamma^2)^3 \; ; \quad G = \frac{1}{\gamma^2 - 1}\left[\frac{\gamma \cosh^{-1}}{(\gamma^2-1)^{1/2}} - 1\right]$$

Here, γ is the ratio of the major to minor semi-axis, σ is the surface tension, P is the pressure within the bubble, r_0 is the radius of the bubble without electric field, ε_1 and ε_2 are relative permittivities of liquid and gas, respectively. Typical results is shown in Figure 20.9.

(2) Interfacial Instabilities. Instability of interface, which leads to bubble disruption, can be obtained from the Kelvin Helmholtz instability analysis (Brunner and Chang, 1980) as follows:

$$(u_g - u_\ell)^2 F_o + \frac{\sigma}{\rho_1}\frac{\partial^2 \xi}{\partial x^2} - g\xi\left(1 - \frac{\rho_2}{\rho_1}\right) \geq \frac{1}{2\rho_1}(\varepsilon_1 - \varepsilon_2)\frac{(\varepsilon_1 + 2\varepsilon_2)}{3\varepsilon_2} E_{int}^2 \qquad (20.13)$$

where ξ is the wave height in gas-liquid interface, u_g and u_ℓ are local gas and liquid velocity, F_o is a constant dependent upon bubble size, g is the gravitational acceleration, E_{int} is the electric field at the interface, ρ_1 and ρ_2 are density of liquid and gas, respectively. The bubble disrupted when ξ equals minor axis of elongated bubble.

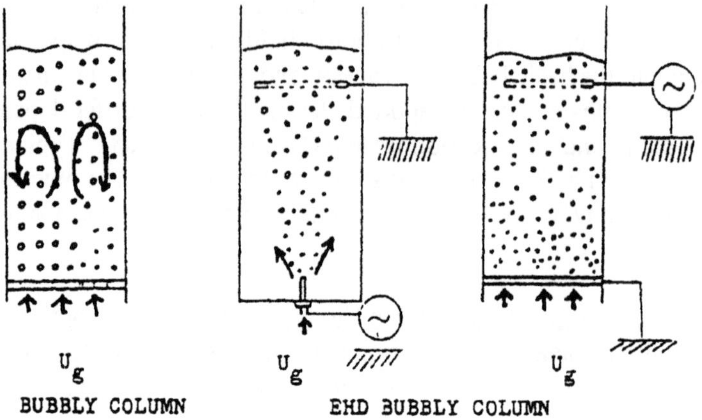

Fig. 20.6 Schematics of EHD bubble column reactors.

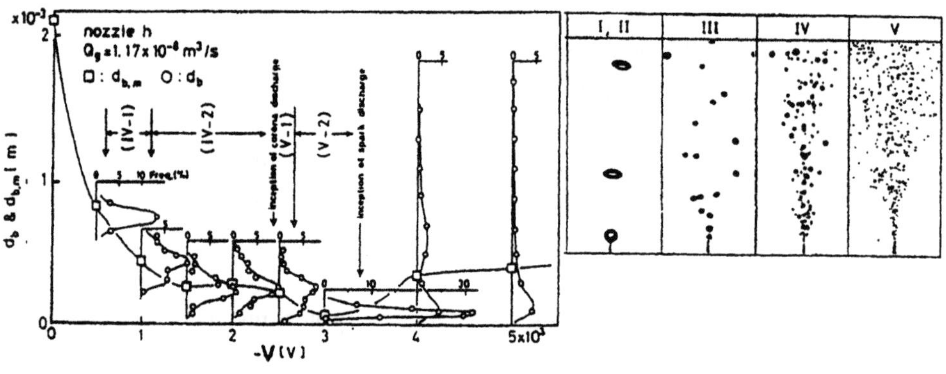

Fig. 20.7 Typical bubble size distribution of EHD bubble column (Sato 1979).

Mechanism of bubble dispersion can be a combination of the two effects, i.e. EHD flow and the enhancement of the interfacial drag forces due to the bubble disruptions. Typical flow velocity enhancement observed is shown in Figure 20.10 (Ogata et al., 1985). By combining these bubble dispersion and disruption effect due to the applied electric field.

The energy efficiency η in current system can be defined as follows

$$\eta = \frac{\text{consumed energy for surface enlargement}}{\text{total energy added to the system}} = \frac{6\,\sigma \cdot M / \bar{d}_B}{V \cdot i + P \cdot M} \times 100 \ [\%] \qquad (20.14)$$

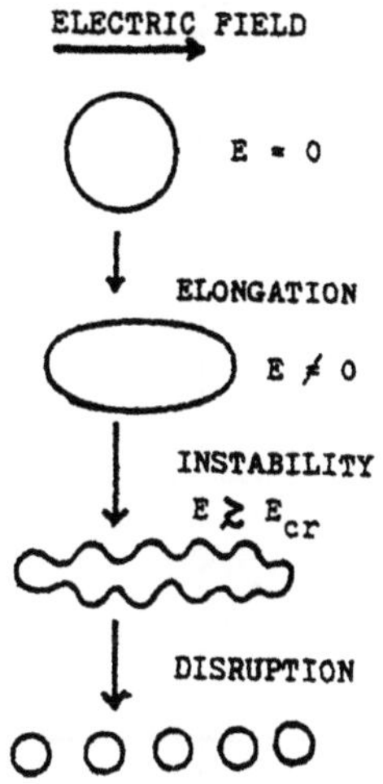

Fig.20.8 Mechanism of bubble disruption processes.

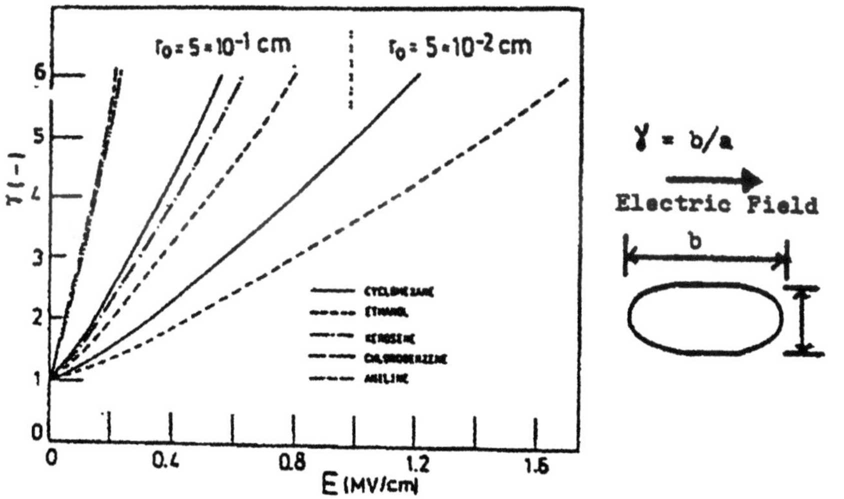

Fig. 20.9 Elongation of gas bubbles in an uniform electric field for various working fluids (Ogata et al., 1985)

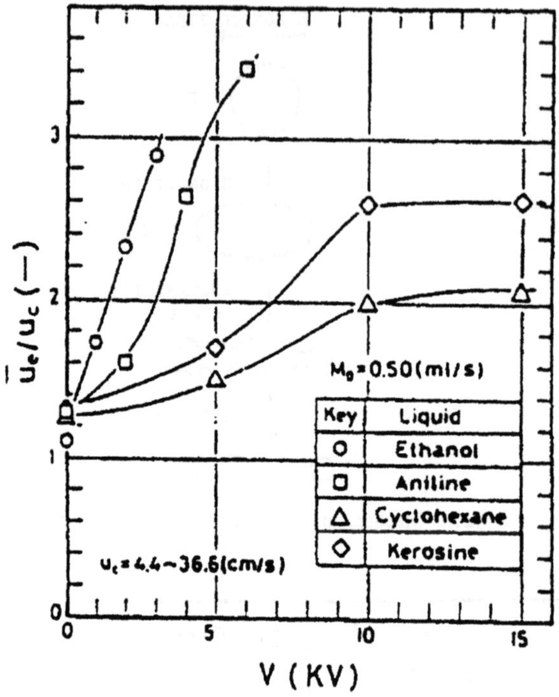

Fig.20.10 Typical flow velocity under electric field (Ogata et al. 1985).

The first term of the denominator shows electrical energy, while the second is mechanical energy. The average diameter d_B (Sauter mean diameter) is

$$\overline{d}_B = \sum_{i=1}^{N} d_{Bi}^3 \Big/ \sum_{i=1}^{N} d_{Bi}^2 . \tag{20.15}$$

Typical comparison with current EHD bubbly column and conventional stirred vessel and orifice mixer results are shown in Figure 20.11. Depend on the working fluid EHD method has a substnatial energy saving compared with existing systems. The other advantages of the EHD effect in a bubbly column reactor is mass transfer enhancement due to the interfacial velocity enhancement in inner bubbles (Chang and Berg, 1985).

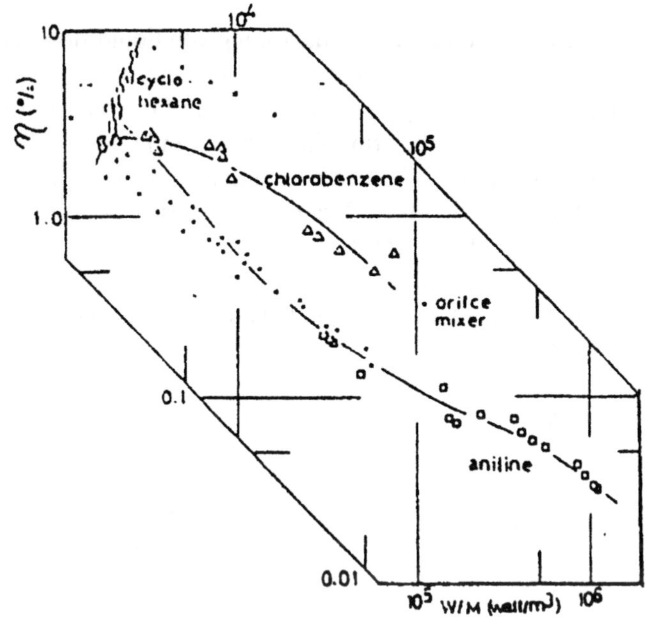

Fig. 20.11 Overall mixing efficiency as a function of power applied for mixing (Ogata et al. 1985).

20.4.2 Cocurrent EHD Two-Phase Flow

For the single phase cases, Saville and Palusinski (1986) presented electrophoretic separation of materials by using a relatively large electrode. Campeau et al. (1986) tried to use a travelling wave electric field for separation and gating of the bubbles in a dielectric fluid. Suzuki (1984) developed the gas-liquid separator based on the force in Eq. (19.29) term 2 for the bubbly column. Recovery of injected gas becomes significantly enhanced after critical applied voltages.

20.4.3 Countercurrent EHD Two-Phase Flow

For the falling film reactor, liquid-gas interfacial area becomes significantly enhanced due to the surface wave enhancement as described by equation (19.29), therefore, overall reaction is expected to be enhanced.

Another advantage of the EHD falling film reactors is control of the flooding phenomena (Revankar and Chang, 1984), since flow rate of the falling film liquid is limited by the gas flow rate as shown in Figure 20.12. In Figure 20.12, the relationship between the liquid and the gas flow rates in a general two-phase countercurrent flow system is shown. Region 1 represents the flow regime where the stable operation of the system can be observed. In region 2, the continuous operation of the two-phase countercurrent flow sysem is not possible. Thus, for a given liquid flow rate, there exists a limit to which the gas flow can penetrate through the channel and vice versa. The locus of these flooding points forms an envelope of all maximum flow rates of the two-phase flow system, and this curve is called the "flooding curve". In Figure 20.13 we have shown the flooding curve for various applied electric field obtained by Revankar and Chang (1984). It is clear from the figure that the effect of the electric field is to reduce the flooding point. For smaller liquid flow rates, i.e., for small normalized liquid flow velocity, K_ℓ, the enhancement of flooding velocity due to electric field predomintes. Therefore, we can flow more liquid and gas reactants for each reaction tubes by applied EHD forces in a falling film reactors.

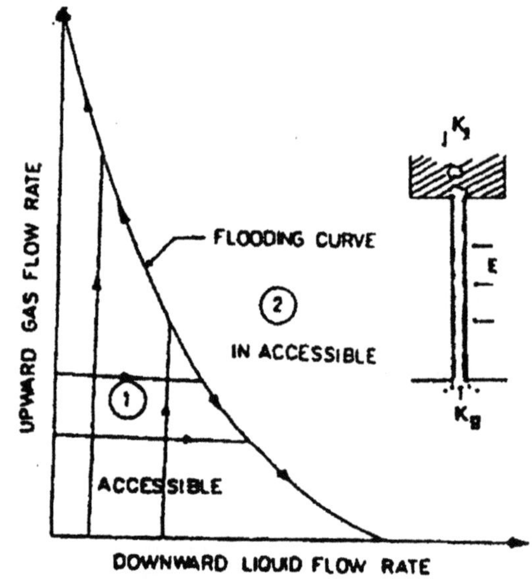

Fig. 20.12 Flow regime maps for counter current EHD flow.

20.5 Boiling Phenomena Under EMF

(i) <u>Boiling Curves Under EMF</u>

Boiling phenomena can be categorized by pool and flow boilings. The study of boiling phenomena is still not completely understood even in simple conditions. However, pool boiling has a so called "boiling curve" as shown in Figure 20.14 for most fluids.

For distinctly different heat transfer modes (i) single phase natural convection; (ii) nucleate boiling; (iii) transition boiling; (iv) film boiling, can be observed in most cases.

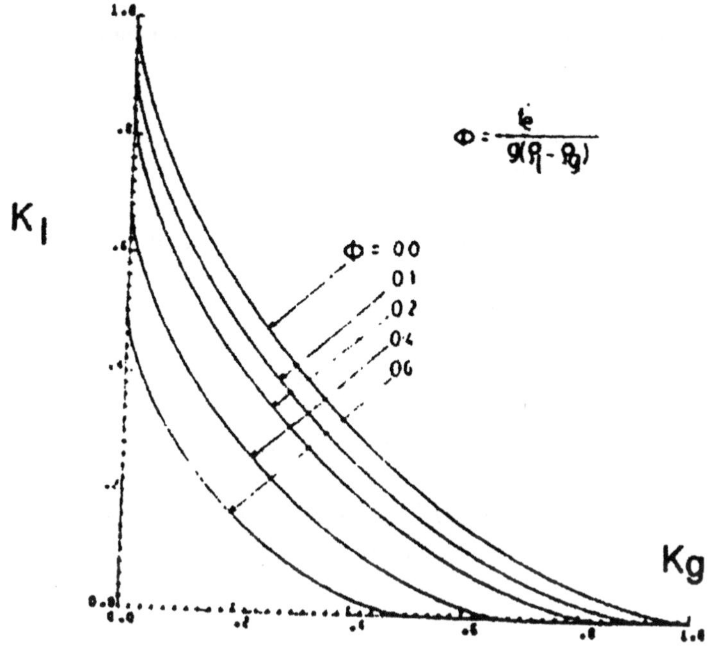

Fig.20.13 Theoretical EHD flooding curves for various dimensionless applied voltages (Revankar and Chang 1984).

For relatively small values of superheat ($T_w - T_{sat}$), the heat flux Q/A is predicted by the relationships for "natural convection" and operation occurs along curve A-B because there is no significant generation of vapour bubbles under these conditions even though the surface temperature is in excess of the saturation temperature. The temperature differences required for sufficient vapour bubble generation to occur that the characteristic curve would be affected depends upon the pressure at which the system is boiling and the type of fluid although the type and condition of the heater surface is significant as well since these factors determine the number of the active nucleation cavities at which the bubbles form.

Once the superheat has exceeded that corresponding to point B, the characteristic curve begins to rise more steeply and operation occurs along curve B-C which indicates a large increase in heat flux q/A with only moderate increase in superheat ($T_w - T_{sat}$). This process is known as "nucleate boiling" heat transfer because sufficient vapour bubbles are formed at discrete nucleation sites on the surface to promote large heat transfer rates. When the liquid temperature is at the saturation temperature, the vapour bubbles formed grow until the inertia and/or buoyancy forces overcome the surface forces holding the bubble to the heater surface at which point the bubble breaks free and rises in the liquid. When the liquid temperature is below the saturation temperature, sufficient heat will be exchanged between the vapour bubble and the liquid through which it is rising that the bubble will get smaller and may completely condense or collapse before arriving at the free surface.

As the heat flux increases, the number of active nucleation sites becomes more numerous and the activity at the nucleation sites increases as well. Discrete vapour bubbles merge into vapour columns at each of the sites and then vapour columns merge to form large globules of vapour. Ultimately a situation is reached in the vicinity of point C where the surface is blanketed with vapour and circulation of the surrounding liquid is prevented so that the ability of the boiling process to remove heat is diminished. A limiting condition referred to as the "critical heat flux" (or maximum heat flux) arises such that any further increase in heat flux leads to a rapid excursion to a new operating condition in the vicinity of point D. Even though the energy input to the heater is the same, the superheat increases several orders of magnitude and the heater surface often fails by melting so that "burnout" is said to have occurred.

Should the heater surface not fail, then operation along curve D-E is possible. A continuous layer of vapour blankets the surface and although the interface between the liquid and vapour is in continuous motion because of the regular formation of vapour globules at discrete points, the liquid never contacts the surface. The process is known as "film boiling" and energy is transferred between the heater surface and the liquid by conduction and radiation. Should the heat flux now be reduced to that corresponding to point F (minimum heat flux), a sudden excursion back to nucleate boiling in the vicinity of point G may occur and further reductions would lead to operation along curve G-B.

The dashed curve describes an unstable phenomenon known as "transition boiling". Operation along curve F-C would be observed if the heater surface temperature were controlled rather than if the heat flux were controlled. In transition boiling, a combination of film boiling and nucleate boiling occurs. The fluid motion is very chaotic and operation under these conditions is avoided whenever possible.

(ii) Nucleate Boiling Region

With the addition of EMF, the experimental results observed to be enhanced heat flux are shown schematically in Figure 20.15. One cause of the enhancement is the enhancement of natural convections observed by Chang et al. (1985, 1995). However, significant enhancement in the two-phase flow region from B to C cannot be explained by an enhancement of natural convection alone, since the enhancement of maximum heat flux by EMF is observed to be an order of magnitude larger than in natural convection region. Possible heat flux enhancement due to EMF can be summarized as follows.

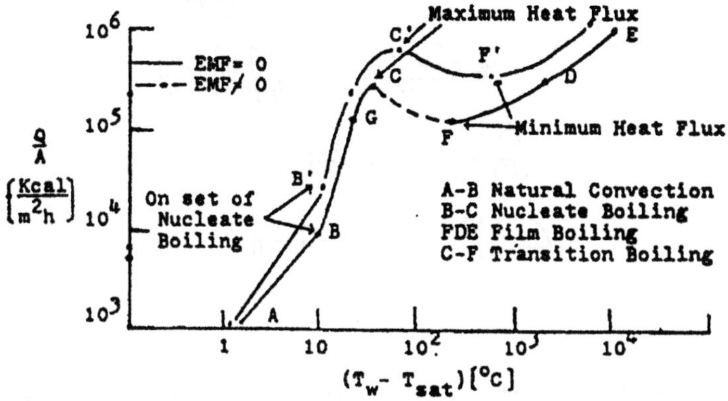

Fig. 20.14 Schematics of boiling curves, the heat flux to the surface of q vs (surface temperature-saturation temperature, $T-T_s$) with and without electromagnetic fields.

1. The enhancement of condensation in a latent heat transport process, since the condensation phenomenon has been observed to be enhanced by an EMF (Choi, 1968) (Figure. 20.15a).
2. The microconvection inside bubbly flow motion may be enhanced by EHD and MHD flows, since bubble convective motions may be enhanced by EHD (Ogata et al., 1985) (Figure 20.15b).
3. The enhancement of microlayer evaporation and thermal capillary flows by an EMF instability at gas-liquid interfaces, since the surface wave of the gas-liquid interface is significantly enhanced by an EMF (Melcher, 1981) (Figures 20.15f and g).
4. The enhancement of natural convection inside bubble columns by EMF as discussed in the sections to follow.
5. The occurrence of bubble disruptions by EMF (Ogata et al., 1985).

(iii) <u>Predictions of the maximum heat flux under EMF</u>

In the maximum (critical) heat flux region, Chang (1984) suggested that the natural convection is enhanced by a gas-liquid interfacial motion as shown in Figure 20.16a. With the EMF forces at interface F_{EBI}, the maximum heat flux becomes

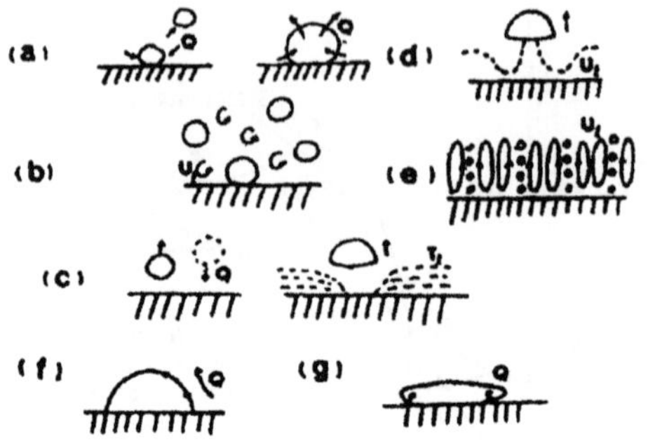

Fig.20.15 Mechanisms of heat transfer in nucleate boiling: a) Latent heat transport; b) Microconvection; c) Vapor-liquid exchange; d) Wake flow; e) Enhanced natural convection; f) Thermal capillary flows; g) Microlayer evaporation.

$$\frac{(Q/A)_{max}}{h_g \rho_g} = c_1(E)\, U_{g\,max}$$

$$= \left\{ \frac{\rho_\ell F_{EBI}}{2\rho_g(\rho_\ell + \rho_g)} \left[\left(1 + \frac{4(\rho_\ell - \rho_g)^2 g \sigma}{F_{EBI}^2} \right)^{1/2} - 1 \right] \right\}^{1/2} c_1(E) \qquad (20.16)$$

where $c_1(E)$ can be obtained from experimental results. For E and H perpendicular to the gas-liquid interfaces, we obtain

$$F_{EBI} = \frac{\varepsilon_g(\varepsilon_\ell - \varepsilon_g)^2}{\varepsilon_\ell(\varepsilon_\ell + \varepsilon_g)} E_g^2 + \frac{\mu_g(\mu_\ell - \mu_g)^2}{\mu_\ell(\mu_\ell + \mu_g)} H_g^2 + \frac{\varepsilon_\ell(\varepsilon_\ell - \varepsilon_g)^2}{\varepsilon_g(\varepsilon_\ell + \varepsilon_g)} E_\ell^2 + \frac{\mu_\ell(\mu_\ell - \mu_g)^2}{\mu_g(\mu_\ell + \mu_g)} H_\ell^2 \qquad (20.17)$$

where E_g, E_ℓ, H_g, H_ℓ are electromagnetic field near interfaces. Equation (20.17) agrees with the results of Johnson (1968) when $H_g = H_\ell = 0$. Equations (20.17) and (20.17) also show that the maximum heat flux is significantly enhanced by the existence of EMF, hence the transitions to the burnout become significantly suppressed.

(iv) Predictions of the minimum heat flux under EMF

In the film boiling condition is shown in Figure 20.16b, the minimum heat flux under EMF (Chang, 1984; Johnson, 1978) as follows:

$$\left.\frac{q(E)}{q(O)}\right|_{min} = \frac{\left[\dfrac{F_{EBI}}{2} + \dfrac{2(\rho_\ell - \rho_g) g\sigma}{F_{EBI} + \left[F_{EBI}^2 + 4(\rho_\ell - \rho_g) g\sigma\right]^{1/2}}\right]^{1/2}}{\left[(\rho_\ell - \rho_g) g\sigma\right]^{1/4}} \qquad (20.18)$$

Typically the ratio of the maximum and minimum heat fluxes with and without applied EMF as a function of F_{EBI} for water at 1 at m is shown in Figure 20.17. Figure 20.17 agrees well with the experimental results within 20% in most of cases (Johnson, 1968).

(v) Convective boiling

For the forced convective boiling, the heat flux without EMF was approximated by

$$q = q\text{ (pool boiling)} + q\text{ (single phase forced convection)} \qquad (20.19)$$

Hence, a similar treatment to that pursued in the preceeding and present sections concerning the EMF effect on the single- and two-phase flow can be applied to the first and second terms, respectively. However, for more detailed analyses, the flow pattern effect must be considered, since the two-phase flow pattern also significantly influenced by an applied electric field (Chang, 1989).

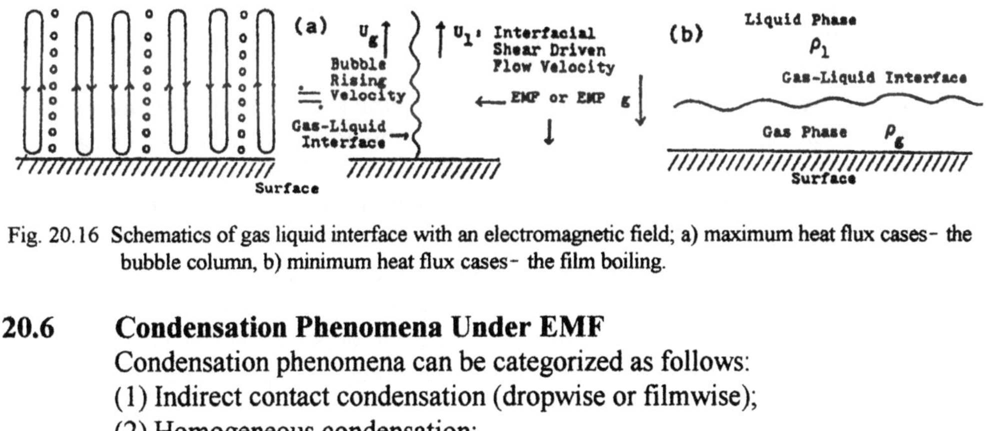

Fig. 20.16 Schematics of gas liquid interface with an electromagnetic field; a) maximum heat flux cases- the bubble column, b) minimum heat flux cases- the film boiling.

20.6 Condensation Phenomena Under EMF

Condensation phenomena can be categorized as follows:

(1) Indirect contact condensation (dropwise or filmwise);

(2) Homogeneous condensation;

(3) Direct contact condensation (bubblewise and drop wise), as shown in Figure 20.18.

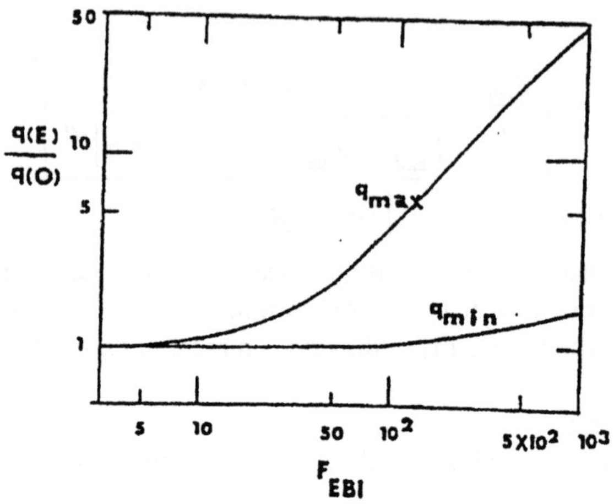

Fig.20.17 The ratios of the maximum or minimum heat fluxes with and without applied electromagnetic field as a function of F_{EBI} for pure water at 1 atm.

(i) Indirect Contact Filmwise Condensation Under EMF

It is well-known from experimental studies that the condensation heat transfer coefficient can be increased significantly by the presence of electric fields on some fluids (Choi, 1968; Bologa et al., 1978, 1980). It has been found that this increase in heat transfer coefficient is related to the appearance of electrohydrodynamic waves on the liquid-vapour interface (Jones and Perry, 1974).

For tube geometries and in the case of a dielectric fluid such as freon, the increase in heat transfer coefficients obtained is from 2 to 10 fold with electric fields of the order of 20-60 kV/cm (Bologa et al., 1978, 1980). Here, to incorporate the effect of condensation under the electric field we multiply the subcooled condensation under the electric field we multiply the subcooled condensation by a parameter $z(\phi)$, which is a function of the applied field and the system configuration, as well as the fluid properties.

Here, z is greater than unity in the presence of the electric field and when $\phi = 0$, i.e. no electric field, $z = 1$. The maximum value z can take will depend upon various factors such as the fluid properties, dielectric constant, tube geometry, strength of the electric field, etc. In the case of freon, experimentally, the rate of enhancement of the condensation in the presence of an electric field is available in terms of heat transfer rates, i.e.

$$(h_E/h_o) = C(E_v/E_{cr})^m \tag{20.20}$$

where h_E, h_o are the heat transfer coefficients with and without the electric field, E_v is the applied electric field and E_{cr} the critical electric field intensity after which enhancement in heat transfer is observed. For the case of Freon-113, $E_c = 26$ KV/cm. The value of m is 1.75. Using this information, the parameter, $z(\phi)$ can be given empirically as

$$Z(\phi) = \eta[\phi/\phi_{crit}]^{1.75} \qquad (20.21)$$

where η is the factor determining the efficiency of condensation similar to the accommodation coefficient in a direct condensation process.

The mechanisms of the heat transfer may be due to the following reasons:

(1) film thickness decrease with EMF
(2) increasing turbulence inside films by EMF,
(3) enhancement of natural convection in both phases by EMF,
(4) enhancement of condensation surface due to the generation of larger amplitude surface waves by EMF
(5) entrainment creations by EMF.

(ii) Dropletwise direct contact condensation

The experimentally observed heat transfer enhancement in a dropletwise direct contact condensation shows that there is a resonance effect arising from the frequency of the applied field as well as experimentally increased applied field effects (Yabe et al., 1985). The cause of these effects is still unknown in this moment, however, strong EHD flow inside and outside of droplets is expected to occur in this system, as well as surface deformations due to the applied fields.

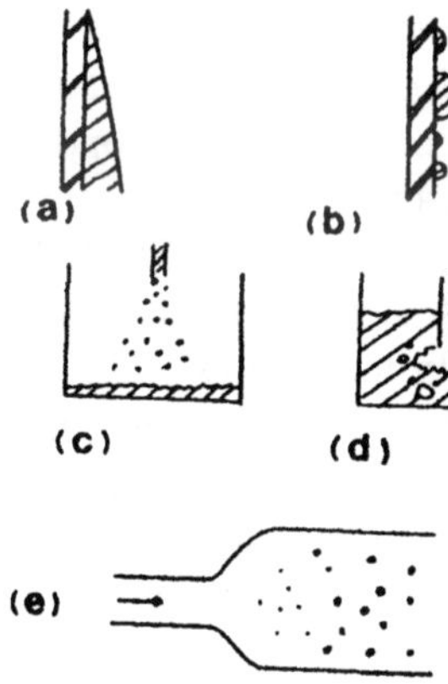

Fig. 20.18 Modes of condensation. a) Filmwise ; b) Dropwise ; c) Spray ; d) Pool ; e) Homogeneous Condensation.

References

Bologa, M.K., I.A. Kozhukhar, O.I. Mardarskiy and V.D. Shkilev (1980), Heat Transfer - Soviet Research, 12, 173.

Brunner, K., P.T. Wan and J.S. Chang (1983), Electrostatics, Inst. Phys. Conf. Ser. No. 66, pp. 215-220.

Brunner, K.S. and J.S. Chang (1980), Proc. 15th IEEE/IAS Conf., pp. 1052-1058.

Campeau, J.C., J.S. Chang and S. Masuda (1986), Japan Electrostatic Soc., Ronbushu. 86, 471-420.

Chang, L.S. and J.C. Berg (1985), AICh.E. J., 31, 149-151 and 551-557.April 26, 1996

Chang, J.S. (1989), Chem. Eng., Japan (Kagaku Kogagu) 53, 100-103.

Chang, J.S. (1984), Proc. 1984 International Electromagnetic Compatibility Conf., D.S. Bem et al., Eds., Tech. Univ. Wrolaw Press, Poland, pp. 719-730.

Chang, J.S. (1988), "Industrial dielectrophoresis and electrohydrodynamics: Its application to multi-phase flow apparatus", Particle Phenomena and Multi-phase Transport", T.N. Veziroglu Ed. Hemisphere Pub. Co. Washington, vol. 5, pp. 35-59.

Chang, J.S. (1989), IEEE Trans. Indust. Appl. Soc. 25, 241-247.

Chang, J.S., S.T. Revankar, R. Raman and F.B.P. Tran (1985), Trans. IEEE/IAS, IA-21, 715-722.

Choi, H.Y. (1968), J. Heat Transfer 90, 98-102.

Didkovski, A.B. and M.K. Bologa (1978), Teplofizika Vysoki Temp., 16, 576.

Garton, C.G. and Z. Krasucki (1964), Proc. Roy. Soc. (London), A280, 211.

Johnson, R.L. (1968), AIAAJ 6, 1456-1460.

Jones, T.B. and M.P. Perry (1974), J. Appl. Phys. 45, 2129-2132.

Mandhane, J.M., G.A. Gregory and K. Aziz (1974), Int. J. Multi-Phase Flow 1, 537-553.

Melcher, J.R. (1981), "Continuum Electromechanics", The MIT Press, Cambridge.

Ogata, S. and J.S. Chang (1985), Proc. Inst. Electrostatics, Japan, 9, 277-280.

Ogata, S., K. Tan, K. Nishijima and J.S. Chang (1985), AIChEJ., 31 62-69.

Revankar, S.T. and J.S. Chang (1984), J. Electrostat., 16, 47-68.

Sato, M., M. Kuroda and T. Sakai (1979), Kaguka Koguka Ronbushu, 5, 380-384.

Saville, D.A. and O.A. Palusinski (1986), AI ChE. J., 32, 207-214.

Shah, Y.T., B.G. Kelkar, S.P. Godbole and W.D. Deckner (1982), AIChE. J., 3(3), 353-379.

Suzuki, M. (1984), "Separating device for an insulating gas-liquid two-phase fluid", U.S. Patent 4472278.

Taitel, Y. and A.E. Dukler (1976), AICh.E.J. 22, 47-54.

Yabe, A., K. Kikuchi, T. Taketani, T. Yamanishi, Y. Mori, K. Hijikata (1985), J. Jpn. Soc. Mech. Eng. B48, 2271-76.

Chapter 21

EHD GAS-POWDER TWO-PHASE FLOW

21.1 Introduction

For the pneumatic conveying of powder in a horizontal pipe flow, due to the slow flow, the flow pattern observed is complex (Zenz, 1960), since under high speed fully dispersed flow, powder may directly impact on the wall and generate electrostatic discharges from triboelectrifications (Sugaya et al., 1989). On the other hand, the optimum pneumatic conveying flow pattern is slug flow with suitable slug lengths, since the maximum total amount of powder transported per pressure drop may be achieved (Chang et al., 1988). On the other hand, many attempts have been made to transport and control powder flow using an EHD force (Masuda et al. 1974; Aoyama and Masuda, 1977). In this chapter, EHD gas-powder two-phase flow phenomena will be introduced.

21.2 Two-Phase Flow Pattern

Typical flow patterns observed and correspondent capacitance void meter waveforms are shown in Fig. 21.1. Four flow patterns, fully packed flow, stratified flow, fully dispersed flow and slug flow are observed, i.e. gas superfacial velocity from 0.5 to 6.5 [m/s] and power flow rate from 0.15 to 0.85 [kg/s] or powder superfacial velocity from 4.6×10^{-2} to 2.0×10^{-1} [m/s]. Unlike gas-liquid two-phase flow, significant amount of unmoving particle bed component has been observed for stratified flow. For the slug flow, several different flow patterns are observed as has been described by Chang et al. (1988). For the fully packed flow, the only moving component is gas flow. Flow regime map obtained for various inclined angles is shown in Fig. 21.2 in terms of superfacial velocities (Brodowicz et al. 1988).

Figure 21.2 shows that the flow regime transition boundary in between stratified to slug flow shifted toward larger superfacial gas velocity and lower powder flow rate directions when inclined angle increases. Similar shift also observed in between stratified flow and fully dispersed powder flow boundaries. These phenomena may be due to the effect of gravity (Irons and Chang, 1983) as has been suggested in the gas-liquid two phase flow case.

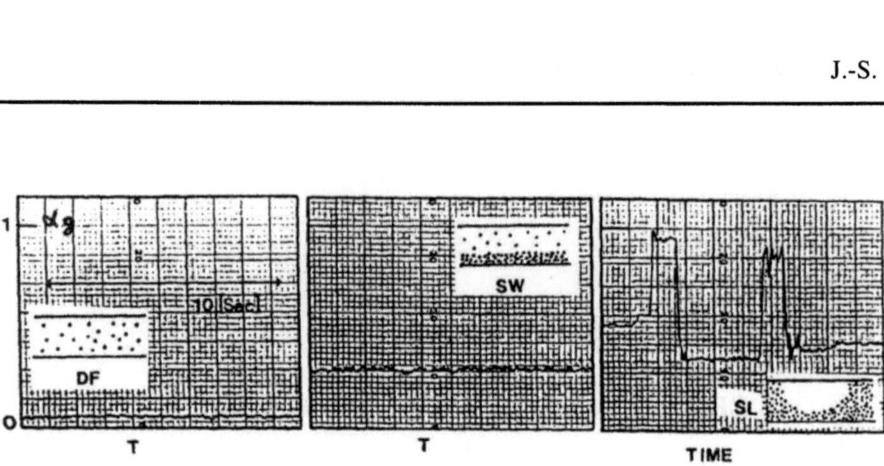

Fig. 21.1 Observed flow pattern and capacitance void transducer waveforms (Chang et al. 1988).

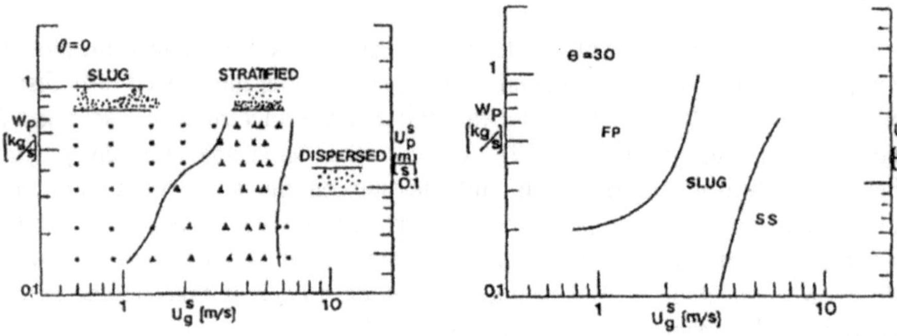

Fig. 21.2 Flow pattern map based on superficial velocities. a) Inclined angle θ = 0; b) θ = 30 (Bodowicz et al. 1988).

For the flow regime transition boundary in between slug and fully packed flow, the transition boundary again shifted toward larger and lower superfacial gas and powder velocities directions, respectively, when inclined angles are increased. Since gravitational forces increases with inclined angles, this flow regime boundary shift may also be better controlled by a superfacial gas velocity than a powder flow rate.

The time-averaged particle fraction is nonmonotonically dependent on both the gas and particle superfacial velocities, and the values of time averaged particle fraction suddenly jumped in the flow regime transition boundaries. Time averaged particle fraction contour maps in terms of superfacial velocity is shown in Fig. 21.3 for $\theta = 0$.

Flow regime map based on the phase velocities is shown in Fig. 21.4, where the phase velocity defined from the mass conservation as $U_k = W_k/A\alpha_R$, W is the flow rate, A is the cross sectional area of pipe, α is the volume fraction, and subscript k refers to each phase. Figure 21.4 shows that significant difference with flow regime map based on the superfacial velocities, and the relatively large slip velocities.

Time averaged pressure drop contour maps are shown in Fig. 21.5. Figure 21.5 shows that the time-averaged pressure drop is less sensitive to the flow pattern transition boundary

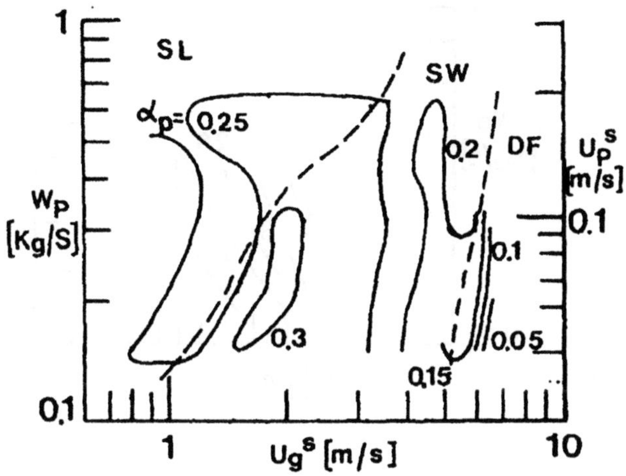

Fig. 21.3 Time averaged particle fraction contour maps in terms of superfacial velocities. θ = 0 (Chang et al., 1988).

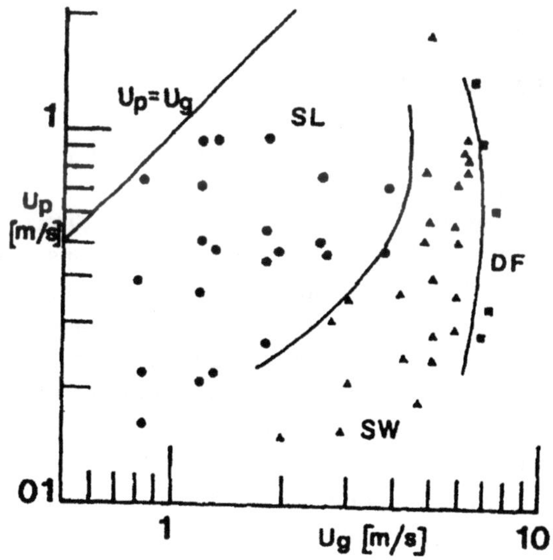

Fig. 21.4 Flow pattern map based on the phase velocities. θ = 0 (Chang et al. 1988).

compound with the particle fractions and the velocity ratios. However, sudden jump of pressure drop due to the flow pattern transition agrees experimental results of the Zenz (1960). The effects of EHD forces on the EHD pressure drop is more complex and time

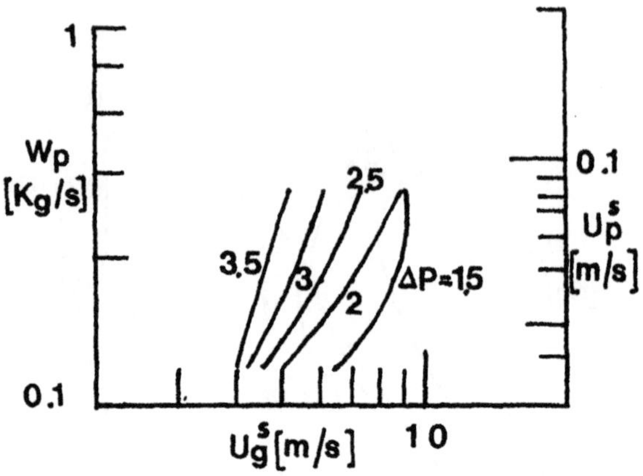

Fig. 21.5 Time averaged pressure drop contour map in terms of superfacial velocities. θ = 0 (Chang et al., 1988).

dependent (Chang and Ichikawa 1995) and no comprehensive study has been conducted.

21.3 Electrofluidized Beds

21.3.1 Fluidization Characteristics

Two different types of electrofluidized beds(Johnson and Mecher 1975) have been investigated as follows:

1. Co-Flow EFB: Gas flow and electric field in same directions.
2. Cross-Flow EFB: Gas flow and electric field directions in 90° angles.

For the cross-flow EFB with cylindrical reaction chambers, Ogata et al. (1980) defined concentric flow EFB. With an external bipolar ion injection, Chang et al. (1986) named Plasma EFB. All of these electrofluidized beds are shown schematically in Fig. 21.6. Typical fluidization characteristics for the co-flow, cross-flow and plasma co-flow type EFB are shown in Figures 21.7, 21.8, and 21.9 respectively.

For total pressure drop as a function of superficial gas velocity is shown in Fig. 21.7 for fluidized bed with and without bipolar ion injections. The location of the first maximum in the total pressure drop is generally called the minimum fluidizing velocity U_{mf} where the bed fluidizes. Further increasing the gas velocity will generate slug flow. In this fluidization region, the total pressure drop ΔP_s is fluctuating with time due to the gas slug motions. The maximum pressure drop in Fig. 21.7 is ΔP_m and the time averaged pressure drop in a saturation region is ΔP_s.

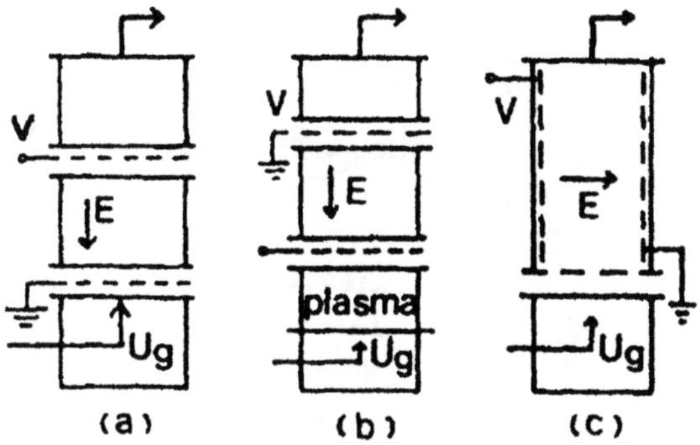

Fig. 21.6 A schematic of electrofluidized beds (EFB): a) coflow; b) plasma; c) crossflow

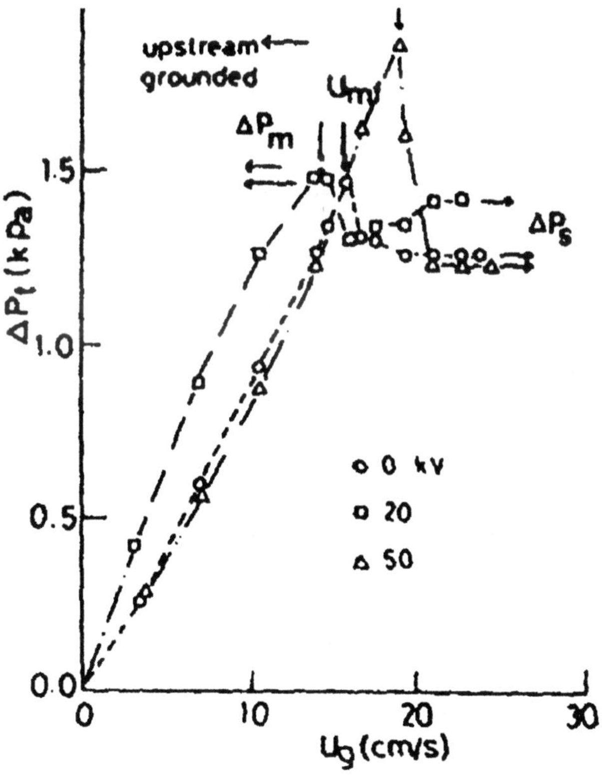

Fig. 21.7 Fluidization characteristics of co-flow type EFB (Chang et al. 1990).

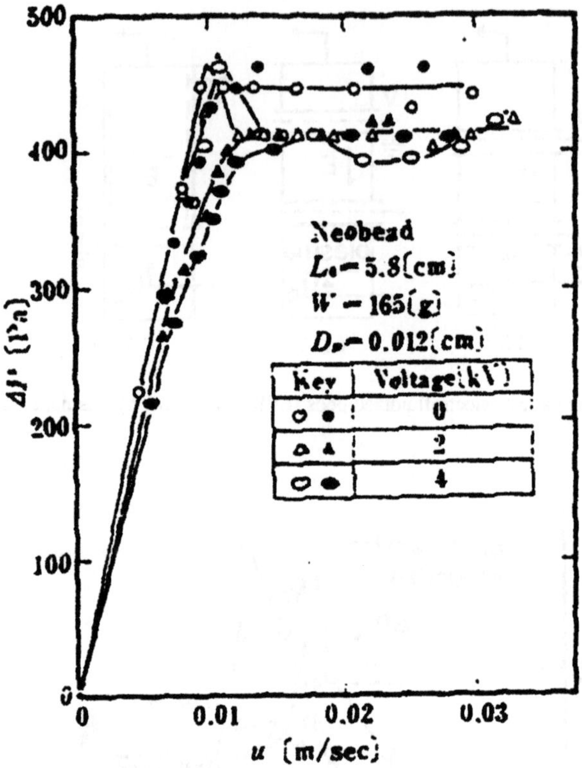

Fig. 21.8 Fluidization characteristics of cross-flow type EFB (Ogata et al. 1984).

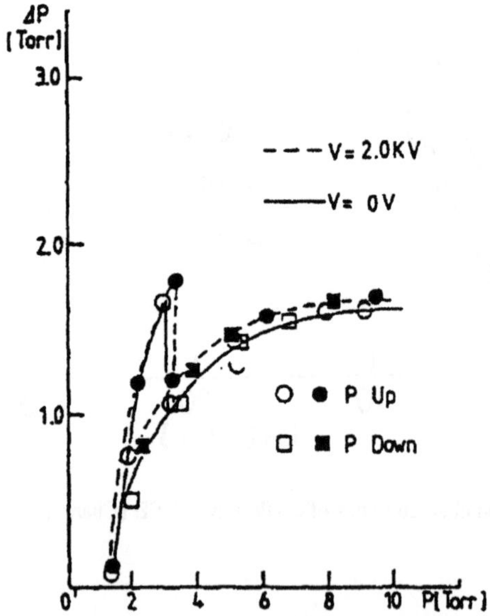

Fig. 21.9 Fluidization characteristics of plasma EFB (Chang et al. 1986).

21.3.2 One-Dimensional EHD Flow Model

From the volume-dimensional mixture momentum conservation model of two-phase flow (Chang et al., 1990 and see Chapter 19), the pressure drops inside the fluidized bed can be expressed as follows:

$$\frac{dp_\phi}{dz} = \tau_{w\phi} - \rho_\phi g + F_\phi(E) - \frac{d(G_\phi^2 V')}{dz} \tag{21.1}$$

$$p_\phi = \alpha_g p_g + \alpha_p p_p \tag{21.2}$$

$$\rho_\phi = \alpha_g \rho_g + \alpha_p \rho_p \tag{21.3}$$

$$G = \alpha_g \rho_g u_g + \alpha_p \rho_p u_p \tag{21.4}$$

$$V' = \frac{(\alpha_g \rho_g u_g^2 + \alpha_p \rho_p u_p^2)}{(\alpha_g \rho_g u_g + \alpha_p \rho_p u_p)} \tag{21.5}$$

where α_g is the void fraction, u is the phase velocity, p is the pressure, τ_w is the wall shear stress, and the subscripts g, p and ϕ are gas-, power- and two-phases respectively. The presence of ions in the system may equally influence all three terms, the ions not only changing net charge density ρ_o but also modifying the electric field as can be expected from Poisson's equations as follows

$$\tilde{\nabla}^2 \chi = -\left(\frac{L}{\lambda_D}\right)^2 (n_i - n_e) = \tilde{\nabla} \cdot \eta \tag{21.6}$$

where χ is the dimensionless electric potential, L is the characteristic length, λ_D is the Debye length, η is the dimensionless electric field and n_i and n_e are the dimensionless ion and electron density, respectively. Equation (21.6) shows that the effect of bipolar ions on the Eq. (21.6) second and third terms is only important for $(L/\lambda_D) \geq 1$. Here we must note that $n_i - n_e \neq 0$ due to the applied field and the on-set of corona discharges near electrodes in the present system.

If we subtracted pressure drops with and without applied field in a fluidization characteristics of Figure 21.7, the roles played by an EHD term in equation (21.1) can be estimated as shown in Figure 21.10. The electrostatic adhesive force in between wall and particle and/or particle and particle (Johnson and Melchor, 1975; Ogata et al., 1980, 1984), clearly dominated to prevent fluidizations. Therefore, the minimum fluidization velocity

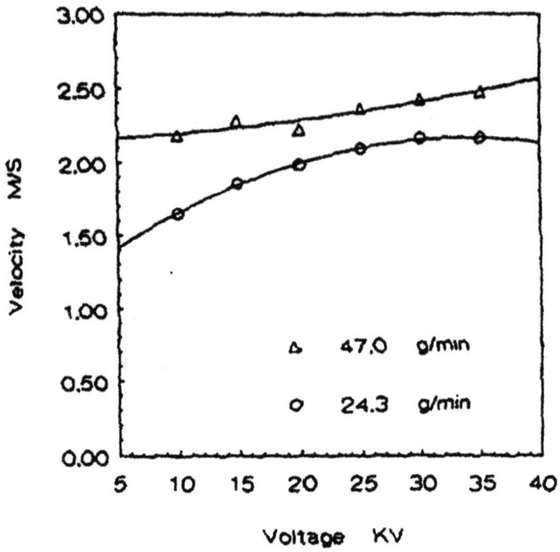

Fig. 21.10 EHD pressure drops for co-flow type EFB (Chang et al. 1990).

becomes larger in spite of the total pressure is only slightly influenced. However, once superfacial velocity is exceeded this minimum fluidization velocity, total pressure drop is substantially reduced due to the enhancement of local particle and gas velocities by EHD flow (Ogata et al., 1982). These opposite influences of the applied electric field method are applied in a particle-to-surface reaction enhancement without fluidizations and particle mixings and segregations (Kiewiet et al., 1978; Ogata et al., 1982).

21.3.3 Multi-dimensional EHD Two-Fluid Model

The modelling of fluidized beds under the fluidization by multidimensional two fluid model with and without applied electric field has been conducted by Soo (1967) and Ogata et al. (1982) respectively. For completeness, we shall repeat briefly some of the theory of the laminar electrostatic fluidized bed developed by Ogata (1982) et al. assumed that the gas flow is laminar and no bubbles inside beds. The bed comprises two sub-beds in a series, in which the lower sub-bed acts as a packed bed with very little solid motion, while the upper sub-bed is in an intensively fluidized state as shown in Figure 21.11. In the upper sub-bed, the solids move upward in the centre and downward near the wall. At the interface between the upper and lower sub-beds, the solids move radially inward. This flow mechanism was confirmed experimentally in ordinary fluidized beds (Leva 1962). For the gas fluidized beds, the experimental results under this operating condition have observed also to be a laminar flow (Soo 1967). In the electrostatic fluidized bed, the electrostatic adhesive force between particles (Ogata et al. 1984) tends to stabilize the bed. Therefore, the flow may be good

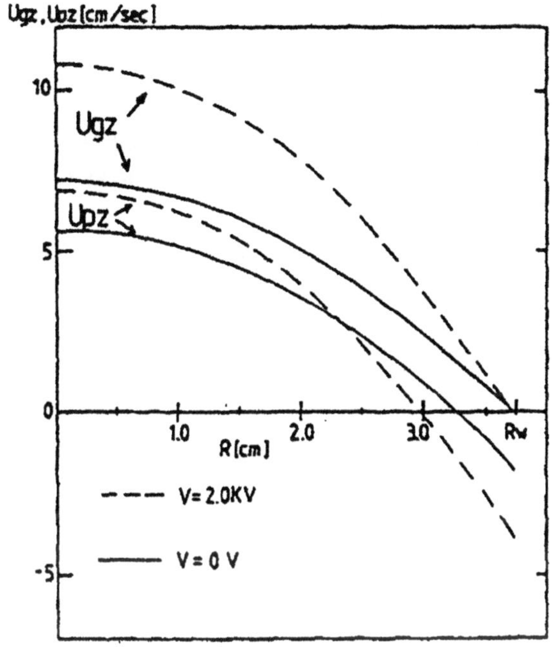

Fig. 21.11 Co-flow type EFB Model and numerical radial velocity profiles

approximation to be treated as a laminar flow if no strong non-uniform electric field drives local EHD flow to onset turbulence. Under these conditions, the steady state momentum equations for each phase can be simplified as follows:

(i) gas phase

$$-\nabla(\alpha_g P_g) + \nabla(\alpha_g \mu_g \nabla u_g) + \rho_p \phi(u_p - u_g) - \alpha_g \rho_g g + f_g(E) = 0 \qquad (21.8)$$

(ii) particle phase

$$-\nabla(\alpha_p P_p) - \rho_p \phi(u_g - u_p) - \alpha_p \rho_p g + \nabla(\alpha_p \mu_p \nabla u_p) + f_p(E) = 0 \qquad (21.9)$$

where P is the pressure, μ the phase viscosity in mixture, u the velocity, ρ the density, ϕ the interfacial drag between two phases, f(E) the momentum change due to the electric field, α_g the local void fraction, θ_p the local mass fraction $(1-\alpha_g)$, $\rho_g \simeq \rho_g/\alpha_g$ and $\rho_p/(1-\alpha_g)$. If we assume fully developed flow in cylindrical coordinate, i.e. dp/dz = const., the continuity condition is satisfied by $u_{gr} = u_{pr} = 0$, and $u_{zg} = u_{zg}(r)$, $u_{zp} = u_{zp}(r)$, ρ_p = const. and ρ_g = const. Further, within the range of fluidization, there is no net transport of particles

$$\int_0^{R_w} u_{zp}\, r dr = 0$$

and $\partial u_{zg}/\partial r = 0$, $\partial u_{zp}/\partial r = 0$ at $r = 0$, $u_{zg} = 0$ at $r = R_w$ (tube radius). We also assumed μ_g = cont. and μ_p = const. The equations now simply reduce to as follows:

$$0 = -\frac{\partial P}{\partial z} + \frac{\mu_g}{r}\frac{\partial}{\partial r}\left(r\frac{\partial u_{gz}}{\partial r}\right) + \frac{\mu_p}{r}\frac{\partial}{\partial r}\left(r\frac{\partial u_{pz}}{\partial r}\right) - (\rho + \rho_p)\, g + f_g(E) + f_p(E) \tag{21.10}$$

$$0 = -\rho_p\, \phi(u_{gz} - u_{pz}) - \frac{\partial P}{\partial z}\left(1 - \frac{\rho_p}{\bar{\rho}_p}\right) - \bar{\rho}_g\, g + \frac{\mu_g}{r}\frac{\partial}{\partial r}\left(r\frac{\partial u_{gz}}{\partial r}\right) \tag{21.11}$$

From equations (21.10) and (21.11), we obtain

$$u_{gz} = -\frac{A^1}{6^2}\left[I_o(6R_w) - I_o(6r)\right] + \frac{\beta R_w^2}{4 6^2}\left(1 - \frac{r}{R_w^2}\right) \tag{21.12}$$

$$u_{pz} = u_{gz} - \left(\frac{A^1}{\phi}\frac{\mu_g}{\rho_p}\right) I_o(6r) + \frac{1}{\rho_p g}\left(r - \frac{\rho_p}{\bar{\rho}_p}\right)\frac{dP}{dz} + \frac{\bar{\rho}}{\rho_p} + \frac{\mu_g \beta}{g 6^2 \rho_p}\frac{f_p(E)}{\phi \rho_p} \tag{21.13}$$

$$A' = \frac{\rho_p g(\mu_g + \mu_p)}{\mu_g \mu_p}$$

$$\frac{\dfrac{\phi \beta R_w^2}{8 6^2 g} + \dfrac{1}{\rho_p g}\left(1 - \dfrac{\rho_p}{\bar{\rho}_p}\right)\dfrac{dP}{dz} + \dfrac{\bar{\rho}}{\rho_p} + \dfrac{\mu_g \beta}{g \rho_p 6^2} - \dfrac{f_p(E)}{\phi \rho_p}}{I_o(6R_w) - 2\left(1 - \dfrac{\mu_g 6^2}{\phi \rho_p}\right)\dfrac{I_1(6R_w)}{6R_W}} \tag{21.14}$$

$$\beta = \left[\frac{dP}{dz} + (\rho + \rho_p)g - f_g(E) - f_p(E)\right]\frac{\phi \rho_p}{\mu_g \mu_p}$$

$$6^2 = \frac{\mu_g + \mu_p}{\mu_g \mu_p}\,\phi \rho_p \tag{21.15}$$

Equations (21.12) to (21.13) show that the maximum velocities for both gas and particle velocity located at centre of tube are significantly influenced by an existing electric field, since A^1 and β are functions of electric field.

Typical numerical radial velocity profile result (Ogata et al. 1982) is shown in Figure 21.11, where experimental observed flow velocity at inlet, total pressure drop, and current voltage characteristics are used. The results show the interfacial friction between particles ϕ and particle viscosity μ_p play an important role in the system and these values observed to be influenced by an applied electric field.

21.4 Motion of Powders Under Travelling Wave Electric Fields

Motion of powders under travelling wave electric field has been studied theroetically by Masuda and Matsumoto (1974) and Aoyama and Masuda (1977). More recently, Chang et al. (1990) used pulsed travelling wave electric field to transport powders under dispersed flow, where a typical electrode arrangement is shown in Figure 21.12. A typical velocity distribution of powders is shown in Figure 21.12. Figure 21.12 shows that particle velocity can be effectively accelerated under the travelling wave electric field and can be reached as fast as 60 cm/s for 2.5 mm particles.

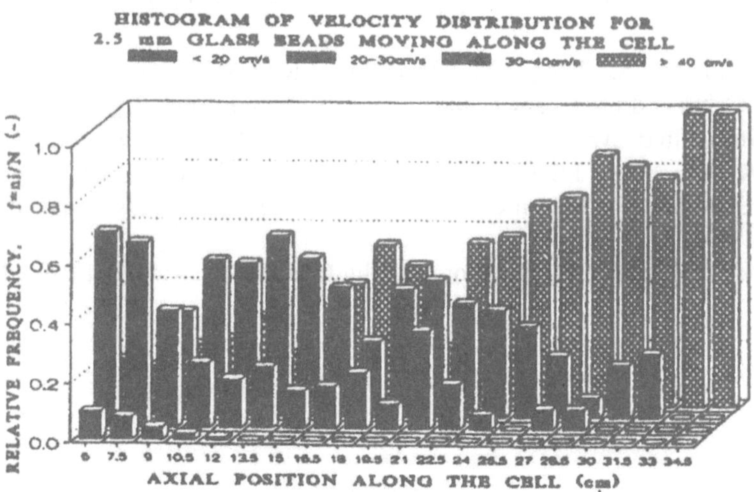

Fig. 21.12 Travelling wave electric field arrangements and typical velocity distribution along test cell (Chang et al. 1990).

References

Aoyama, M. and S. Masuda (1977), Proc. Inst. Electrostatics Japan, 1, 175-185.

Brodowicz, K., J.S. Chang, B. Doneveski, I. Sekovanic and W. Tofluk (1988). Proceedings of Experimental Heat Transfer, Fluid Mechanics and Thermodynamics, R.K. Shah et al., Ed. Elsevier, New York, pp. 1338-1342.

Chang, J.S., W. Tofluk, T.A. Myint, N. Hayashi and K. Brodowicz (1988), Proc. Int. Symp. Multi-phase Transport and Particulate Phenomena, T.N. Veziroglu, Ed., Hemisphere Pub., New York, Vol. 4, pp. 47-56.

Chang, J.S., Witte, M.H.I. Baird and S. Masuda (1990), Material Sciences, 14, 9-14.

Chang, J.S., S. Mielke, R. Scott and S. Ogata (1990), J. Electrostatics, 24, 135-144.

Chang, J.S., S. Mielke, S. Matsumura (1986), Proc. Int. Conf. Plasma Sci. and Tech. Science Press, Beijing, pp. 191-196.

Chang, J.S. and K. Ichikawa (1995), 1995 Conf. Elect. Insulation & Dielectric Phenomena, Ann. Report, pp. 392-395.

Irons, G.A. and J.S. Chang (1983), Int. J. Multiphase Flow, 9, 289-297.

Irons, G.A. and J.S. Chang (1983), Powder Technology, 34, 233-242.

Ishii, M. (1975), "Thermo-Fluid Dynamic Theory of the Two-Phase Flow", Eyrolles Press, Paris.

Kiewiet, C.W., M.A. Bergougnou, J.D. Brown and I.I. Inculet (1978), Trans. IEEE/IAS, JA-14, 526-529.

Johnson, T.W. and J.R. Melcher (1975), Trans. Ind. Eng. Chem. Fund., 14, 146-153.

Masuda, S. and Y. Matsumoto (1974), Inst. Electrical Engineer in Japan, 1994, 515-522.

Ogata, S., K. Tagama, K. Yamada, M. Fujino and H. Shinohara (1980), Powder Eng. (Jpn.), 17, 620-630.

Ogata, S., Fujino and J.S. Chang (1982), Proc. Inst. Electrostatic Japan, 6, 180-185.

Ogata, S., T. Oshio and J.S. Chang (1984), Trans. IEEE Ind. Appl., IA-20,, 1584-1590.

Soo, S.L. (1967), Fluid Dynamics of Multiphase Systems, Blaisdell Pub. Co., Waltham.

Sugaya, M., N. Hayashi and J.S. Chang (1990), The Effect of Highly Charged Powder Flow on the Solid Fraction Measurement by Capacitance Transducers, Trans. IEEE Japan, 110A, 197-205.

Zenz F.A. and D.F. Othmer, Fluidization and Fluid Particle Systems, Beinhold, New York, 1960.

Chapter 22

EHD LIQUID-LIQUID/LIQUID-SOLID FLOW

22.1 Introduction

The effects of electric fields upon fluid motion have found several applications in chemical engineering. The earliest and best-known is that upon the coalescence of particulates in the gas phase. Within the past 25 years, the effects of electric fields upon liquid-liquid systems have been increasingly the focus of research. Conventionally, efficient liquid-liquid mass transfer requires mechanical agitation in order to reduce droplet size and increase specific interfacial area. However, excessive agitation can produce drops so small that they form stable emulsions which are hard to coalesce; also, the mass transfer coefficients are usually very small for small drops because surface forces prevent circulation and they behave like rigid spheres.

Stewart and Thornton (1967) showed that beneficial effects could also be achieved by means of high voltage electric fields, avoiding some of the disadvantages associated with mechanical agitation. Since water and aqueous solutions are moderately to highly conductive it is always necessary to form the field in a continuous organic phase with the water dispersed. Although so far there is no industrial scale application of electrically-aided extraction, research on industrially important systems is increasing. For example, extraction of copper and uranium have been studied by Bailes (1981) and Bailes and Kalbasi (1981). Recently, Millar and Weatherley (1989) have reported on the electrically-aided extraction of products from whole broth fermentation.

The benefits of electrical fields can be attributed to several mechanisms. Firstly, in the drop formation process a charge may be induced on the growing drop resulting in a force which tends to increase the rate of droplet detachment, so that drop sizes are reduced. Hence, the interfacial area per unit volume in the system is increased. The second effect is the acceleration of the drop velocity through the continuous phase under the influence of an electric field. A third factor is the reduction in the effective interfacial tension and the existence of an electrical potential gradient at the interface, promoting circulation and interfacial turbulence despite the small drop size. Finally coalescence is enhanced under the influence of the electric field.

More recently, EHD liquid-liquid systems are used in biological applications and other industrial applications, and details can be found in *Handbook of Electrostatic Processes* (1995).

22.2 EHD Dispersed Liquid-Liquid Flow

A typical still photograph of a drop without EHD forces at the point of detachment (He et al., 1991) is shown in Figure 22.1(a). The spherical drop shape is consistent with the generalized graphical correlation of Clift et al. (1978), where the Eotvos number Eo is in the range 1-3 and the Morton number M is 2.2. The diameters of the droplets, d, were found from still photographs or obtained from the dispersed phase flow rate, Q, and drop formation frequency, f, assuming that drops were spherical.

$$d = (6V'/\pi)^{1/3} = (6Q/(\pi f))^{1/3} \tag{22.1}$$

Droplet size and velocity were also measured by the ultrasonic method (Morala et al., 1984). At high frequencies of drop formation, the drop velocity near the nozzle tip could be calculated from the measured separation distance W between two successive drops.

$$U_{tip} = Wf = 6QW/(\pi d^3) \tag{22.2}$$

The droplet velocities were generally measured over a distance of 1-3 cm from the nozzle tip.

Typical still photographs under EHD effects are shown in Figure 22.1 (He et al., 1991). Figure 22.1(a) shows drop behaviour in the absence of electric field (Regime I). The drops moved vertically downwards from the nozzle, with each drop following the same path. Regime II (Figure 22.1(b)) featured smaller droplets and higher detachment frequencies than Regime I. The detached drops did not move off vertically and showed alternating diverging paths. A multi-droplet dispersion regime (III) occurred in which many small droplets moved off from the nozzle at high velocities, then decelerated a few centimetres away from the nozzle where the field strength was lower. The result, shown in Figure 22.1(c), was a fine dispersion of drops which was slow to settle.

The effects of electric force on drop formation have usually been represented by an equilibrium balance of forces acting upon the drop at the moment of detachment as shown in Table 22.1 (Stewart and Thornton, 1967; Ogata et al., 1976; Takamatsu et al., 1981; Carleson and Berg, 1983; Vu and Carleson, 1986).

At very high applied voltages, the buoyancy forces and the interfacial forces are negligible compared to the electric and drag forces. Hence, the distances between electrode and drop centre, S, becomes (He et al., 1991)

$$S = kd^4V^2/24Q\mu \tag{22.3}$$

where V is the applied voltage, μ is the viscosity of continuous phase and k is the constant. At detachment, S is equated to d/2 so that the diameter of the formed drop is given by

$$d = (12Q\mu/kV^2)^{1/3} \tag{22.4}$$

Thus, in the limiting case of negligible gravity and surface effects, the drop diameter varies as the −2/3 power of the applied voltage, where the effect of flow rate upon k corresponds to the relationship (He et al., 1991)

$$k \propto Q^{-0.57} \tag{22.5}$$

The general equation for the electrohydrodynamic (EHD) force per unit area is (Landau and Lifshitz, 1960):

$$\underline{\tau}_e = \rho_i \underline{E} + \frac{1}{2} E^2 \nabla \varepsilon \tag{22.6}$$

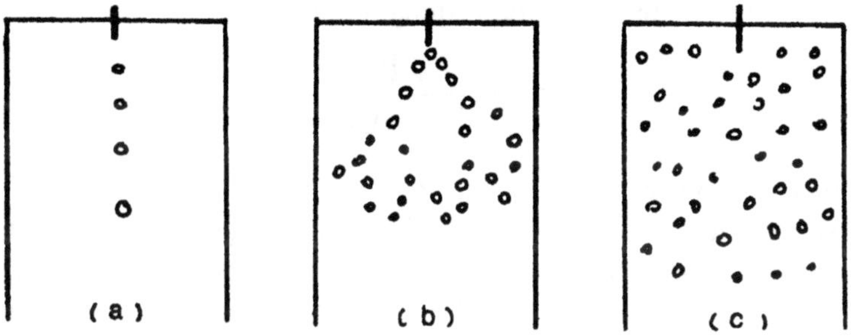

Figure 22.1 Typical photographs of moving drops. (a) Voltage 0 to 2 kV (regime I); (b) Voltage 2 to 7 kV (regime II); (c) Voltage > 7 kV (regime III) (He et al., 1991).

Source	Experimental Conditions	Expression for F_e
Ogata et al. (1976)	Water or ethanol/water drops formed in air at nozzle with positive voltage.	$2\pi\varepsilon_o V^2/\ln(8L/d_o)$
Takamatsu et al. (1981)	Water or furfural drops formed at nozzle with positive voltage, 5 cm from grounded electrode.	$4\pi\varepsilon\,\beta\, r^2 E_o^2$
Carleson and Berg (1983)	Water drops formed in gas phase with SO_2 absorption. Field strength up to 8 kV/cm.	$\pi d^2 CqE$
Vu and Carleson (1986)	Glass column, 0.45 m height. Continuous phase; pentane, hexane.	εKAE^2

Table 22.1 Previous work on electric force (F_e) acting upon a detaching drop

The final term is negligible if a phase with uniform dielectric constant ε is being considered under electric field E.

The current per unit area is determined by charged particle convection, U, mobility, μ, and charged particle diffusion, D, as follows:

$$\underline{J} = \rho_i \underline{U} \pm \rho_i \mu_i \underline{E} - D \nabla \rho_i \tag{22.7}$$

Charged particle diffusion can be ignored, hence Eq. (22.6) can be simplified and rearranged to give

$$\rho_i = \frac{\underline{J}}{\underline{U} \pm \mu_i \underline{E}} \tag{22.8}$$

Substitution of the charged particle density ρ_i in Eq. (22.6), ignoring the dielectric term, gives the force per unit area in terms of the field strength E as

$$\tau_e = \frac{\underline{J}\,\underline{E}}{\underline{U} \pm \mu_i \underline{E}} \tag{22.9}$$

In the situation of a growing drop, it is assumed that $\underline{\tau}_e$ is proportional to F_e/d^2, the current per unit area is proportional to V/d^2, and the electric field strength $\underline{E}$ is proportional to V/L where L is a linear dimension of the apparatus. The velocity $\underline{U}$ is proportional to Q/d^2. When these proportionalities are substituted in Eq. (22.9) the following expression is obtained.

$$F_e \propto \frac{V^2 d^2}{Q L \pm \mu_i V^2 d^2} \tag{22.10}$$

This indicates that the force F_e would vary inversely with the flow rate if the mobility term is small. Figure 22.2 shows that U_{tip} increases sharply with applied voltage, despite the decrease in drop size. A surprisingly good correlation of the results is obtained by a double logarithmic plot of velocity against the corresponding drop diameter, as shown in Figure 22.3 by He et al. (1991). It appears that velocity varies as the inverse square of drop diameter. The limiting case of zero applied voltage (gravitational effects only) is represented by the Hadamard-Rybczinski and Stokes expressions, with velocity proportional to the square of diameter.

The inverse square correlation between velocity and diameter can be derived subject to several assumptions. First a steady state is assumed, with the electric force on the moving drop assumed to be a balancing drag force expressed by Stokes' law as follows

$$F_e = k V^2 d^2 = 3 \pi \mu d U_{tip} \tag{22.11}$$

The drag force may be overestimated by Stokes' law since the drops follow each other in a closely spaced succession and because of electrical effects. The final assumption is that drop size is that drop size is the simple function of applied voltage given in Eq. (22.4). With these

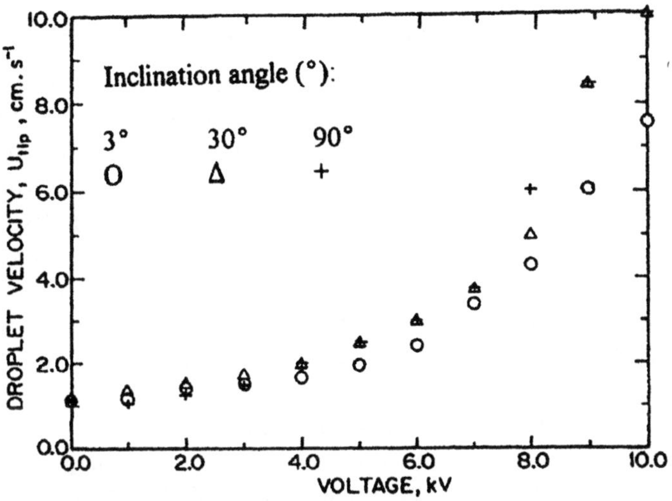

Figure 22.2 Effect of applied voltage on average drop velocity at Q = 26 μL.s^{-1} and d_o = 0.71 mm (He et al., 1991).

Inclination angle (°):	3°	30°	90°
Symbol:	O	Δ	+

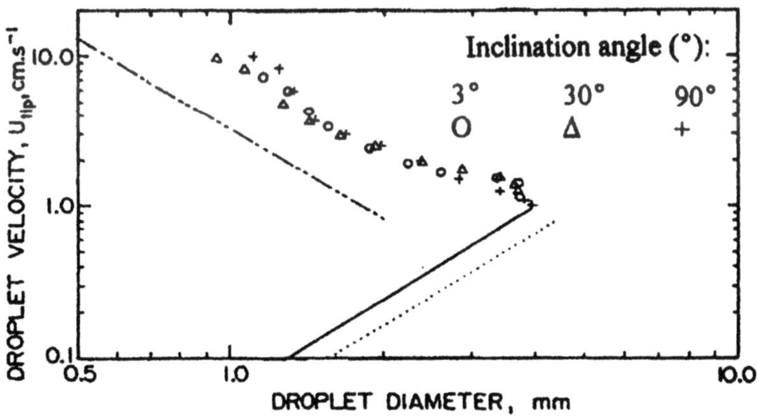

Figure 22.3 Relation between average drop velocity and drop diameter at Q = 26 μL.s^{-1} and d_o = 0.71 mm. ——— Hadamard-Rybczinksi Equation; ------- Stokes Equation; --·-- Equation (28.12)(He et al., 1991)(He et al., 1991).

Inclination angle (°):	3°	30°	90°
Symbol:	O	Δ	+

simplifying assumptions, the kV^2 term and the viscosity can be eliminated between Eqs. (22.4) and (22.11) to give the result

$$U_{tip} = 4\,Q/(\pi\, d^2) \tag{22.12}$$

This simplified derivation has been plotted on Figure 22.3 (He et al., 1991) and it is seen to run parallel to but below the band of data points by a factor of about 2.5. One factor in this difference is the wake effect of the drops following each other closely, whereby the drag is reduced below the value expected from Stokes' law (Clift et al., 1978). However, this effect is only in the order of 10% at the drop separations observed. A more likely reason for the discrepancy is an increase in electrical force due to drop-nozzle repulsion which would be greatest just after the charged drop detaches from the nozzle. This is consistent with the observed rapid acceleration of the drops just after detachment. Another possible factor in increasing the drop velocity is electrically induced circulation (He et al., 1991).

There do not appear to be many publications on droplet charge in liquid-liquid two-phase systems and most of the literature related to droplet charge only considers gas-liquid systems. It has been noted that both the Rayleigh (1882) and Vonnegut and Neubauer (1952) models apply to the maximum charge that a droplet can hold before it will split into smaller droplets in air. In the case of one dielectric liquid being dispersed into a second dielectric liquid medium due to the effect of an applied electric field, these models should be modified before they can be used to predict the maximum charge limit for the dispersed droplets in a liquid-liquid system:

the dielectric constant or permittivity of free space in these models should be replaced by an effective dielectric constant, ε_{eff} which is defined below;
instead of using the surface tension of dispersed liquid, the interfacial tension between dispersed liquid phase and continuous liquid phase should be used.

The behaviour of a droplet in another dielectric liquid which is subject to an electric field is quite complex due to the electric stresses created at the interface. The local electric field with the droplet is also dependent on the droplet shape. The electric field, E_d, in a dielectric spherical droplet immersed in a continuous phase of another immiscible dielectric liquid which is in an imposed non-uniform electric field, $\mathbf{E}_c$, is (Kao, 1961):

$$\mathbf{E}_d = \frac{3\,\varepsilon_c}{2\,\varepsilon_c + \varepsilon_d}\,\mathbf{E}_c \tag{22.13}$$

The electric field in the droplet is obviously different from the applied electric field at the location of the droplet, except for the case where $\varepsilon_c = \varepsilon_d$. It is noted also from Equation (22.13) that when ε_d is very large, $\mathbf{E}_d$, the electric field in the droplet is very small, corresponding to the case where the droplet is electrically conductive.

We then define an effective dielectric constant, ε_{eff} as (He et. al., 1995):

$$\varepsilon_{eff} = (\varepsilon_d - \varepsilon_c)\cdot\left(\frac{3\,\varepsilon_c}{\varepsilon_d + 2\,\varepsilon_c}\right) = \frac{3\,\varepsilon_c(\varepsilon_d - \varepsilon_c)}{\varepsilon_d + 2\,\varepsilon_c} \tag{22.14}$$

A. Correlations and comparisons of charge acquired on detached droplet

The maximum charge acquired during the droplet formation in the present work may be correlated by the modified Rayleigh instability model, or modified Vonnegut and Neubauer model, which is similar to the modified Rayleigh instability model with the only difference being a factor of 1/2 in the constant, after a more rigid theoretical analysis. These modified correlation models are not only useful as a criterion to correlate and check the consistency of the experimental results but also useful to investigate the mechanisms of EHD droplet formation and dispersion. The different forms of the modified Rayleigh instability model are listed below:

expressed as the total charge on droplet in terms of droplet diameter,

$$q = (8\pi^2 \varepsilon_{eff} \sigma d^3)^{1/2} = \left\{ 8\pi^2 \left[\frac{3\varepsilon_c(\varepsilon_d - \varepsilon_c)}{2\varepsilon_c + \varepsilon_d} \right] \sigma d^3 \right\}^{1/2} \tag{22.15}$$

expressed as the volume charge density in terms of the droplet volume,

$$\frac{q}{V_d} = \left(\frac{48\pi^2 \varepsilon_{eff} \sigma}{V_d} \right)^{1/2} = \left\{ \frac{48\pi\sigma}{V_d} \cdot \left[\frac{3\varepsilon_c(\varepsilon_d - \varepsilon_c)}{2\varepsilon_c + \varepsilon_d} \right] \right\}^{1/2} \tag{22.16}$$

where V_d is the droplet volume.

or expressed as droplet charge-to-mass ratio which is quite often used in gas-liquid systems:

$$\frac{q}{m_d} = \frac{12}{\rho_d} \left(\frac{2\varepsilon_{eff} \sigma}{d^3} \right)^{1/2} = \frac{12}{\rho_d} \left\{ \frac{\sigma}{d^3} \left[\frac{6\varepsilon_c(\varepsilon_d - \varepsilon_c)}{2\varepsilon_c + \varepsilon_d} \right] \right\}^{1/2} \tag{22.17}$$

where ρ_d is the density of the drop phase and m_d is the mass of one droplet.

For the modified Vonnegut and Neubauer model, the numerical values for liquid-liquid systems are each divided by 2 from gas-liquid system.

The experimental results indicate that droplet can acquire relatively higher electric charge initially from the hollow electrode possibly due to unipolar charge injection (Atten and Moreau, 1972) and induction charging (Cross, 1987) mechanisms under high electric field intensity, and then may gradually lose most of its acquired charge as it travels downwards, probably due to recombination with other charged species in the bulk dielectric continuous phase (Kitahara, 1984) or by discharging to the grounded planar electrode wall.

As shown in Figure 22.4 (He et al., 1996), the modified Rayleigh instability model and the modified Vonnegut and Neubauer model have been included for comparison with the experimental results. The modified Vonnegut and Neubauer model agrees with most of the experimental results obtained at hollow electrode where the electric field intensity is

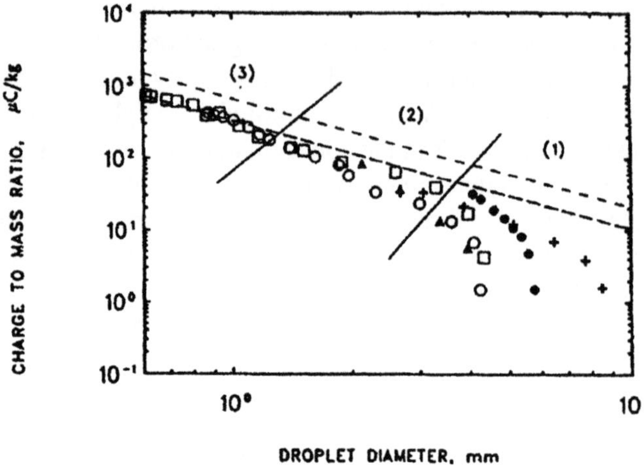

Figure 22.4 Comparison of droplet charge-to-mass ratio with models and literature present work (He et al., 1991).
- - - - - Modified Rayleigh instability limit model; and - - - - Modified Vonnegut & Neubauer
(1) Single droplet regime controlled by gravitational force;
(2) Single droplet regime controlled by EHD force;
(3) Multi-droplet regime controlled by EHD force.

high, except for the large drop size which is equivalent to the single droplet regime with low applied voltages as seen in the figure. Therefore based on the modified Vonnegut and Neubauer model, the mechanism of droplet break up at hollow electrode may be postulated: the EHD instability generated at high applied voltages probably due to over-charging of the droplet at hollow electrode is dominated by charge injection and induction charging mechanisms as already discussed. Figure 22.4 also indicates that the modified Vonnegut and Neubauer model is also better than the modified Rayleigh model for the prediction of maximum droplet charge density as well as the maximum charge-to-mass ratio in the liquid-liquid system.

The mechanisms of EHD droplet formation and dispersion from the hollow electrode have been proposed by He et al. (1996) for liquid-liquid two phase system. This can be explained in terms of three different droplet formation modes.

B. Single droplet regime controlled by gravitational force

Droplets formed at applied voltages less than 2 kV would behave like those formed in the absence of electric fields, i.e. a single droplet detaches from hollow electrode vertically downward. The gravitation is still the dominant body force and the EHD forces are negligible. Therefore the droplet could not acquire enough induced charge to reach the maximum value before it detaches from hollow electrode, and the charge induced on droplet could be mainly from charge induction in dispersed phase or possibly from the effect of how electrification (Cross, 1987). As shown in Figure 22.4, the area marked (1) which covers the data scattered below the modified Vonnegut and Neubauer model represent this kind of droplet formation mode. It is noted that most of the data from reported previous work are located in this mode, since all these work was conducted at single droplet region. After the droplet detaches from hollow electrode, its initial velocity is very close to the droplet terminal velocity in the absence of applied electric field.

C. Single droplet regime controlled by EHD force

Between 2-7 kV, smaller droplets and higher detachment frequencies are observed than those being observed in the first formation mode. The detached drops do not move off vertically but show alternating diverging paths which indicate the EHD force effect on both droplet size reduction and subsequent repulsion between drops due to the acquired charge. The charge acquired on a droplet could be mainly from charge induction in dispersed phase.

As shown in Figure 22.4, the area marked (2) which covers the data very close to but still below the modified Vonnegut and Neubauer model represent this type of droplet formation mode. This means that the charge acquired on the forming droplet does not reach the maximum charge limit. After droplet detaches from hollow electrode, its initial velocity is higher than the terminal velocity because of the droplet acceleration caused by charged droplets repulsion (He et al., 1991).

D. Multi-droplet regime controlled by EHD force

A multi-droplet formation would occur in which a fine dispersion of drops bursts from the hollow electrode at very high velocities, then gradually decelerates to reach their terminal velocities. The fine droplet size and very high initial droplet velocity indicate the dominant EHD force at high electric field intensity. The charge induced on these dispersed droplets could be mainly from unipolar charge injection and charge induction in dispersed phase based on the postulated mechanism made on current flow.

As shown in Figure 22.4, the area marked (3) which covers the data coincide with the modified Vonnegut and Neubauer model. This means that the charge acquired on these fine dispersed droplets has reached the maximum charge limit which determines and controls the droplet size. Then the EHD force generated at high electric field would contribute to the dispersion and acceleration of the droplets as observed. Thus the initial velocities of these droplets are significantly enhanced.

22.4 Control of Liquid-Solid Particle Two-Phase By Travelling Wave Electric Field

Liquid-solid flow is observed in many industrial processes. However, unlike other two-phase flow systems, only dispersed solid particles in liquid phase are most commonly observed and used.

Flow regime observed under liquid-solid flow is similar to gas-solid flow. However, due to the larger viscosity of liquids compared with gases, two-phase flow phenomena are more complex. Normally, a small or low density solid particles do not affect buffer liquid flow unless the two-phase interaction factor $\Sigma_\phi > 1$ (Chang, 1984), where $\Sigma_\phi = K_\phi L \rho_p/\rho_\ell U_s$, K_ϕ is the interfacial drag, ρ_p/ρ_ℓ is the density ratio between solid particles and buffer liquids, and L is the characteristic length of particles.

Control of solid particle motions in liquids is proposed by Pohl (1951, 1978) using a dielectrophoresis and by Masuda et al. (1986) using a non-uniform travelling wave electric field. Masuda et al. (1986) observed various modes of particle motion in a non-uniform travelling wave electric field as shown in Figure 22.5, where three phase applied voltage

$$\phi(x, y, t) = f(x, y) \cos \omega t + f(x, y - 1/3\lambda) \cos(\omega t - 1/3\pi) + f(x, y - 2/3\lambda) \cos(\omega t - 4/3\pi)] V_p \tag{22.18}$$

is applied on the three phase electrodes denoted by U.V.W. in Figure 22.5. Here, f(x, y) is the potential when unit voltage is applied to the U-phase electrodes while putting other electrodes to grounded potential, V_p is the peak value of the applied balanced three phase voltage, λ is the distance between two U-phase electrodes. Depending on applied voltage, operating frequency and electrode distances, forward, backward, spinning and circular motions of particles and their velocity can be controlled as shown in Figure 22.5. A typical mode diagram of particle motions corresponding to Figure 22.5a is shown in Figure 22.5b in terms of applied voltage and operating frequencies (Masuda et al., 1986).

22.5 Atomization of Electrohydrodynamic Solid-Liquid Slurry Spray

Liquid spraying is important for many industrial processes such as material processing, spray painting, fuel injection, medicinal spraying, agriculture spraying, etc. More recently, limestone-water slurry in-furnace injection was used for controlling NO_x and SO_x emission from coal combustion flue gas. Atomization of limestone-water slurry is presently driven by a compressed air, which requires extra air to be introduced into the combustion region of the furnace. This makes it difficult to control the furnace temperature and air-fuel ratio as well as increasing the amount of a flue gas. In the present work an electrohydrodynamic (EHD) atomization method has been investigated to produce limestone-water slurry spray for replacing compressed gas atomization.

Electrohydrodynamic spray process was discovered many years ago. The effect of electric charge on the rate of cavitation of liquid in a narrow capillary was discovered in the seventeenth century and this phenomenon was later explained by the action in a liquid drop of electric forces opposing those of surface tension. When the charge on the drop reaches a certain value, the electric forces become stronger than those of surface tension and the drop becomes unstable. A criterion of stable spherical droplet for ideally conducting liquid was first established by Rayleigh (1882) and is known as Rayleigh limit criterion. The first attempt to investigate the mechanism of formation of an aerosol particle by electrohydrodynamic atomization of liquid was done by Zeleny (1920). He observed that a liquid filament was drawn-out from a narrow capillary and then broke down into separated drops. The practical experimental results to obtain a highly uniform monodisperse aerosol particles by an electrohydrodynamic method were first published by Vonnegut and

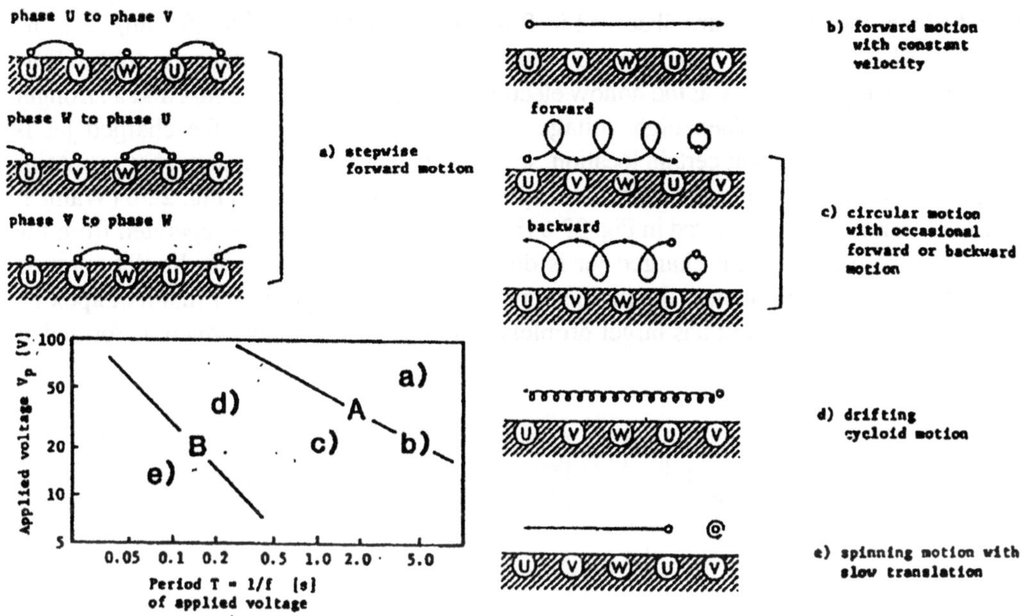

Fig. 22.5 Various modes of particle motions (a) and mode diagram (b) in liquid-solid particle dispersed flow under a non-uniform travelling wave electric field (Masuda et al., 1986).

Neubauer (1952). Since then, this type of atomization has been investigated by many authors in connection with various applications, such as surface coating, agricultural spraying, fuel spraying, etc. All these investigations have demonstrated the complexity of electrohydrodynamic spraying phenomena and the process is still not fully well understood except at lower flow rate. On the other hand, EHD atomization of two phase slurry has not yet been well studied.

22.5.1 Functioning Mode

As mentioned above, all previous investigations have demonstrated the complexity of electrohydrodynamic atomization phenomenon and the diverse morphology of spray jet patterns. Many parameters are involved to affect the atomization results, the most important of which are the liquid physical properties (such as electrical conductivity, surface tension and viscosity), the liquid flow rate, the applied voltage, the geometry of the system used and the dielectric strength of the ambient medium. Depending on the values of these parameters different functioning modes or regimes are obtained. Cloupeau and Prunet-Foch (1990) proposed a classifications for these modes, the main ones are: dripping, microdripping, cone jet, spindle, simple jet and ramified jet. As indicated by Cloupeau and Prunet-Foch, those modes are observed in flow rate range from 10^{-6} to 10^{-2} ml/s. In the present work, much higher flow rate (400-800 mm^3/s) has been investigated. In this flow rate range, the liquid velocity at the hollow electrode exit is about 0.6-1.2 m/s and a stronger injection is obtained. When high voltage is imposed to the liquid, the charged jet is disintegrated into droplets at certain location, so that the electrohydrodynamic spray contains two parts: continuous filament column and spread droplets as shown in Fig. 22.6 (Wang et al., 1993). It is also was observed in Fig. 22.6 that many fine droplets are separated from the jet surface before the main charged jet is disintegrated. The dispersed droplet spray is composed of two components: the main spray which is formed at liquid filament apex by disruption and emission and has larger droplets, and the minor spray which is formed at

Fig. 22.6 A typical liquid-solid particle slurry atomizations for various electric field (Wang et al., 1993).

liquid filament surface by emission and has fine droplets. This atomization mechanism is quite different from those modes appeared at lower flow rate conditions. It is reasonable to refer the present atomization mode as the emission mode.

Huebner (1990) observed a so-called fan mode when he investigated the disintegration of charged liquid jets of isopropyl alcohol on 0.448 ml/s flow rate condition. In this mode several secondary jets are developed from the primary jet. Fan mode and the emission mode observed in the present work are similar, but there are two major differences. One is that a number of fine droplets are separated from liquid filament surface in emission mode instead of secondary jet in a fan mode. Another difference is that the fan mode appears only at higher applied electric potential conditions (higher than 35 kV), however, emission mode can be observed in a wide voltage range (from 10 to 35 kV). Huebner noted that the fan configuration mode does not occur with pure water spray, however, the emission mode appeared in water spraying experiments.

While the electric potential increases, the liquid filament column length decreases, the spread angle increases and the apex of the liquid filament moves closer to the location where the fine droplets are emitted as shown in Fig. 22.7 (Wang et al., 1993). It is also illustrated from this figure that the spread angle of surface emitted droplets is always larger than that of the main spray.

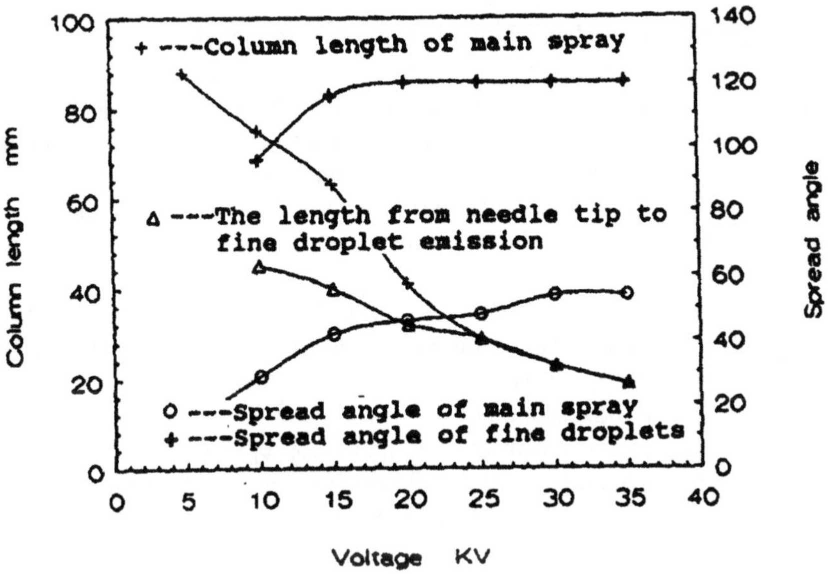

Fig. 22.7 A typical slurry column length and spread angles as a function of applied voltage under EHD atomizations (Wang et al., 1993).

22.5.2 Droplet Size

Three mean diameters of droplets are defined as following:

Length mean diameter

$$D_{10} = \frac{\Sigma D_i}{N} \tag{22.19}$$

Volume mean diameter

$$D_{30} = \left(\frac{\Sigma D_i^3}{N} \right)^{1/3} \tag{22.20}$$

Volume-to-surface mean diameter

$$D_{32} = \frac{\Sigma D_i^3}{\Sigma D_i^2} \tag{22.21}$$

where D_i is the diameter of the ith droplet and N is the sample number. The mean diameter as functions of electrical potential are shown in Fig. 22.8. It is obvious that $D_{32} > D_{30} > D_{10}$. Fig. 22.8 indicates also that the length mean diameter decreases with increasing the electric potential, however, the volume-to-surface mean diameter has a minimum value around 30 kV. Volume-to-surface mean diameter is more important than that of length mean diameter for investigations of mass and heat transfer phenomena and chemical reactions. Based on the variation of volume-to-surface mean diameter, it is found that improving the atomization by increasing applied voltage has a limited value. Similar phenomenon was reported by Ogata et al. (1978).

One of the characteristics of electrohydrodynamic atomization is the monodispersity of droplets. By using combined AC and DC electric potentials, Sato (1984) obtained very uniform droplets. However, monodispersion of charged liquid appears only for smaller flow rate. A bimodal size distribution was observed by Huebner (1970) for the disintegration processes occurring with charged jet of isopropyl alcohol at flow rate 0.488 ml/s. Fig. 22.9 shows a typical droplet size distribution of electrohydrodynamic atomization of limestone-water slurry with 0.78 ml/s flow rate (Wang et al., 1993). It is observed from Fig. 22.9 that the droplet diameter distribution characteristics have two peaks as has been observed by Huebner earlier. The first peak is located at 0.3 mm droplet diameter for all three voltages. The second peak appears at 1.1 mm diameter for applied voltages of 25 kV and 35 kV but appears at 1.3 mm diameter for 15 kV. Bimodal droplet distribution usually discussed in terms of an interplay of molecular, gravitational and electric forces. Assuming the double emission mode, a fine droplet peak is attributed to the surface emission and larger droplet peak originates from emission of the main jet.

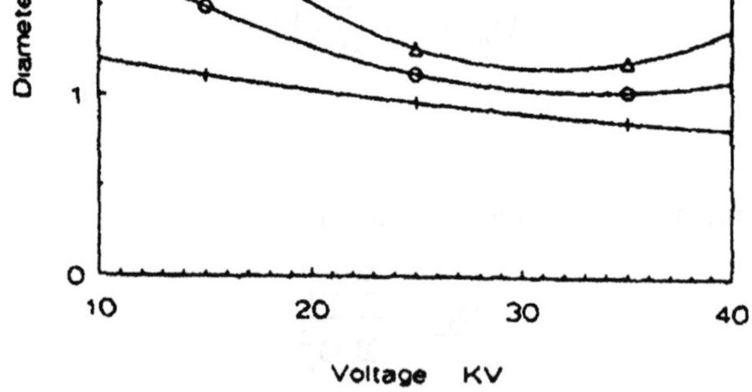

Fig. 22.8 A typical mean droplet diameter based on length, volume and volume-to-surface area as a function of applied voltages (Wang et al., 1993).

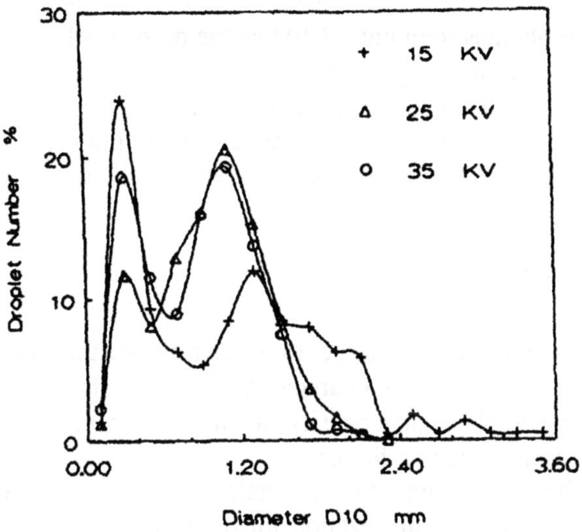

Fig. 22.9 Droplet diameter distributions for Fig. 29.4 (Wang et al., 1993).

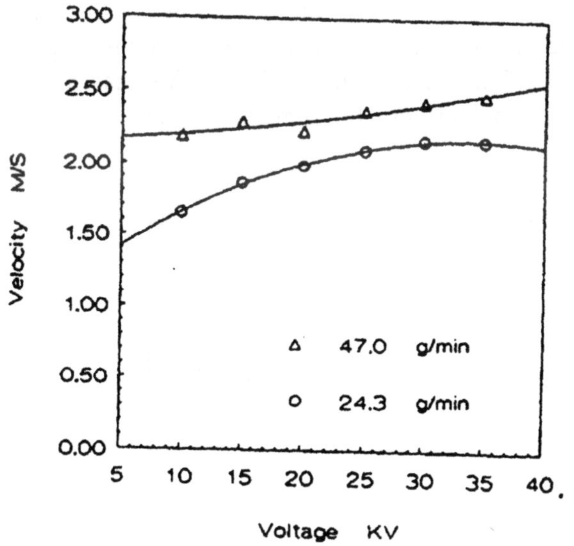

Fig. 22.10 A typical EHD atomization droplet velocity as a function of applied voltage for liquid-solid particle slurry flow (Wang et al., 1993).

22.5.3 Droplet Velocity

Fig. 22.10 shows a typical average velocity as a function of applied voltage for two flow rates (Wang et al., 1993). It is indicated that the droplet velocities increase with increasing voltage and change from 1.6 to 2.5 m/s at both flow rate used in the experiment (0.4 ml/s and 0.8 ml/s). It is obvious from Fig. 22.10 that the droplet velocities increase with increasing flow rate at same applied voltage, since higher flow rate means higher superficial liquid velocity (ratio of flow rate to the outlet area) at the hollow electrode exit. Considering how the ratio of average droplet velocity to the exit superficial liquid velocity changes with the applied voltage, it is found that the droplet velocity ratios under the lower flow rate are larger than those for higher flow rate.

22.5.4 Charge-To-Mass Ratio

Droplet charge density or charge-to-mass ratio are the basic parameters to affect the electrohydrodynamic atomization characteristics. Generally speaking, the droplet size decreases with increasing droplet charge density. It may be interpreted as the facilitation of droplets rupture by electrical repulsion forces.

Fig. 22.11 shows the charge-to-mass ratio of limestone-water slurry droplet produced by an electrohydrodynamic atomization for the electrode with 0.84 mm inside diameter and 25 mm length (Wang et al., 1993). Fig. 22.11 shows that the charge-to-mass ratio increases with increasing applied voltage. If the applied voltage is less than 25 kV, the droplet charge-to-mass ratio increase slowly. When the voltages are larger than 25 kV, the charge-to-mass ratio increases sharply and the droplet charge density distribution becomes wider, as indicated by band in Fig. 22.11. As mentioned above the droplet diameter do not increase when the applied voltage is beyond 30 kV. In this case higher voltage do not contribute to improve the liquid dispersion but it may enlarge the charge density.

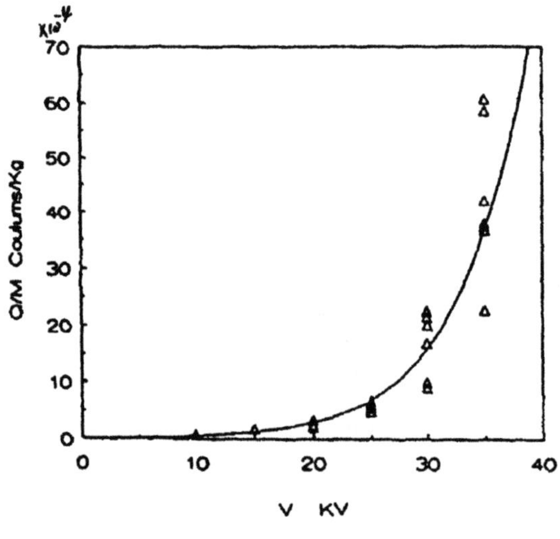

Fig. 22.11 A typical EHD atomized droplet charge-to-mass ratio as a function of applied voltage for slurry flow (Wang et al., 1993).

References

Atten, P. and R. Moreau (1972), J. Mécanic, 11, 471-520.
Bailes, P.J. (1981), Ind. Eng. Chem. Process Des. Dev., 20, 564-570.
Bailes, P.J. and M. Kahlbasi (1981), J. Electrostatics, 10, 81-88.
Carleson, T.E. and J.C. Berg (1983), Chem. Eng. Sci., 38, 871-876.
Chang, J.S., A.J. Kelly and J.M. Crowley (1995), "Handbook of Electrostatic Processes", Marcel Dekker Inc., New York.
Chang, J.S. (1984), "CANDU Thermalhydraulics", Ch. 11, R. Bonalumi et al., Eds., MIES Press, Hamilton.
Clift, R., J.R. Grace and M.E. Weber (1978), "Bubbles, Drops and Particles", Academic Press, New York.
Cloupeau, M. and B. Prunet-Foch (1990), J. Electrostatics, 25, 165.
Cross, J.A. (1987), "Electrostatics: principles, problems and applications", Adam Hilger, Bristol.
He, W., M.H.I. Baird and J.S. Chang (1991), Can. J. Chem. Eng., 69, 1174-1183.
He, W., M.H.I. Baird and J.S. Chang (1996), IEEE Trans. Industry Applications, 32, 146-152.
Huebner, A.L. (1970), Science, 168, 118.
Kao, K.C. (1961), Brit. J. Appl. Phys., 12, 629-632.
Kitahara, A. (1984), "Chapter 5: Nonaqueous systems", in Electrical Phenomena at Interfaces, (A. Kitahara and A. Watanabe eds.), Marcel Dekker, Inc,. New York.
Landau, L.D. and E.M. Lifshitz (1960), "Electrohydrodynamics of Continuous Media", Addison Wesley, Reading, MA.
Masuda, S., M. Washizu and I. Kawabata (1986), Conf. Record of IEEE IAS 1986 meeting, pp. 1361-1365.
Millar, M.K. and L.R. Weatherley (1989), Chem. Eng. Res. Dev., 67, 227-231.
Morala, E.C., D. Cheong, P.T. Wan, G.A. Irons and J.S. Chang (1984), "Multi-phase Flow and Heat Transfer III, Part A: Fundamentals, Elsevier Science Publishing, Amsterdam, pp. 501-511.
Ogata, S., Y. Hara and H. Shinohara (1978), Inter. Chem. Eng., 18, 482.
Ogata, S., T. Kawashima, O. Nakaya and O. Shinohara (1976), J. Chem. Eng. Japan, 9, 440-444.
Pohl, H.A. (1951), J. Appl. Phys. 22, 869-871.
Pohl, H.A. (1978), Dielectrophoresis, Cambridge University Press, New York.
Rayleigh, L. (1882), Phil. Mag., 14, 184.
Rayleigh, J.W.S. (1882), Phil. Mag. (Ser. 5), 14, 184-186.
Sato, M. (1984), J. Electrostatics, 15, 237.
Stewart, G. and J.D. Thornton (1967), Symp. Ser. No. 26, Inst. Chem. Engrs., 29, 29-42.
Takamatsu, T., Y. Hashimoto, A. Yamaguchi and T. Katayama (1981), J. Chem. Eng. Japan, 14, 178-182.
Vonnegut, B. and R.L. Neubauer (1952), J. Colloid Sci., 7, 616-622.

Vu, N. and T.E. Carlson (1986), AIChE J., 32, 1739-1742.
Vonnegut, B. and R. Neubauer (1952), J. Colloid Sci., 7, 616).
Wang, S.H., J.S. Chang and A.A. Berezin (1993), J. Electrostatics, 30, 235-246.
Zeleny, J. (1920), Phys. Rev., 16, 102.

Chapter 23

EHD HEAT EXCHANGER AND CHEMICAL REACTORS

23.1 Introduction

The application of high voltage electric field to enhance the rate of mass transfer in liquid-liquid extraction has been an active subject of investigation for past 25 years. Among the large amount of published work, there have been several review papers and short articles which have summarized most of the work having been done to date (Thornton, 1968; Scott, 1989; Weatherley, 1992; Ptasinski and Kerkhof, 1992). The general idea of directly using an electric field to improve the performance of mass transfer in liquid-liquid extraction was initially described by Stewart and Thornton (1967). The improvement of interfacial mass transfer may be achieved in several ways due to the additional electrohydrodynamic (EHD) forces on the droplet-continuum interface: (1) reducing the effective interfacial tension; (2) increasing interfacial area for mass transfer and (3) enhancing interfacial disturbance and drop circulation.

The basic techniques to achieve electrically enhanced liquid-liquid extraction are briefly summarized in Table 23.1. Although there is so far no industrial scale application of electrically enhanced mass transfer in liquid-liquid extraction (Scott, 1989; Weatherley, 1992), investigations on some industrially related systems have been carried out. For example, extraction of nickel, copper, uranium and rare earths (Bailes, 1977, 1981; Martin et al., 1984; Hund and Lancelot, 1986), and extraction of penicillin G and other products from whole fermentation broths (Millar and Weatherley, 1989; Weatherley et al., 1990) have been reported. Table 23.1 shows that most of the work conducted with a uniform electric field was either confined to the interpretation of total mass transfer performance or restricted to the discrete droplet regime; and all systems selected had low viscosities. In most cases, the mechanisms of electrically enhanced mass transfer were still not clearly understood and the reported results were very much system dependent.

On the other hand, two-phase heat transfer enhancement applications by EHD forces are conducted manily for single component cases such as boiling and condensing two-phase flow, where two-phase EHD heat exchangers developed are listed in Table 23.2. In this moment, several industrial scale proto-type heat pumps are tested (Yabe 1995).

Table 23.1. Previous work on electrically aided mass transfer in liquid-liquid systems.

Source	Apparatus	Confined range	Mass transfer system
Thornton (1968)	glass column: 10 cm (i.d.) d.c., 0–17 kV, electrode: single nozzle, 1.1 mm (i.d.) 2 parallel plates (gap: 2.5 cm, 7.0 cm)	discrete single droplet droplet size: 2.1–4.2 mm	c: toluene d: water s: benzoic acid (c → d) high interfacial tension low viscosity
Bailes & Thornton (1971)	glass column: 9 cm (i.d.), 19 cm high d.c., 0–20 kV, electrode: single nozzle, 1.1 mm (i.d.) 2 parallel plates	discrete single droplet drop phase: 91 mL/h; drop size: 1.5–2.2 mm	binary, partial misible c: n-heptane d: furfuraldehyde low interfacial tension, low viscosity
Bailes & Thornton (1974)	glass column: 9 cm (i.d.) d.c., < 9.6 kV, electrode: 19 nozzle-plate, 1.1 mm (i.d) 2 parallel plate (gap: 4–12 cm)	multi-droplet drop size: 1.7–2.2 mm	binary, partial misible c: n-heptane d: furfuraldehyde low interfacial tension, low viscosity
Bailes (1977)	glass column: 6 cm i.d., 35 cm high d.c., 0–5 kV, electrode: single nozzle, 1.0 mm i.d.(for nickle) single nozzle, 2.0 mm i.d.(for copper) single planar plate (statinless steel) down below 30 cm for nickel extraction down below 5 cm for copper extraction	discrete single droplet drop size: 3.1–3.7 mm drop phase: 79 mL/h (for nickel extraction) 93 ml/h (for copper extraction)	(1) nickel extraction c: D2EHPA with toluene d: nickel sulphate solution s: nickel (d → c) (2) copper extraction c: LIX64N in Escaid 100 d: copper sulphate solution s: copper (d → c) high interfacial tension low viscosity
Kowalski & Ziolkowski (1981)	glass column (coated with polyethylene) 5 cm i.d., 45 cm high multi-drop distributor a.c. (50 Hz), < 11 kV electrode (4 metal rods): two 8 mm dia. coated with polyethylene (to H.V. source), two 6 mm dia. without coating (to ground)	multi-droplet drop phase: 6.4–38.6 L/h continuous phase: 30.8 L/h	(1) c: xylene d water s: acetic acid (c → d) (2) c: carbon tetrachloride d: water s: acetic acid (c → d)
Martin et al. (1983)	glass column: 3.9 cm i.d., 120 cm high multi-drop distributor plate with nine 1 mm dia. holes d.c., < 15 kV, electrode: 8 stages of statinless steel rods, 10 cm long each coated with teflon	multi-droplet drop phase: 0–18 L/h continuous phase: 0–0.45 L/h	c: LIX65N in Escaid 100 d: copper chloride solution s: copper (d → c) high interfacial tension low viscosity
Yamaguchi et al. (1985)	glass column. 30 cm i.d. d c., 0–36 kV, electrode: single nozzle, 0.6 mm i d.(to H.V. source) 2 parallel plates, 30 cm dia. gap: 5 cm, 15cm	discrete single droplet drop phase: 8–40 mL/h drop size: 3–5 mm dia.	c: cyclohexane d: water s: iodine (d → c) high interfacial tension low viscosity

Freeman & Laurizio (1986)	glass column d.c., < 10 kV electrode: single nozzle & two parallel plates	discrete single droplet drop phase: 480 mL/h drop size: 2.72–5.17 mm dia.	c: cyclohexane d: water s: toluene (d → c) high interfacial tension low viscosity
Wham & Byers (1987)	tapered glass contactor (0.4 L capacity) d.c., 0–20 kV; pulse (0–50 Hz), 10 kV electrode: two stainless steel rods, 0.32 mm dia. positioned 2.5 cm apart	discrete single droplet drop size: 3 mm dia.	binary, partially miscible c: 2-ethyl hexane d: water low interfacial tension low viscosity
Yang & Carleson (1987)	glass column for single drop; 3 stage column for multi-drop d.c., electrode: single nozzle & two prallel plates (single drop) gap: 2–12 cm distributor plate, 3.175 mm dia. orifices (multi-drop)	single droplet & multi-droplet drop phase: 18–30 mL/h	(1) ternary system c: toluene d: water s: methnol (d → c) high interfacial tension low viscosity (2) binary, partial miscible c: n-heptane d: furfural low interfacial tension low viscosity
Yamaguchi et al. (1988,1989)	spray column, 9 cm i.d., 30 cm hight, drop distributor: 31 nozzles, 0.3 mm i.d. d c., 0–23 kV, electrode: 4 pole vertical copper rods, 0.5 cm dia., 30 cm long (2 to H.V. source, 2 to ground)	multi-droplet drop phase: 6.6 L/h continuous phase: 22.5 L/h drop size: 2.5–4.0 mm dia.	c: cyclohexane d: water s: iodine (d → c) high interfacial tension low viscosity
Scott & Wham (1989)	tapered glass contactor (1 L capacity) a.c. (60 Hz), 0–20 kV; pulse (0–200 Hz), 0–25 kV, electrode: 2 parallel stainless steel plates, 5 cm × 5 cm, positioned vertically, 2.5–5.0 cm apart	emulsion phase contactor drop size: 1–10 μm dia. drop phase: 19.8 mL/h continuous phase: 35.4 mL/h	c: methyl isobutyl ketone d: water s: acetic acid (d → c) low interfacial tension low viscosity

Table 23.2 EHD Heat Exchanger Developments

Investigators (year)	Orientation	Type of Heat Exchanger	Comments
Loerke & Debes (75)	Vertical	Heat Pipe	
Jones & Perry (74)	Horizontal	Heat Pipe	R-113
Kikuchi et al. (83, 84)	Horizontal	Heat Pipe	R113, R11, R13 Perfluorocarbon
Bologa et al. (80)	Inclined	Heat Pipe	R-113
Revankar & Chang(84)	Vertical	Reflux Condenser	
Yabe et al. (85, 92)	Vertical	Shell Side Condenser	
Bergham et al. (86)	Horizontal	Shell Side Condenser	
Ogata & Yabe (93)	Horizontal	Boiler	
Allen & Cooper (87)	Horizontal	Evaporators	
Ohadi (91)	Vertical/Horizontal	Condenser/Boiler	
Chang (86)	Vertical	Pressurizer	

23.2 Mass Transfer Under Electric Fields

23.2.1 Mass Transfer without Electric Fields

Considerable experimental and theoretical work has been done on mass transfer to or from an isolated drop moving through a continuous phase without any external force imposed except gravity. In general, mass transfer occurs in three different stages: (1) during formation at a nozzle; (2) during free rise or free fall; and (3) during coalescence at an interface. In order to investigate the mechanisms of drop mass transfer during its free rise or free fall period, the proportion of the mass transfer that occurs during drop formation and coalescence periods, the so-called "end effects", must be evaluated.

The magnitude of the "end effects" has usually been estimated by varying the height for steady risk or fall of the drop and extrapolating the amount of mass transfer to zero height (Slater et al., 1988), and in the mean time to minimize the effect of droplet coalescence. An experimental technique in which drops are formed and withdrawn at the same nozzle, without release (Popovich et al., 1964; Streicher and Schiigerl, 1977), provides data on the rate of mass transfer while a drop is growing at a nozzle. The third approach is of using various plotting techniques for combining "end effects" analysis and the examination of the mass transfer mechanism (Johnson and Hamielec, 1960). It is this approach that has been used in this work to investigate the drop mass transfer mechanism for both charged and uncharged drops.

Mass transfer from the drop phase to a continuous phase may be expressed as the changes in solute concentration inside the drop based on the three stage mechanism with the assumption of concentration homogeneity within the drop:

$$\begin{array}{ccccccc} & \text{forming} & & \text{falling} & & \text{coalescing} & \\ C_0 & \rightarrow & C_1 & \rightarrow & C_2 & \rightarrow & C_3 \end{array}$$

The fractional approaches to each concentration equilibrium (mass transfer efficiency) during drops formation, free fall and coalescence, respectively have been defined by Johnson and Hamielec (1960) as follows:

$$E_{df} = \frac{C_0 - C_1}{C_0 - C^*} \tag{23.1}$$

$$E_{ff} = \frac{C_1 - C_2}{C_1 - C^*} \tag{23.2}$$

$$E_{dc} = \frac{C_2 - C_3}{C_2 - C^*} \tag{23.3}$$

where E_{df}, E_{ff} and E_{dc}, are the efficiencies for drop mass transfer at different stages respectively and C^* is the solute concentration in a drop which is in equilibrium to that in the continuous phase. It is assumed that C^* is the same for each of the three mass transfer steps. in practice only C_o and C_3 are directly measurable, and the corresponding overall (total) efficiency is related to the three contributing efficiencies:

$$E_t = \frac{C_0 - C_3}{C_0 - C^*} = 1 - (1 - E_{df})(1 - E_{ff})(1 - E_{dc}) \tag{23.4}$$

Generally, the overall mass-transfer coefficient based on the dispersed phase is defined as K_{od}, K_{df} is defined as overall mass-transfer coefficient during drop formation stage and K_{ff} is defined as overall mass-transfer coefficient during drop free fall stage. Obviously, both K_{df} and K_{ff} are based on the dispersed phase.

Various mechanisms proposed in the literature for mass transfer during droplet formation (Popovich et al., 1964) were based on the unsteady-state diffusion theory with different assumptions about the hydrodynamics of the drop formation process. All these models can be considered as modifications of the Higbie's penetration theory (1935), according to which mass transfer is determined only by the molecular diffusivity and the contact time, t_{df}:

$$K_{df} = Z\left(\frac{D_c}{\pi\, t_{df}}\right)^{1/2} \tag{23.5}$$

where Z is a dimensionless constant which depends upon the assumptions made for diffusion, D_c, combined with the growth rate of drop volume and interfacial surface area.

Higbie's model (1935) assumes that the drop surface is mobile with surface renewal occurring at time interval of $t = d/v_t$. This corresponds to

$$K_{ff} = 2\left(\frac{D_c\, v_t}{\pi\, d}\right)^{1/2} \tag{23.6}$$

23.2.2 Mass Transfer in the Presence of Electric Field

The observed behaviour of charged droplets formed under the applied electric field during mass transfer experiments was similar to the liquid-liquid flow works described in Chapter 19. Under the effect of electric field both drop size reduction and enhancement of drop initial velocity were significant, especially at higher applied voltages. Therefore larger interfacial areas were generated and interfacial disturbance might be enhanced as well, compared with the results obtained in the absence of electric field. These phenomena provide the possibilities for enhancing the mass transfer rate in liquid-liquid extraction processes.

Figure 23.1 shows the total mass transfer efficiency, E_t, for two different drop phase flow rates is significantly enhanced in the presence of high voltage electric field (He et al. 1993). As shown in Figure 23.1 (a), the total mass transfer efficiencies, E_t measured at 10 kV applied voltages are much higher than those measured in the absence of electric field for various drop phase flow rates . As seen in Figure 23.1 (b), the total mass transfer efficiency increases approximately by 2-4 times in the present range of investigation, where the relative total mass transfer efficiency, R has been defined as the ratio of E_t at applied voltages divided by the E_t without electric field applied at the same drop phase flow rate.

From charge to mass ratio measurements and visual observations during mass transfer experiments, it was seen that in the presence of electric field the drop formation period took much less time compared with the duration of its free fall. The droplets formed were of uniform size at a fixed applied voltage, even in the drop dispersion regime at very high applied voltages. Therefore, the mass transfer enhancement during drop formation stage may be primarily due to the effect of increased specific interfacial area associated with reduced drop diameter. Which means that droplets formed of the same drop size have the equivalent mass transfer characteristics during formation for a given setup geometry and system, as postulated by Bailes (1981). The drop coalescence effect on mass transfer under the applied voltage is still considered negligible as in the absence of electric field. Then Equation (23.5) may still be used to evaluate K_{df} for the charged drop mass transfer during drop formation provided the drop size be considered as a variable as well as the time duration for drop for

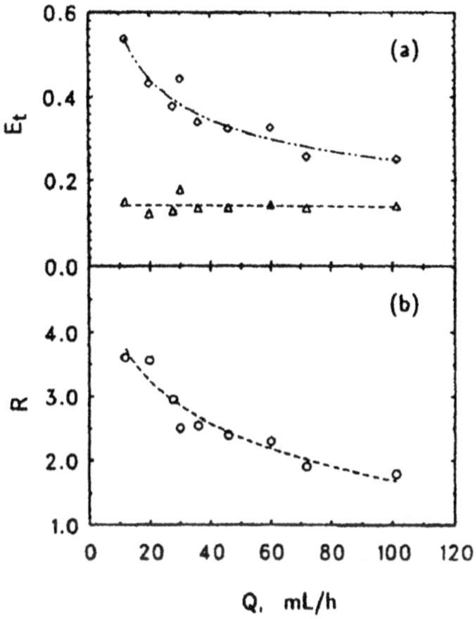

Figure 23.1 Total mass transfer efficiency as the function of drop phase flow rate at 10 kV applied D.C. voltage (He et al. 1993).
(a) ◇ voltage: 10 kV; Δ voltage: 0 kV;
(b) relative ratio of total mass transfer efficiency
(c) Total mass transfer efficiency as the function of D.C. applied voltage
O 30 mL/h; Δ 60 mL/h

mation. In the case of imposed electric field, different drop diameters at a fixed drop phase flow rate indicate the implicit values of different applied voltages. Equation (23.5) may be written in terms of droplet diameter and drop phase flow rate with the substitutions of $t_{df}\left(=\frac{\pi d^3}{6Q}\right)$ as follows:

$$K_{df} = C'\left(\frac{Q}{d}\right)^{1/2} \tag{23.7}$$

where $C' = C\left(\frac{5}{36}\right)\left(\frac{6}{\pi}\right)^{1/2}$. Therefore the modified model of Equation (23.7) can be used to estimate mass transfer coefficient during drop formation for both uncharged and charged droplet mass transfer cases. Once K_{df} can be estimated for charged drop mass transfer using Equation (23.7), the mass transfer efficiency for charged drop formation stage, E_{df}, can also be calculated.

Figure 23.2 shows the results of the extended model of Equation (23.7), where the mass transfer coefficient, K_{df} is plotted as the function of Q/d together with the results for uncharged droplets. It is noted from Equation (23.7) that larger K_{df} values can be obtained by either reducing drop size or increasing drop phase flow rate. In the case for uncharged drops, the drop size is constant; while for the charged drops, drop size becomes smaller when increasing the applied voltages. Obviously, the higher the applied voltage in the present mass transfer system, the larger the K_{df} is obtained.

The mass transfer efficiency for charged drops in the free fall stage, E_{ff}, is readily estimated using Equation (23.4) from the total measured mass transfer efficiency and from the values of E_{df} evaluated as described above. Finally the mass transfer coefficient for charged drops in free fall, K_{ff}, can be calculated using Equation (23.6), where the terminal velocity for charged droplets is evaluated using Stokes equation, since the electric field intensity down stream of the nozzle was so weak that the induced charges on the reduced drops were negligible based on the experimental results.

Both charged and uncharged drop mass transfer coefficient during drop free fall (found from Equation (23.7)) arc plotted as the function of drop diameter as shown in Figure 23.3. Figure 23.4 shows the correlations between Sherwood numbers and Reynolds numbers while Figure 23.5 shows the correlations between Sherwood numbers and Peclet numbers for the mass transfer results obtained by He et al. (1993). Both figures give linear correlations on the log-log plot and are compared with two well-known existing models as shown in these two figures.

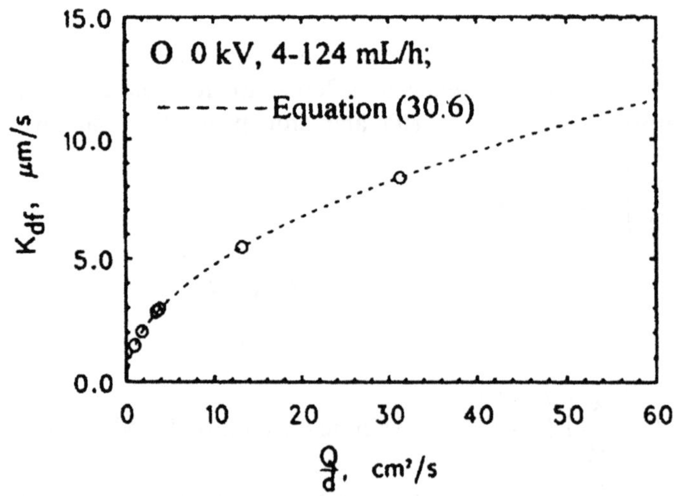

Figure 23.2 Mass transfer coefficient during drop formation stage as a function of the ratio of drop flow rate to drop diameter (He et al. 1993).
O 0 kV, 4-124 mL/h; ------Equation (22.6)

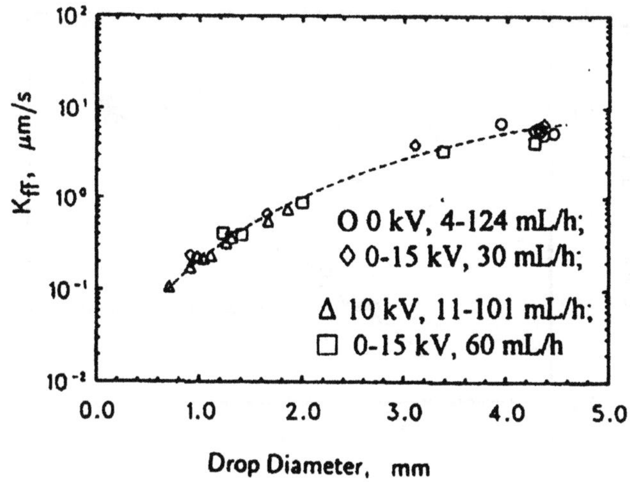

Figure 23.3 Mass transfer coefficient during drop free fall stage as a function of drop sizes (He et al. 1993).
O 0 kV, 4-124 mL/h; Δ 10 kV, 11-101 mL/h;
◇ 0-15 kV, 30 mL/h; □ 0-15 kV, 60 mL/h

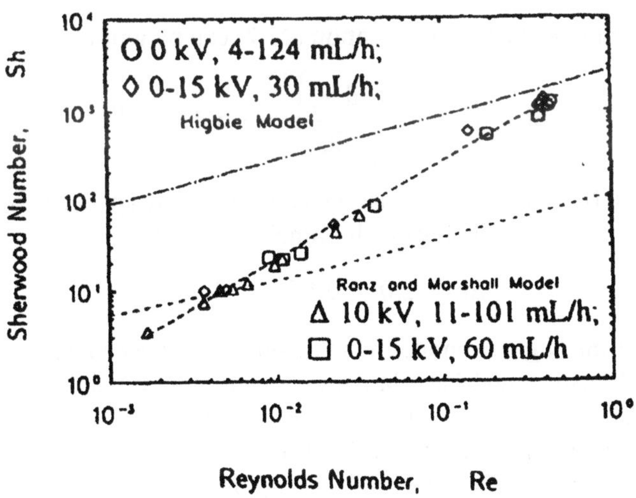

Figure 23.4 Correlations of Sherwood Numbers as a function of Reynolds Numbers (He et al. 1993).
O 0 kV, 4-124 mL/h; Δ 10 kV, 11-101 mL/h;
◇ 0-15 kV, 30 mL/h; □ 0-15 kV, 60 mL/h

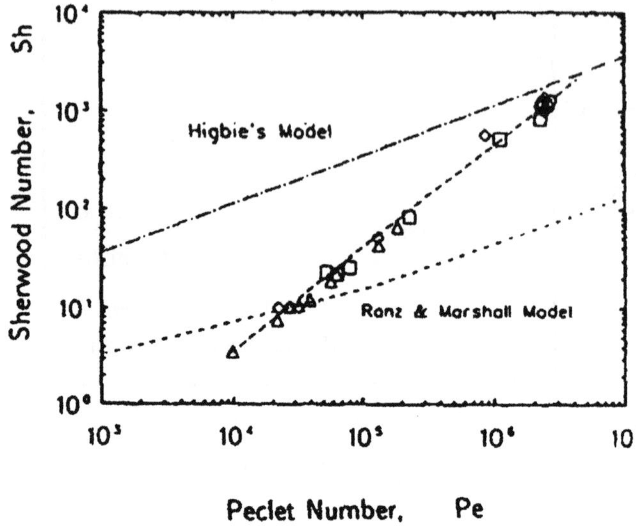

Figure 23.5 Correlations of Sherwood Numbers as a function of Peclet Numbers (He et al. 1993). O 0 kV, 4-124 mL/h; Δ 10 kV, 11-101 mL/h; ◊ 0-15 kV, 30 mL/h; □ 0-15 kV, 60 mL/h

The Higbie's model (1935) of Equation (23.7) can be written

$$Sh = \left(\frac{2}{\pi^{1/2}}\right)\left(\frac{d_c v_t}{D_c}\right)^{1/2} = 1.13\ Pe_c^{1/2} \tag{23.8}$$

The Rant and Marshall (1955) equation has been widely applied to mass transfer from solid spheres or from drops or bubbles with immobile surfaces:

$$Sh = 2.0 + 0.6\, Re_c^{1/2}\, Sc_c^{1/3} \tag{23.9}$$

The shift from Higbie's model at relatively higher Reynolds numbers or Peclet numbers to Rant & Marshall model at relatively lower Reynolds numbers or Peclet numbers suggests that the drops were circulating during drop free fall when the drop size were relatively large in the absence of electric field or at low applied voltages; as the applied voltage is increased, the drops became gradually rigid due to the reduced drop size and the increasing relative importance of trace interfacial contaminants (He et al. 1993).

The mechanisms of mass transfer during the formation of charged droplets and during the free fall of these dispersed droplets are the same as those obtained in the absence of the applied electric field. The enhancement of mass transfer achieved in this system in the presence of electric field is mainly due to the significantly increased specific interfacial area. This in turn is due to strong local electrical forces promoting rapid detachment of drops from the nozzle.

23.3 Two-Phase Transfer Under Electric Fields

Fundamental principle of heat transfer enhancements are already discussed in Chapter 20. However, the advantages of EHD techniques in heat transfer application are not limited to heat transfer enhancement. They have been used to control the critical heat flux (CHF) in boiling

systems and the flooding limits in counter-current flow as summarized in Table 23.3 (Chang, 1988).

For EHD condenser applications, three different type condensers are studied as follows:

1. Horizontal tube side condensers (Jones and Perry 1974, Fig. 23.6a);
2. Vertical tube side reflux condensers (Revankar and Chang, 1984; Chang et al., 1985, Fig. 23.6b); and
3. Vertical shell side condensers (Yabe 1995, Fig. 23.6c).

Heat transfer enhancement in vertical shell side EHD condenser is based on the two principles as follows:

1. Generation of thin liquid films by a continuous removal of condensate by high voltage helical, since the heat transfer rate nearly linear proportional to liquid thin film thicknesses wires; and

2. Transformation of thin liquid thin films to a dropwise condensation by high voltage plate electrodes, since dropwise condensation heat transfer rate is much higher compared with falling film condensation heat transfer rates.

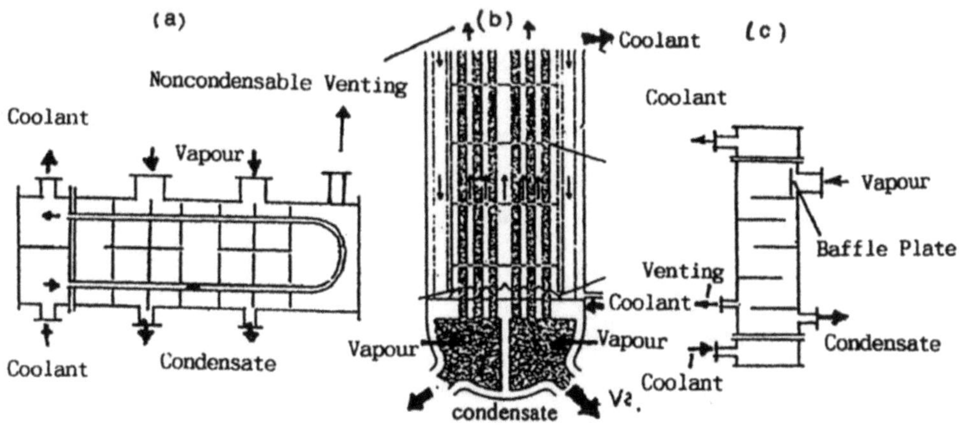

Figure 23.6 EHD condensers; a) horizontal tubeside; b) vertical tube side; and c) vertical shell side condensers.

Table 23.3 Advantages of EHD Techniques for Various Heat Exchangers

Type of Components	Operating Phases	Advantages
Steam Generators	Two-Phase	Boiling heat transfer enhancement, CHF control
Boilers	Two-Phase	Boiling heat transfer enhancement, CHF control
Condensers	Two-Phase	Condensing heat transfer enhancement, control of flooding
Preheaters	Single or Two-Phase	Heat transfer enhancement
Pressurizers	Two-Phase	Control of pressure transient CHF control, Boiling and Condensing heat transfer enhancements
Heat Pipe	Two-Phase	Horizontal two-phase heat transport and heat transfer enhancements
Heat Pump	Two-Phase	Heat transfer enhancements

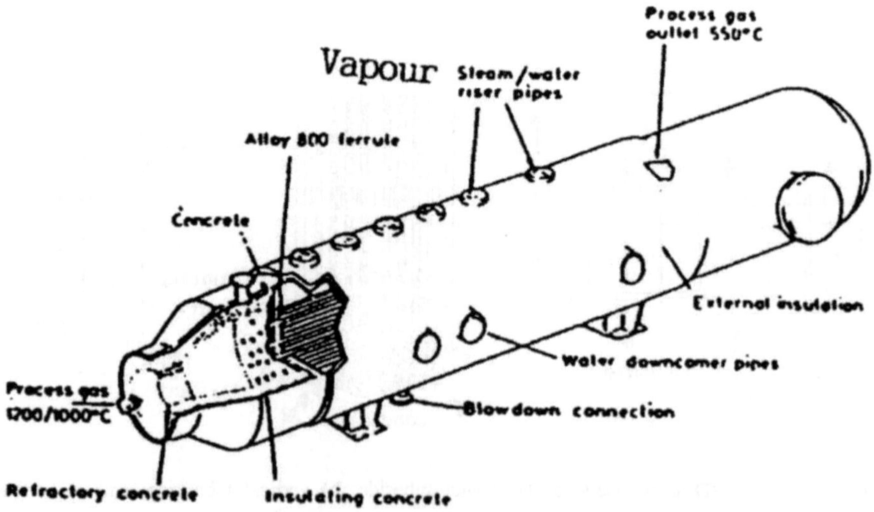

Figure 23.7 EHD shell side boiler.

For EHD boiler applications, only horizontal shell side boiler is shown in Fig. 23.7 is studied. These boilers and condensers are also integrated and applied to EHD heat pumps, heat pipes, electrical transfer and transmission line coolings as shown in Figures 23.8a), b), c) and d) respectively. The main objectives of the EHD heat pipe/heat transformation in a horizontal direction and application of heat pipe under reduce gravity environments. EHD two-phase heat transfer enhancement techniques are also proposed apply for a nuclear power plant emergency core cooling system as summarized in Table 23.4, since no moving components are required and also be able to operate by emergency electrical power supply systems such as batteries.

Components	**Control and Advantages**
Fuel Channel	Heat Removal to Calandria CHF Control
Steam Generator	Flooding Control Heat Removal to Secondary Side
Emergency Water Spray	Atomization Direct Contact Heat Transfer
Flow Loop	EHD Pumping-Enhancement of Natural Circulation
Pressurizer	Pressure Control

Table 23.4 EHD Emergency Core Cooling Systems

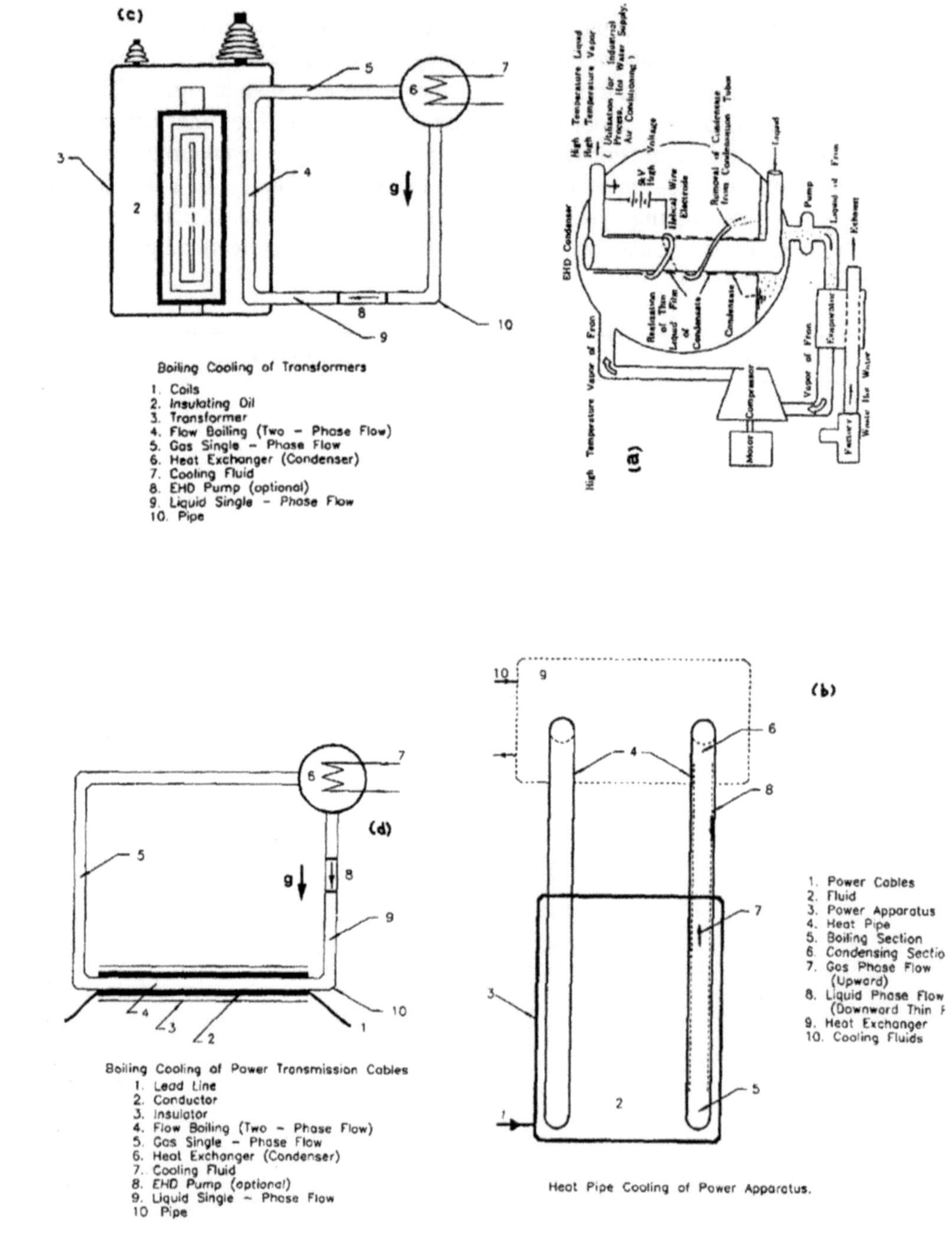

Figure 23.8 EHD integrated heat exchangers; a) heat pump; b) heat pipe; c) transformer cooling; and d) transmission line cooling systems.

References

Bailes, P.J. and J.D. Thornton (1971), Proc. ISEC'71, pp. 1431-1439.
Bailes, P.J. and J.D. Thornton (1974), Proc. ISEC'74, pp. 1011-1027.
Bailes, P.J. (1977), Proc. ISEC'77, pp. 233-241.
Bailes, P.J. (1981), Ind. Eng. Chem. Process Des. Dev., 20, pp. 564-570.
Chang, J.S., R. Raman, S.T. Revankar and F.B.P. Tran. (1985), IEEE Trans. Industry Appl. Soc., IA21, 715-723.
Chang, J.S. (1989), Chem. Eng. Japan, 53, 100-103.
Chang, J.S. (1988), Particulate Phenomena and Multi-Phase Transport, T.N. Veziroglu Ed., Hemisphere Publ. Co., Washington, Vol. 5, pp. 35-62.
Freeman, R.W. and J. Laurizio (1986), Chemical Separations (C.J. King et al., Eds), pp. 295-301.
He, W., M.H.I. Baird and J.S. Chang (1993), J. Can. Chem. Eng., 71, 366-376.
Higbie, R.Z. (1935), Trans. Am. Inst. Engrs., 31, p. 365.
Hund, M. and F. Lancelot (1986), Proc. ISEC'86, pp. 1-510-517.
Johnson, A.I. and A.E. Hamielec (1960), A.I.Ch.E. J., 6, pp. 145-149.
Jones, T.B. and M.P. Perry (1974), J. Appl. Phys., 45, 2129-2132.
Kowalski, W and Z. Ziolkowski (1981), Int. Chem. Eng., 21, pp. 323-327.
Martin, L., P. Vignet, C. Fombawlet and F. Lancelot (1983), Separation Sci. & Tech., 18, pp. 1455-1471.
Millar, M.K. and L.R. Weatherley (1989), Chem. Eng. Res. Des., 67, pp. 227-231.
Ogata, J. and A. Yabe (1993), Int. J. Heat and Mass Transfer, 36, 775-791.
Popovich, A.T., R.E. Jervis and 0. Trass (1964), Chem. Eng. Sci., 19, pp. 357-365.
Ptasinski, K.J. and P.J.A.M. Kerkhof (1992), Sep. Sci. and Tech., 27, pp. 995-1021.
Revankar, S.T. and J.S. Chang (1984), J. Electrostatics, 16, 69-77.
Scott, T.C. and R.M. Wham (1989), Ind. Eng. Chem. Res., 28, pp 94-97.
Slater, M.J., M.H.I. Baird and T.-B. Liang (1988), Chem. Eng. Sci., 43, pp. 233-245.
Stewart, G. and J.D. Thornton (1967), I.Chem.E. Symposium Series No. 26, pp. 29-42.
Streicher, R. and K. Schiigerl (1977), Chem. Eng. Sci. 32, pp. 23-33.
Thornton, J.D. (1968), Rew. Pure and Appl. Chem., 18, pp. 197-218.
Weatherley, L.R., I. Campbell, D. Kirton and J.C. Slaughter (1990), J. Chem. Tech. Biotechnol., 48, pp. 427-438.
Wham, R.M. and C.H. Byers (1987), Separations and Technology, 22, pp. 447-466.
Yabe, A. (1995), Chapter 25 in "Heat Engineering", Handbook of Electrostatic Processes, J.S. Chang et al., Eds., Marcel Dekker Inc., New York, pp. 555-580.
Yamaguchi, M., T. Takamatsu, F. Yoshida and T. Katayama (1985), J. Chem. Eng. Japan, 18, pp. 325-330.
Yamaguchi, M., H. Sugaya and T. Katayama (1988), Proc. ISEC'88, pp. 210-213.
Yamaguchi, M., H. Sugaya and T. Katayama (1989), J. Chem. Eng. Japan, 22, pp. 25-29.
Yang, W. and T.E. Carleson (1987), Conf. Rec. IEEE IAS 1987 Annual Conference, pp. 1464-1470.

MIX
Papier aus verantwortungsvollen Quellen
Paper from responsible sources
FSC® C105338

If you have any concerns about our products,
you can contact us on
ProductSafety@springernature.com

In case Publisher is established outside the EU,
the EU authorized representative is:
Springer Nature Customer Service Center GmbH
Europaplatz 3, 69115 Heidelberg, Germany

Printed by Libri Plureos GmbH
in Hamburg, Germany